Contents

Preface

The science of radio astronomy began in the 1930s with Karl Guthe Jansky's serendipitous discovery of radio emission from the galactic nucleus. Subsequent research in radio astronomy led to the discovery of an incredibly broad range of phenomena, from molecules in space to the discovery of the cosmic microwave background radiation that provides verification of the hot big bang theory. Whole new classes of objects have been discovered at radio frequencies including giant molecular clouds (GMCs), cosmic jets, active galactic nuclei (AGNs), radio nova, and rapidly rotating neutron stars called pulsars. Research in radio astronomy has greatly enhanced scientists' models of other cosmic objects like the structure and dynamics of stars, the Milky Way and other galaxies, supernova remnants (SNRs), the interstellar medium, and suspected black holes by revealing previously unobserved structure. The contributions of research in radio astronomy to the present understanding of celestial phenomena are considerable. The use of radio telescopes in research has greatly enhanced fundamental astrophysics by allowing astronomers to observe a wide variety of objects and astrophysical processes. The achievements in astrophysics of radio astronomy techniques have been widely appreciated. Researchers using the National Radio Astronomy Observatory (NRAO) instruments have made tremendous contributions to the evolution of the science.

The purpose of this book is to examine the sequence of events that led to the establishment of the NRAO and the construction and development of instrumentation, to describe the contributions and discovery events, and to relate the significance of these events to the evolution of the sciences of radio astronomy and cosmology. A brief discussion of the early days of the science sets the stage. The developmental and construction phases of the major instruments including the 85-foot Tatel telescope, the 300-foot telescope, the 140-foot telescope, and the Green Bank interferometer are examined. The technical evolution of these instruments is traced and their relevance to scientific programs and discovery events is discussed.

The history is told in a narrative format interspersed with technical and scientific explanations, providing an interpretive discussion of selected programs, events, and technological developments that epitomize the contributions of the NRAO. Scientific programs conducted with the NRAO instruments that were significant to galactic and extragalactic astronomy are discussed. NRAO research programs presented include continuum and source surveys, mapping, a high precision verification of general relativity, and SETI programs. Modern NRAO instruments including the VLA and the VLBA and their scientific programs are presented in the final chapter as well as a discussion of NRAO instruments of the future such as the GBT.

An attempt has been made to use units consistent with the instrument discussed. For example, the NRAO 140-foot radio telescope is not referred to in meters; similarly, the 25-meter Dutch radio telescope at Dwingeloo is not referred to by its equivalent size in English units. Metric/English equivalents are provided the first time the units appear in the text. In some cases, antiquated units are replaced with the modern version. Cycles per second (cps), for example, is replaced with Hertz. This strategy is intended to provide the reader with a sense of continuity of the wavelengths, frequencies, energies, and dimensions discussed. It is my hope that this book will give the reader some insight into the subtlety and complexity of research in radio astronomy and into the fantastic universe revealed.

Acknowledgments

To the NRAO staffs, past and present, and astronomers who have used the NRAO instruments over the years, I wish to extend grateful acknowledgment. The study was made possible through the use of NRAO archives and data which were made available by Dr. G. A. Seielstad, former director of Green Bank operations. I wish to express appreciation to the following for providing invaluable data, information, advice and counseling, without which the project would not have been possible: James F. Crews, Richard Fleming, J. Richard Fisher, Jay Lockman, Ron Maddalena, Ron Monk, Carl Chestnut, Sue Ann Heatherly, Patricia J. Smiley, and the NRAO staff. I wish to acknowledge the work of previous researchers and historians: John Findlay, Lloyd Berkner, David Heeschen, Allan Needell, W. T. Sullivan, and Owen O. Gingerich. I also wish to extend grateful acknowledgment to Dr. Patricia Obenauf, James F. Crews, Dr. Gregory Good, Dr. Emory Kemp, Dr. Ron Iannone, Dr. Perry Phillips, Dr. Gerrit Verschuur, Melanie R. Culbertson, Travis E. Hoyt, Sonya C. Matthews, and Mary Roberts, without whose guidance and assistance the project could not have been accomplished. I also acknowledge the very important work of previous researchers and historians: John Findlay who wrote a series of Annual Reports published in the *Bulletin of the American Astronomical Society,* beginning in 1969 and was a very active astronomer at NRAO; Lloyd Berkner, who wrote the 1969 Annual Report and was a key player in the observatory's establishment; David Heeschen who wrote the 1970 through 1979 reports and directed the NRAO at Green Bank from 1962 to 1979; and M. S. Roberts and R. J. Havlen for the 1980 report. These reports provided valuable information to aid in tracing the technological and intellectual developments fostered by the NRAO. Also invaluable to the project was the work done by Allan Needell in investigating the federal role in the astropolitics leading to the establishment of the NRAO, resulting in his classic publication, *Lloyd Berkner, Merle Tuve, and the Federal Role in Radio Astronomy.* Other works that were extremely informative include two compilations by W. T. Sullivan, *Classics in Radio Astronomy,* and *The Early Years of Radio Astronomy;* Owen O. Gingerich's *Astrophysics and Twentieth-Century Astronomy to 1950, Part 4A;* and the proceedings of a conference for the history of radio astronomy held at the NRAO in 1983, *Serendipitous Discoveries in Radio Astronomy,* edited by K. Kellerman and B. Sheets. It is my hope that this project extends the work done by these researchers into a relatively comprehensive history of the evolution of instrumentation in radio astronomy at the NRAO. Finally, I am deeply grateful to my family Royce, Jean, Royce V., Lizzie, Steve, Neil, Patsy, and Junior who encouraged my endeavors in the long years of research, preparation, and writing that have culminated in this project.

Introduction

The history of radio astronomy presents a unique opportunity to study in detail the evolution of a science from its inception to the present day. Radio astronomy has evolved from serendipitous beginnings into big science in only 50 years. Numerous first- and secondhand accounts of historical events in radio astronomy are therefore available; indeed many of the early pioneers are currently actively involved in research.

Several ingredients add intrigue to the story. In 50 years a unique science has evolved with a history steeped in serendipity, the development of a war technology, and intense international competition. This period has witnessed the construction of the largest scientific instruments ever conceived and with them the discovery of the most dramatic and energetic phenomena in the universe. In attempting to reveal and understand these phenomena, radio astronomers have incorporated the diverse fields of astronomy, electrical engineering, and theoretical physics; radio astronomy therefore has a rich interdisciplinary history.

Radio astronomy is a highly technique-oriented science due to the elaborate hardware necessary to observe the farthest reaches of the known universe. The energies involved with radio signals from cosmic objects are inconceivably small. An analogy radio astronomers like to use is that the combined energy collected by all radio telescopes since the beginning of the science in 1932 would be roughly equivalent to the kinetic energy contained in a falling snowflake. The subtlety of the human ability to acquire knowledge about the universe is epitomized in this analogy. It has been necessary to develop elaborate instruments and observation techniques to collect and analyze this weak cosmic radio frequency radiation. The evolution of technology involved with observation techniques and instrumentation in radio astronomy has been tremendous. In the short history of the science, the sensitivity and resolution of radio telescopes have each increased by an amazing factor of 10^{10}.

The history of radio astronomy includes simultaneous invention, international rivalry, secrecy over a much desired war technology, and in the end, the evolution of a pure, abstract science with no conceivable malevolent purpose. Few sciences or fields of inquiry rival the accomplishments made by radio astronomers in enhancing our view and understanding of objects in the universe, and the universe itself. Their research has led to the discovery of an incredibly broad range of phenomena—from molecules in space to the discovery of the cosmic microwave background radiation that essentially verifies the hot big bang theory. Whole new classes of objects have been discovered at radio frequencies: giant molecular clouds (GMCs), cosmic jets and active galactic nuclei (AGNs), radio nova, and rapidly rotating neutron stars called pulsars. Radio astronomy has also greatly enhanced our views of known objects like the Milky Way's spiral arm structure, supernova remnants (SNRs), and suspected black holes by revealing previously unobserved structures. The National Radio Astronomy Observatory (NRAO), through a fortunate series of circumstances, is an institution that was conceived, developed, and constructed at an ideal time in the formative history of the science.

The NRAO has played a major role in the evolution of radio astronomy into big science. The body of published research and discovery events made by astronomers using NRAO instruments is enormous. The body of technical papers reporting developments in instrumentation at the NRAO is equally extensive. The many accomplishments attest to the integral role of the NRAO in the evolution of radio astronomy. Key accomplishments include a precise experimental verification of general relativity at radio frequencies and the discovery of radio nova with the Green Bank interferometer, which represents the discovery of a whole new class of celestial objects. The papers on the discovery of giant molecular clouds, interstellar masers, and the beginnings of molecular radio astronomy also focus on research performed at the NRAO. The NRAO was conceived and developed as a national user institution for radio astronomers from the United States and all over the world. The amount of important scientific research occuring at the NRAO is a testimony to the soundness of the initial concept of a national user institution for research in radio astronomy.

The intent of this project is to examine the sequence of events that led to the establishment of the NRAO, the development and construction phases, the contributions and discovery events, and to relate the significance of these events to the world research community and to the evolution of the sciences of radio astronomy and cosmology. A report of the technical and intellectual development alone does not represent a holistic approach to the history of a science or even a scientific institution. A more accurate picture must include an investigation of the social history and human dimension as well. The general social structure that researchers work within provides much information about attitudes and processes in scientific research. The pervasive teamwork structure of the user institution is an important aspect of the evolution of science at the NRAO and therefore requires investigation. A writer of the historical sociology of science acknowledges that the human element has much to offer to the history of science and technology. Interaction with larger social structures, the human interaction, and exchange of information are integral components in the evolution of a scientific idea. The human dimension entails the politics, day-to-day research, ideals, interpretations, occasional mistakes, sometime happy accidents and other human interest stories underlying the scientific process. A certain "humanization" of science is perpetuated by the inclusion of social history.

The establishment of the NRAO represents a unique way of doing modern science via user institutions. The NRAO represents the establishment of a national laboratory that was the first international user institution for re-search in astronomy. The concept of a user institution is one in which telescope time and services are awarded based purely on the scientific merit of the investigation. The use of the NRAO's instruments is open to all scientists, regardless of university affiliation or even nationality. Historically, scientists from all over the world have used the NRAO telescopes. The concept of the user institution for the pursuit of science is a trademark of postwar science. The vision and eventual establishment of the NRAO as a user institution and the subsequent implications for the evolution of radio astronomy warrant investigation.

The accomplishments of the NRAO are considerable. The number of the resulting scientific and technical papers is copious. It is therefore necessary to apply an interpretive selection of events and developments in the history of the NRAO. The sequence of events leading to the establishment of the NRAO is examined along with an overview of the protohistory and developments of the early pre-NRAO years of the science to place the present study in proper historical context. The major focus, however, is to investigate the evolution of instrumentation at the NRAO and the relation of these instruments to scientific discovery in radio astronomy. Investigations in radio astronomy have radically altered our view of the universe, its evolution and its contents. Researchers using the NRAO instruments have provided us with a dramatic view of a previously invisible universe.

Chapter 1

The Beginnings of Radio Astronomy

Looking at the stars always makes me dream,
as simply as I dream over the black dots
representing towns and villages on a map.
Why, I ask myself, shouldn't the shining dots
of the sky be as accessible as the black dots
on the map of France?

> Vincent Van Gogh
> 1889

Since prehistory humans have gazed into the sky and pondered the nature of the universe. Astronomers had been confined to the study of the visible light emitted by stars, which comprises only a minute portion of the electromagnetic spectrum of radiation naturally emitted by objects in the universe. The electromagnetic spectrum consists of radiation which has wavelengths one thousand millionth as long as light in the case of high frequency gamma rays to a hundred million times longer for low frequency radio waves. Only a limited knowledge could be achieved with the vast majority of the electromagnetic spectrum effectively invisible. An analogy astronomers frequently use is that astronomy prior to the science of radio astronomy was an effort to understand the symphony of the universe by only being able to hear the piano's middle C and the two adjacent notes.

Many objects that are completely invisible to optical astronomers emit immense amounts of energy at radio frequencies, yet were completely unknown until the twentieth century. In 1932, the story began to change as radio waves from outer space were accidentally discovered. The advent of radio astronomy not only produced many exciting discoveries, but opened the door to reveal the cosmic radiation over the entire electromagnetic spectrum. X-ray astronomy, infrared, ultraviolet, and gamma ray astronomy evolved as an indirect consequence of the accidental discovery of radio astronomy. Figure 1.1 shows the electromagnetic spectrum, the energy of each type of radiation, typical cosmic emitters and their temperatures, as well as atmospheric penetration. The relationships between energy, frequency, wavelength, and temperature are given.

Whole new classes of objects have been discovered at ra-

dio frequencies. Detailed views of areas of the universe previously completely invisible, such as the nucleus of the Milky Way, have been provided by radio frequency observations. Information has been revealed that gives insight into the early stages of evolution of the universe. Objects too far away to be seen optically have been discovered, thus extending the size of the known universe. The cosmological implications of the discoveries of radio astronomy are enormous. A well-ordered universe where change rarely occurred has been replaced by a universe of unimaginable beauty and violence. The story of radio astronomy is truly the story of the invisible universe revealed.

The protohistory of radio astronomy can be traced to Heinrich Hertz and the first production of radio waves in 1886. Hertz serendipitously verified J. C. Maxwell's theory of electromagnetic radiation during an experiment he performed with high frequency oscillating currents. Maxwell had unified electricity and magnetism and explained the physical nature of light in the theory of the electromagnetic field. Hertz was the first to generate and detect the Maxwellian-type electromagnetic pulses. In 1886 Hertz showed that electromagnetic waves generated by an oscillating electric current travel at the speed of light, and thus made an association of these waves with light. Hertz invented a generator of high frequency electrical oscillations and a simple detector for these high frequency oscillations. With these devices Hertz generated and detected meter-length electromagnetic radiation and thereby experimentally proved Maxwell's theory. He later went on to show that these waves exhibited the same physical properties as light: reflection, refraction, interference, and polarization.[1] The known electromagnetic spectrum of radiation had been enormously extended.

The scientific advances made by work with the Hertzian waves turned into a technological revolution when Guglielmo Marconi and others realized the implications of these Hertzian waves for the transmission of information across great distances. Marconi improved Maxwell's generator and developed the "aerial" or "antenna" to replace the dipole. In 1901 Marconi established the first wireless communication across the Atlantic Ocean.[2] A revolution in communications technology had begun.

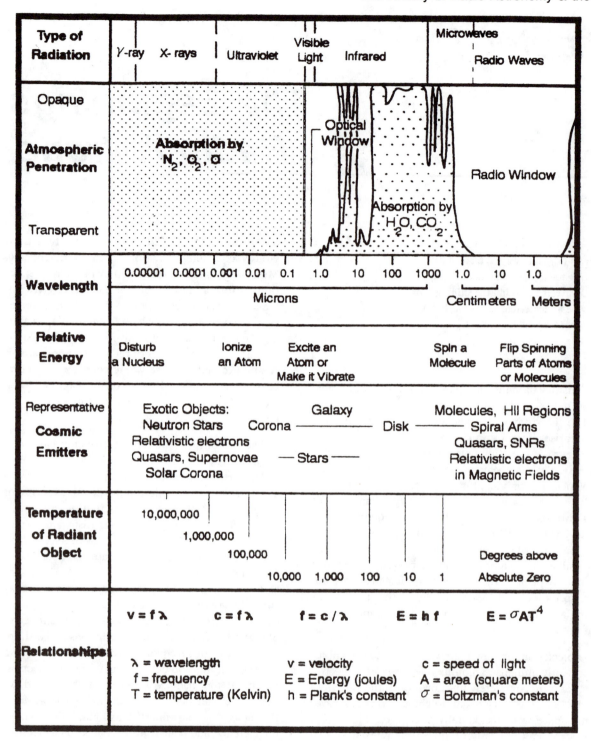

Figure 1.1   The electromagnetic spectrum of naturally emitted radiation.

Several attempts to detect Hertzian waves from the sun were made as the technology developed. Thomas A. Edison provided the first known design for such an attempt in 1890. Edison and his associate A. E. Kennedy proposed the construction of an antenna made of a cable draped on poles around a mass of iron ore to detect solar Hertzian waves. Thomas Edison very nearly became the father of radio astronomy. The idea, however, never materialized. Other unsuccessful attempts were made during the early 1900s in England by Oliver Lodge and by J. Wisling and J. Scheiner. An attempt was made in 1901 with a 175-meter antenna by C. Nordmann in the French Alps. Although the antenna in retrospect appears suitable, Nordmann was observing at a time of sunspot minimum and failed to detect a signal.[3]

Karl Guthe Jansky and the Serendipitous Origin of Radio Astronomy

The revolution caused by radio astronomy did not actually begin until 1932, when a serendipitous discovery occurred—cosmic radio waves were detected by Karl Guthe Jansky. Jansky was working for Bell Laboratories where many discoveries in radio astronomy have since taken place. By 1927 Bell Labs had introduced the first transatlantic radiotelephone. The system operated at the extremely low radio frequency of 60 kHz which corresponds to a long wavelength of 5 kilometers. Figure 1.1 gives the relationship used to perform the conversion. By 1929 Bell Labs had converted to shorter waves in the 10 to 20 MHz range. Electrical disturbances created significant noise at these frequencies, and scientists at Bell Labs were very interested in reducing the consequent noise levels during telephone conversations. Jansky was assigned to the Holmdel, New Jersey, site in 1929 to investigate the source of this "unwanted" noise. Jansky was, of course, interested in the signal-to-noise ratio of long distance signals. To investigate the origin of the noise, Jansky constructed a rotatable, Bruce array-type antenna (Figure 1.2) and a simple superheterodyne receiver (Figure 1.3) which operated at 20.5 MHz (corresponding to a wavelength of 14.6 meters). Little was known about long distance propagation effects at these frequencies. The properties of the antenna and associated hardware were not well understood at these shortwave frequencies. The instrument, however, had four major components essential to modern radio telescopes: (1) a directional antenna, which was prob-

Figure 1.2 Karl Guthe Jansky and the Jansky antenna. The original antenna was constructed at Bell Labs, Holmdel, New Jersey. A replica was constructed in the original workshop and relocated at Green Bank. *Courtesy of AT&T Archives.*

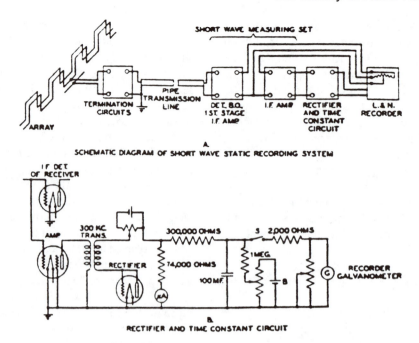

Figure 1.3 Schematic diagram of Jansky's antenna and receiver. A detailed schematic diagram of the rectifier and time constant circuit is shown. These schematic diagrams were published in Jansky's first paper published in the Institute of Radio Engineers in 1932. © *IRE, now the Institute of Electronics and Electrical Engineering, Inc.*

ably the largest antenna in existence at the time; (2) a receiver that was as quiet as the state of the art permitted in which the noise level was limited only by the electron noise of the vacuum tubes; (3) a receiver that was responsive to a wide bandwidth of wavelengths which was far wider than in conventional receivers of the period; and (4) an averaging arrangement which had a long time-constant circuit to smooth out the readout on the recorder chart.[4] Jansky had inadvertently provided the basic technology for the tools of a new science.

Jansky was making regular observations by 1930 and discovered that the "unwanted" noise was due to three sources: local thunderstorms, distant thunderstorms, and "a steady noise of unknown origin."[5] Eventually Jansky observed that the third type of data was directional. It apparently came from a fixed position in the sky. He also observed that the signal rose in the sky 4 minutes earlier each day. All that is left of the 3 years of Jansky's original data is shown in Figure 1.4.[6]

Jansky had the insight to realize that the signals were not of terrestrial origin. Objects in the sky rise 4 minutes earlier every day because the earth's position (revolution) changes relative to the stars. He later pinpointed the origin of these radio waves to an area in the constellation Sagittarius which is the center of the Milky Way. Jansky had discovered the radiation emitted by the galactic nucleus. This was the first detection of nonoptical radiation from outer space. A reconstruction of Jansky's data shows structural features of the Milky Way. Figure 1.5 shows a map of the brightness distribution of radio waves derived from Jansky's drift scans.[7]

Structural details of the Milky Way are shown including a concentration of radiation toward the galactic plane and the galactic nucleus.

NBC (National Broadcasting Company) Blue Network of New York City broadcasted an interview with Jansky on May 15, 1933, regarding the discovery of "radio impulses from among the stars." A direct hookup to Holmdel allowed the public all across the nation to listen to an audio version of the hiss-like impulses.[8] The American public as well as the scientific community was almost immediately aware of the discovery of radio waves from space.

Jansky published his findings in the *Proceedings of the Institute of Radio Engineers* (October 1932) under the unauspicious title, "Directional Studies of Atmospherics at High Frequencies."[9] Two more articles followed in the IRE in 1933 and 1935 in which he interpreted the star noise as coming from the entire Milky Way. Jansky was not sure if the radiation was coming from individual stars in the antenna's beam or the interstellar medium.[10] The quandry lay in the fact that he could not detect radio waves from the sun. The basic characteristics of the emission and the explanation of the emitters would not be understood until the 1950s. A basic question of the nature of the radiation, whether it was thermal or nonthermal in origin, remained unanswered.

It was a matter of chance that Jansky had chosen a frequency at which the galactic center emits copious amounts of radiation and at which the earth's atmosphere is transparent. It was also a matter of chance that he was working at a time of sunspot minimum (which occurs every 11

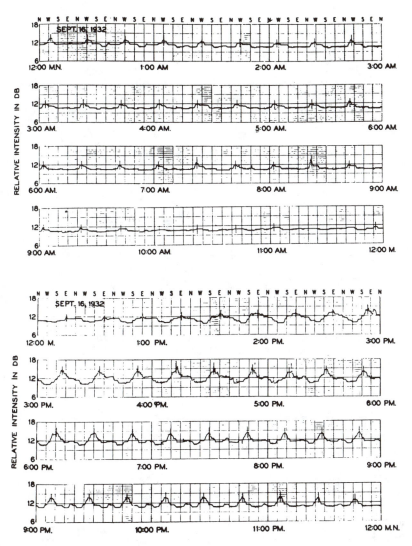

Figure 1.4 Jansky's original data. A signal of the Milky Way, the first discovery of radio waves from space, is observed as a peak as the antenna sweeps around the sky. The 1932 data is all that remains of the three years of Jansky's research on radio waves at Bell Laboratories. The data was published in Jansky's first paper published in the Institute of Radio Engineers in 1932. © *IRE, now the Institute of Electronics and Electrical Engineering, Inc.*

years). At sunspot maximum, the ionosphere would have blocked all radio waves from space at 20 MHz and signals from the Milky Way would have therefore been undetectable. Grote Reber, an early pioneer of radio astronomy, eloquently described the often chance nature of modern science (in relation to Jansky's discovery): "Major discoveries are often the product of being in the right place, at the right time, with the right equipment, asking not necessarily the right questions."[11]

The full significance of this landmark discovery was not realized by the scientific community. Ironically few astronomers appreciated the significance of Jansky's discovery. Astronomers did not have the training or skills to fully understand or appreciate the radio frequency observations. Perhaps those that understood the significance were not in a position to act. There was, however, at least one attempt to experimentally verify Jansky's discovery. Potapenko and

Folland, physicists from Caltech, confirmed Jansky's results 2 years later and observed the galactic nucleus from a field station in the Mojave Desert. A Caltech radio antenna (Figure 1.6) was planned by the group in 1936 but was never constructed.[12] Although there was some pre-World War II interest in radio astronomy, expressed by American astronomers, Whipple, Struve, Greenstein, Baade, and Minkowski,[13] there were no other immediate experimental efforts to pursue astronomical observations at radio frequencies. Jansky's discovery was not even published in an astronomical journal. The pursuit of radio astronomy was interrupted by the World War II and the recruitment of scientists for the wartime development of radar.

Karl Jansky accidentally opened a new window on the universe in detecting radiation from space. His identification of the radio emission with the structure of the Milky Way galaxy was a discovery that would eventually revolu-

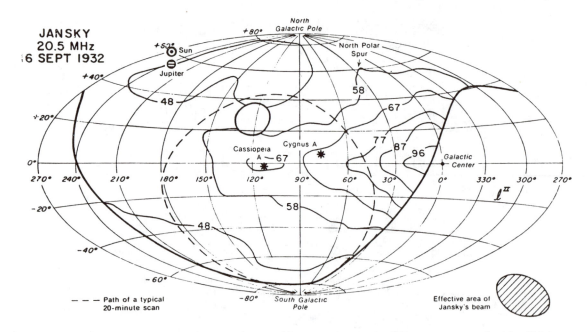

Figure 1.5   Radio brightness temperature (in kelvins) derived (Sullivan, 1978) from the drift scans at a wavelength of 14.6 meters taken by Jansky in 1935. The approximate beam size is indicated. Structural features of the Milky Way are evident such as the concentration of radiation toward the galactic plane and the galactic nucleus. *Courtesy W. T. Sullivan.*

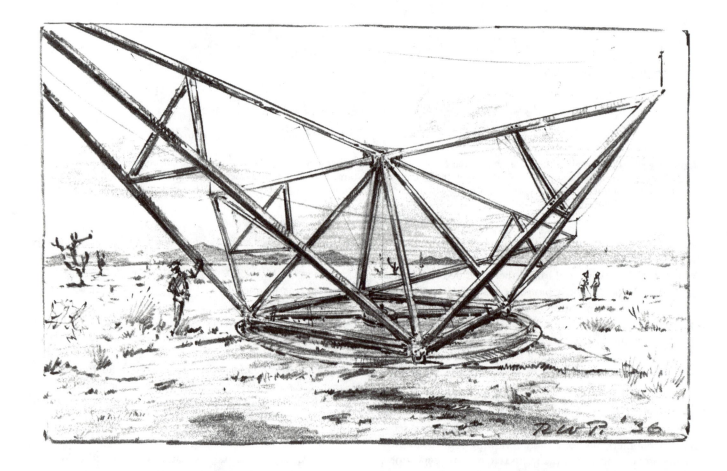

Figure 1.6   Proposed Caltech radio telescope. This 1936 sketch is by Russell Porter who aided in the design. The instrument was never constructed. *Courtesy the California Institute of Technology.*

Figure 1.7   Grote Reber and a model of the 31-foot antenna. *Photo by H. Richard Bamman, initially published in the* Wheaton Daily Journal.

tionize astronomy. The revolution, however, was slow to take place.

Grote Reber and the Early Years of Radio Astronomy

Very little interest developed over Jansky's discovery, particularly in the astronomy and physics community. Jansky proposed building a larger and improved antenna to investigate the "star noise" but the proposal was apparently lost in the bureaucracy at Bell Labs.[14] A 25-year-old engineer, Grote Reber, however, had read Jansky's findings published in the IRE and realized the great significance of the discovery. Reber (Figure 1.7) was determined to pursue the research, convinced of its tremendous significance. He went to "all the top research centers for astronomy in America" to enlist support to continue the research, but was turned down. Reber could not find a research center willing to pursue the research,[15] but evidently was not discouraged. With a great sense of commitment, he resolved to continue the research of this "star noise" himself. Reber decided to construct a radio telescope in his backyard in Wheaton, Illinois.

Using his own funds and a radical new approach to the design, he began construction in 1937. Reber decided on a parabolic design, with the feed horn and single dipole antennas suspended at the focus. This innovation was the first time a paraboloid was used to concentrate the weak incoming radio signals. Reber constructed the 31-foot (9.45-meter) paraboloid out of parts that he purchased, borrowed, or acquired from colleagues (see Figure 1.8). He later admitted that the size was determined not by a multiple of the wavelength observed (as astronomers had since theorized), but because "in building the supporting superstructure the longest lumber available at the hardware stores in Wheaton was twenty feet, dictating a maximum diameter of thirty-one feet for the dish."[16] Reber operated the telescope through WWII and made many significant discoveries. He made the first radio maps of the sky, discovered radio waves emitted by the sun, and made rather detailed radio maps of the Milky Way (see Figure 1.9). His "homemade" radio telescope was the most sensitive, innovative, and productive radio telescope in operation until the end of WWII. In constructing the first true radio telescope, Reber began the deliberate science of radio astronomy.[17] Grote Reber had singlehandedly confirmed Jansky's findings and made further significant discoveries that supplied the impetus to establish radio astronomy as a science.

The design of Reber's telescope was visionary. The parabolic, central focus dish was placed on a meridian-transit

mount. A meridian-transit telescope is pointed in declination and scans the sky as the earth rotates. Pointing in declination was achieved using the differential gear from a Ford Model T truck (see Figure 1.10). The reflector surface was made of galvanized sheet metal screwed to 72 wooden rafters cut to a parabolic shape. Figure 1.11 shows the wooden superstructure and a drain used to avoid collecting rainwater. An overall surface tolerance of 0.5 centimeter was maintained. The radio waves were focused at the central feed where a dipole converted the radiation to an electric current. Reber constructed the first receiver to work at 9 centimeters, the shortest wavelength practical which was dictated by the surface precision and which would provide high resolution. The construction of a receiver for this short wavelength required obtaining custom-made vacuum tubes and a great deal of experimental work testing the components. Reber chose a centimeter wavelength because he reasoned that if the radiation were thermal, this wavelength would be very intense according to Planck's law of blackbody radiation.

Reber's first observations were systematic surveys of the galaxy, the sun, planets, and some of the bright stars. Thermal radiation at this wavelength would be 26,000 times stronger than at Jansky's wavelength and would be easily detectable.[18] He failed to detect any signal. Even this investigation, however, was a success in the sense that Reber concluded that the radio wave radiation was nonthermal. This

Figure 1.8   The first true radio telescope, Grote Reber's 31-foot (9.45-meter) paraboloid. The telescope was constructed single-handedly by Grote Reber in his backyard in Wheaton, Illinois, in 1937. The original Reber telescope is displayed at NRAO, Green Bank, West Virginia. *Photo by B. Malphrus.*

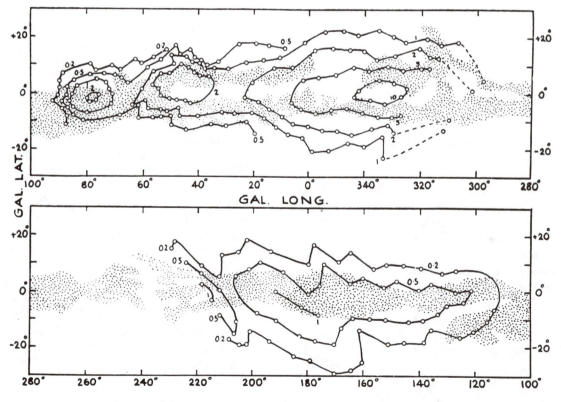

Figure 1.9   Radio brightness distribution map of the Milky Way at 1.9 meters wavelength (Reber, 1944) superimposed on the optical Milky Way in Norton's Star Atlas (Northcott and Williamson, 1948).

Figure 1.10 Declination drive of Reber's telescope. The telescope was originally a transit design, with pointing in declination achieved by using the differential gear from a Ford Model T Truck. *Photo by B. Malphrus.*

Figure 1.11 Superstructure of the Reber telescope. The reflector surface of galvanized sheet metal was attached to 72 wooden rafters cut to a parabolic shape to form the superstructure. The rainwater drain may be seen. *Photo by B. Malphrus.*

later proved to be a landmark conclusion in the science of radio astronomy.

Reber converted to a wavelength of 33 centimeters after concluding that the radio radiation was nonthermal. No signal was detected at this wavelength even with an improved receiver. Reber then tried an even longer wavelength of 1.87 meters. For this wavelength he designed a cylindrical cavity waveguide fabricated from an aluminum sheet. The dipole antenna was placed at the "cavity resonator" at the dish's focus. By 1939 Reber had detected radio waves from the galaxy at 1.87 meters wavelength.[19] Reber was convinced that the radiation was nonthermal because the intensity at this wavelength was far less than predicted for thermal radiation based on Jansky's original data. He suggested that the production mechanism of these radio waves could be explained as "free-free" radiation from electrons.[20] The implications of this theory of nonthermal radiation are far-reaching in the science of radio astronomy.

Reber published his initial results in the *Proceedings of the Institute of Radio Engineers* in February 1940.[21] Reber broke the barrier with the astronomical community by publishing a paper in the *Astrophysical Journal* in June 1940.[22] The editor of the journal found it difficult to publish because of the radio terminology that was unfamiliar to astronomers. Ironically, the editor, Otto Struve, would become the first director of NRAO Green Bank in 1959.[23]

For nearly a decade Reber was the world's only active radio astronomer. He independently conducted an extensive sky survey. While working as an engineer by day, he kept a lonely vigil of the heavens at night. The results of this survey were published in preliminary form in 1941[24] and a more extensive account in 1944 in the *Astrophysical Journal*.[25] These two papers along with Reber's 1940 paper represent the first publications of research in radio astronomy

in an astronomical journal. Radio astronomy had begun the long road to acceptance as valid science by the astronomical community.

The radio maps of the Milky Way that Reber produced during his sky survey revealed some very interesting features. The waveguide allowed Reber to "tune" his telescope to a much shorter wavelength than Jansky's instrument, which provides higher resolution (resolution is given by the relationship: $\alpha \sim \lambda/d$, where d is the diameter of the aperture times a factor of 1.22). The resolution angle of Reber's telescope was approximately 12° as compared to 30° for Jansky's instrument. A detailed picture of the sky could be provided with this degree of resolution. A detailed structure of the galaxy was revealed. The strongest radiation came from the galactic center and a concentration toward the galactic plane of the radiation was well defined. "Hot spots" of radiation in the constellations of Cygnus and Cassiopeia were also discovered. These sources were later found to correspond to extremely distant energetic objects. Reber also tried unsuccessfully to detect radiation from the planets, stars, and nebulae. In 1943, he detected radio waves from the sun and published the findings in 1944.[26] Transient surges of solar radio waves had been accidentally detected in 1942 by J. S. Hey as he investigated noise that jammed British radars during the Blitz on London. Because of WWII, Hey's discovery was kept secret and was not published until after the war. Solar radio waves were also discovered independently in 1942 by a third researcher, George Southworth of Bell Laboratories, who was operating at a wavelength of 3 centimeters.[27]

Reber's ambitious project began in 1937 at a time when the threat of WWII pervaded public thought. The public reaction over the construction of Reber's dish in his backyard in Wheaton, a suburb of Chicago, was that of great skepticism. After observations at Wheaton ended in 1947, the radio telescope was moved to the United States Bureau of Standards. In 1960, the telescope was relocated to the NRAO Green Bank, West Virginia to join other historical radio telescopes on the site.

Radio Astronomy During World War II

After Jansky and Reber the young science somewhat stagnated during WWII, although the war would eventually become a major catalyst for the advancement of radio astronomy. During the war years, the international scientific community could not interact as much as during peacetime. J. S. Hey's 1942 discovery of the solar radiation, for example, was not circulated in the scientific community due to wartime security. The exchange of information was at a minimum with one notable exception. A copy of Grote Reber's *Astrophysical Journal* paper eventually reached the Leiden Observatory in Nazi-occupied Holland. A famous Dutch astronomer, Jan Oort (who predicted the existence of the Oort Comet Cloud), upon reading Reber's research, became interested in using radio techniques to observe the interstellar medium. He suggested the possibility to a student of his, Hendrick van de Hulst, who then suggested three possibilities: observing thermal emission, recombination lines, and the 21-centimeter spectral line. The significant implication was that van de Hulst predicted that "line" radiation from atoms and molecules in space may exist.[28] He suggested searching for 21-centimeter radiation from neutral hydrogen. Van de Hulst's theoretical work on neutral hydrogen revealed a very powerful tool for probing the depths of the galaxy as well as a method for searching for atoms and molecules in space.

Although developments were impeded due to the preoccupation with the war effort, technological developments that would dramatically improve instrumentation and techniques in radio astronomy occurred. One important technological development was radar. The more technologically advanced civilization becomes and the more communications improve, the greater the likelihood of simultaneous invention. In the wartime atmosphere of WWII, radar was simultaneously invented in Germany, Holland, Britain, and the United States.[29] It was a technological development whose time had come. The technologies involved in radar systems are essentially very similar to those involved in radio astronomy instrumentation.

The English, perhaps out of necessity, produced the most highly developed and extensive radar technology. The constant fear of air raids by the Luftwaffe during the Blitz and the threat of German V1 and V2 missiles necessitated a distant early warning system. The system was developed by the Army Operational Research Group (AORG). In the course of development and use of the radar system important astronomical discoveries took place. J. S. Hey and his colleagues at AORG made three important astronomical discoveries: the discovery of the sun at radio frequencies, the

detection of radar echoes from meteor trails, and detection of the first discrete source of radio emission in the constellation Cygnus.[30]

In February 1942, severe interference jammed the AORG distant early warning system. Hey observed from the reports of widely separated radar stations that the source of interference was coming from a direction within a few degrees of the sun. Hey concluded that the interference was associated with a large sunspot that happened to be crossing the disk of the sun. The second discovery occurred during the latter part of the war as the distant early warning system was used to detect the arrival of German V2 missiles. The system was very effective at detecting the V2 missiles, but transient echoes occasionally occurred that led to false alarms. Investigation of these echoes led Hey to the discovery of meteor trails at radio frequencies. Hey's third discovery, that of the first discrete radio source, was more directly related to the work of his American predecessors. Reber had observed two "hot spots" but it was not until the work of Hey at AORG that one of these sources, the source in Cygnus, was pinpointed to a small intense source as opposed to general areas of the sky. Cygnus was therefore the first discrete source of radio waves observed other than the sun.[31] The "fluctuations" observed by Hey in the source that allowed him to pinpoint it to a small area were later found to be associated with fluctuations in the ionosphere and not intrinsically with the source.

Emergence of Radio Astronomy in Great Britain

After the war, scientists in Britain and elsewhere began to devote time to the purely scientific aspects of the radio technology. Physicists and engineers highly trained in radar and radio techniques began to conduct peacetime research. Radio astronomy groups formed in many countries including Britain, Australia, the Netherlands, and, to a lesser extent, the United States. Two distinct groups formed in Britain that would do much to transform radio astronomy into an established science.

Many British physicists and radar engineers returned to or were recruited by universities. The Department of Physics at Manchester University under P. M. S. Blackett and the Radio Physics Group at the Cavendish Laboratory at Cambridge under J. A. Ratcliffe were the primary recruiters of these wartime scientists. Major centers at Jodrell Bank and Cambridge University eventually evolved to pursue research in radio astronomy. Research at the two centers distinctly differed from the beginning, both in original intent and instrumentation. This early differentiation in instrumentation, the development of interferometric arrays at Cambridge as opposed to the use of single-dish paraboloids at Jodrell Bank, represents a long-standing differentiation in instrumentation approaches in radio astronomy. The nature of the problem investigated dictates the best approach

Figure 1.12   The Kootwyk telescope, Netherlands. Many German Wurzburg radar reflectors were adapted to radio astronomy after WWII. The 7.5-meter Kootwyk telescope is a classic example. *Courtesy NRAO/AUI.*

to instrumentation, or vise versa, the instrumentation dictates the nature of the problem investigated.

After the war, research centers around the world acquired surplus army radar equipment from their respective governments and German radar equipment captured during the war. A major example is the appearance of Wurzburg reflectors in Britain, Holland, Australia, and the United States. The Wurzburg reflector pictured in Figure 1.12 is the 7.5-meter Dutch Kootwyk telescope. Similar instruments were used at Cambridge to provide accurate positions of radio sources that eventually allowed the first optical identifications of radio sources to be made.

Early Years of Radio Astronomy at Jodrell Bank

The observatory at Jodrell Bank emerged as a result of the wartime work of Bernard Lovell and P. M. S. Blackett. They had theorized that it might be possible to detect cosmic ray showers using radar techniques. Lovell acquired surplus army equipment and installed these instruments at a site at Jodrell Bank with the help of Hey and Parsons from AORG.[32] Consequently cosmic ray work dominated the early research.

The beginnings of Jodrell Bank in mid-December 1945 were indeed humble. An army surplus 4.2-meter radar sys-

tem and a few Yagi aerials were the only resources. By early 1946 the researchers at Jodrell Bank were beginning to realize that the radar echoes they were investigating were coming from meteors and not from cosmic ray showers. Much successful work was done on meteors. The direction, velocity, and orbit of meteors could be detected with the instrumentation (see Figure 1.13). The scope of research at Jodrell Bank began to change. Under Lovell, the research group grew, new lines of inquiry were pursued and new instruments were constructed for new investigations. Around 1951, the Jodrell Bank group began to investigate "radio stars," which is more closely aligned with current pursuits in radio astronomy.[33] The Cambridge group, on the other hand, initially began to investigate radio emissions from the sun and other discrete cosmic sources.

The group at Jodrell Bank did much to advance the technology involved in radio astronomy instrumentation by building several large telescopes, beginning with a 218-foot (66.5-meter) fixed paraboloid (seen in Figure 1.14).[34] By 1955 several discrete sources had been identified with their optical counterparts. In 1955 a small catalog of radio sources was produced by the group at Jodrell Bank. None of the sources was identified with normal stars. It is now realized that this phenomenon is due to the fact that normal stars typically emit only about one-millionth of their total energy in the radio spectrum. Thus, the brightest radio

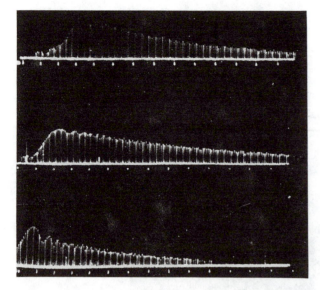

Figure 1.13 The radio echo from a meteor trail. The time markers on the X-axis are in intervals of 1/50th second. The amplitude variations are Fresnel diffraction patterns from which the velocity of the meteor may be determined.

Figure 1.14 The 218-foot (66.5-meter) fixed paraboloid at Jodrell Bank circa 1949. The telescope was constructed of wires stretched from supporting posts into a shallow paraboloid. *Photo courtesy R. Hanbury Brown.*

sources correspond to extremely energetic objects such as quasars, radio galaxies, and unusual stars at the endpoint of stellar evolution.

Radio astronomy at Jodrell Bank by 1951 had become well established. The success of the 218-foot telescope represented the advantage of large paraboloids in radio observations, the advantages of a pencil-beam view of the sky. The ability to see sources of large angular size with a pencil beam is ideal for sky surveys and mapping. The angular size of an extended object is the apparent size of that object in the sky. A limitation of interferometric techniques is the inability to see sources of large angular extent. The success of the 218-foot caused the subsequent planning and construction of large aperture telescopes at Jodrell Bank.

The 218-foot fixed shallow paraboloid was constructed of wire stretched across the ground from supporting posts. A 126-foot mast suspended the feed horn at the central focus. At a wavelength of 1.87 meters, a pencil beam of 2° was produced that could provide detail in radio sources. Some pointing ability was provided by tilting the central mast by adjusting 18 supporting guy wires. A transit telescope that could scan the sky as the earth rotated was cleverly, if laboriously, operated. A swath of sky roughly 30° wide could be observed by tilting the central mast.[35] An instrument that worked very well for surveys was thereby produced.

An early 1950 survey by Hanbury Brown and Cyril Hazard resulted in the detection of the Andromeda galaxy, the first detection of a "normal" galaxy which demonstrated the existence of two types of galaxies: weak radio emitters and "radio" galaxies which are intense radio emitters. The radio emission from the Andromeda galaxy was shown to be very similar to that of the Milky Way. The first measurements of brightness distribution of a radio source resulted from this investigation (see Figure 1.15).[36] This technique evolved into an extremely powerful tool for examining the structure of extragalactic radio sources.

The Brown and Hazard survey investigated the most powerful "radio stars." By the end of 1952 a survey of 23 sources within the field of view had been cataloged. A strong concentration toward the plane of the Milky Way was observed from the data. Observations of the discrete source in Cygnus revealed that the fluctuations previously observed were ionospheric in origin. The intensity of the source, averaged over time, was the same whether the source was fluctuating or not. Measurements derived from widely spaced receivers later confirmed that the source of the fluctuations was ionospheric.[37] A record of the transit of Cygnus A (Figure 1.16) also reveals an interesting feature, multiple deflections, now realized to be associated with a background object, Cygnus-X, a suspected black hole. A comparison of this data with data taken in 1986 with the NRAO modernized 40-foot radio telescope, shown in Figure 1.17, attests to the sensitivity and accuracy of the early Jodrell Bank 218-foot paraboloid.[38]

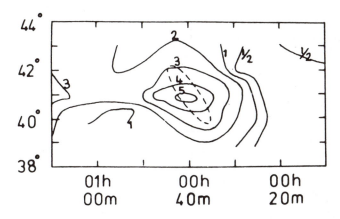

Figure 1.15 Radio brightness distribution of the Andromeda galaxy (Brown and Hazard, 1951). This map of Andromeda at 1.9 meters wavelength represents the first measurement of the brightness distribution of any radio source. *Courtesy R. Hanbury Brown.*

The success of the 218-foot paraboloid motivated an ambitious program under Lovell's direction at Jodrell Bank to construct a series of large paraboloids beginning with an enormous 250-foot (76.25-meter) steerable paraboloid. Construction of the 250-foot Mark I steerable telescope began in 1953 and was completed in 1957 (see Figure 1.18). For many years it was the world's largest and most sensitive telescope. During construction the surface was redesigned from wire mesh to solid aperture because of the 1951 detection of the 21-centimeter hydrogen line by Ewen and Pur-

cell in the United States. A solid surface would focus the radiation better and provide higher aperture efficiency at the shorter wavelength of 21 centimeters. The huge paraboloid was constructed of a steel superstructure covered with 7,000 sections of sheet steel reflecting surface weighing about 700 tons and was alt-azimuth mounted for automatic tracking of radio sources. A 280-foot diameter stabilizing wheel secured to the superstructure provides for motion in altitude (visible in Figure 1.18). Elevation bearings from a battleship gun turret (the *Royal Sovereign*) allow the whole dish to be turned completely upside-down to easily reach the feed point from the ground. The superstructure is supported by two 170-foot (51.9-meter) towers. The entire assembly is mounted on a circular railway track to turn in azimuth. The primary aerial feed is mounted at the focus on a single steed tower rising 62.5 feet (19 meters) from the apex of the paraboloid. The cross-section of the tower is very small near the focus to avoid obscuration of the primary feed.[39] The engineering plans and designs were contracted to H. C. Husband of London, a famous bridge builder. Husband was the engineer that designed and built the bridge over the river Kwai for the film and planned and oversaw its subsequent demolition.[40] Husband & Company became immediately noted as builders of large radio telescopes with the success of the Mark I. Efforts in the construction of radio astronomy telescopes of Husband & Company included the submission of detailed design plans for the proposed NRAO 140-foot (42.7-meter) telescope which were not

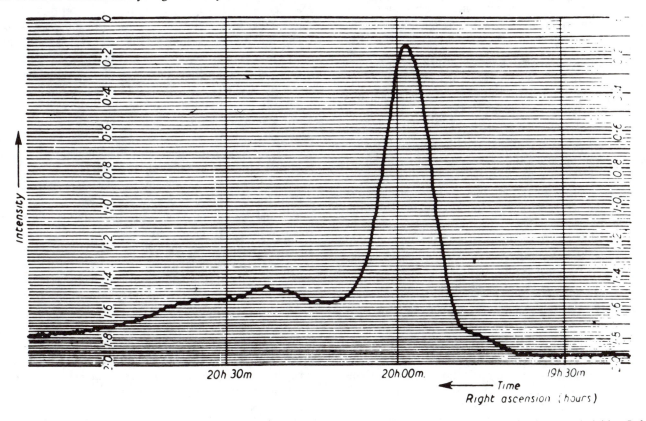

Figure 1.16 Transit observation of Cygnus A (Brown and Hazard, 1949). Cygnus A is observed using the 218-foot paraboloid at Jodrell Bank at a wavelength of 1.89 meters. *Courtesy R. Hanbury Brown.*

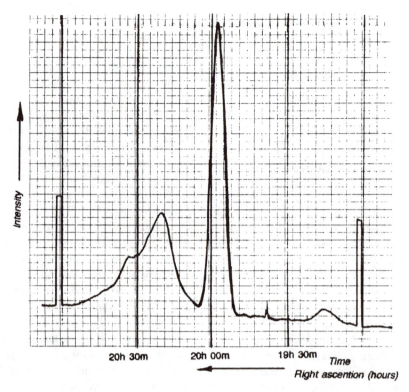

Figure 1.17 Transit observation of Cygnus A (B. Malphrus, 1989). Cygnus A is observed with the refurbished NRAO 40-foot (12.2-meter) radio telescope at 21 centimeters wavelength. A comparison of this data with the Brown-Hazard transit data illustrates the accuracy and sensitivity of the 218-foot paraboloid at Jodrell Bank; all major structural features of the Cygnus A area are shown in the 1949 transit data.

used.[41] Husband of London represented the beginning of a trend in which bridge builders would design and construct major radio telescopes.

Shortly after its completion the scientific (and other) potentials of the great telescope were made obvious to the whole world. In October 1957 when the Russians launched the Sputnik satellite, the Mark I was the only telescope in the world large enough to initially detect it.[42] The Mark I tracked the booster rocket and satellite by radar. Figure 1.19 shows ionospheric fluctuations observed in the signal from Sputnik. This event was a great accomplishment for British radio astronomers and soundly established them as leaders in the new science. Astronomers around the world, particularly in the United States, began to plan the construction of large aperture telescopes. This would prove to be factor in the unfolding of events that led to the establishment of a National Radio Astronomy Observatory to construct a large aperture telescope in the United States.

The Mark I telescope, and the upgraded and resurfaced version redesignated Mark IA, subsequently has been an excellent tool for many aspects of radio astronomy research. Radar investigations of meteors, aurorae, the moon, and inner planets as well as radio investigations of the sun, the Milky Way galaxy, and discrete sources have provided valuable information for astronomers. Other instruments were envisioned by Lovell and his colleagues in the early 1960s. A 125 foot × 83 foot 4 inch (38.1 × 25.5 meter) oval parabo-

loid steerable telescope was constructed as a prototype for much larger instruments. The Mark II prototype was constructed and in operation by 1964. Britain and the United States were considering the construction of gargantuan aperture dishes at this time. Several studies proposed the construction of telescopes 1,000 feet (305 meters) in diameter, but it was discovered that paraboloids much larger than the Mark I were not feasible. Physical engineering limitations prohibit the construction of dishes larger than 100 meters as the cost soars for larger apertures. Plans for construction of larger telescopes in Britain—the 400-foot (122-meter) Mark V (Figure 1.20) and the 375-foot (114-meter) Mark VA at Jodrell Bank—were consequently abandoned.[43] Construction of the 600-foot (183-meter) U.S. Naval Sugar Grove instrument was also abandoned for similar financial and engineering reasons, even though the base with the azimuth track was already under construction. A physical limit to the feasible size of enormous aperture telescopes had essentially been realized by the mid-1960s.

As early as the late 1950s angular resolution had become a major problem of the radio astronomer. Angular resolution, the smallest angle of sky that a telescope can observe (and therefore the apparently smallest object), is a ratio of the wavelength used to the diameter of the aperture, or dish (as described by the equation provided earlier). Radio waves are extremely long (expressed in millimeters, centimeters, meters, and even kilometers) compared to light,

Figure 1.18 Two views of the 250-foot (76.25-meter) Mark I A radio telescope of the University of Manchester's Nuffield Laboratory at Jodrell Bank, England. *Photos by B. Malphrus.*

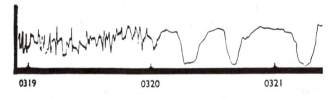

0319   0320   0321

Figure 1.19 Signal from Sputnik I taken at 40 MHz with the 250-foot Jodrell Bank radio telescope in 1957. Scintillations on the left are the result of ionospheric irregularities.

which is typically measured in nanometers (10^{-9} meters). Radio telescopes would therefore have to be enormous to have the same resolution as even a small optical telescope. As an example, for a radio telescope operating at a frequency corresponding to a wavelength of 21 centimeters to have the same resolving power as the 5-meter Hale telescope (which in practice is on the order of 1 arcsecond), the diameter of the reflecting dish would have to be 43 kilometers (26.7 miles). Sensitivity to sources of low flux density was also a problem, but one that large apertures helped to alleviate. In general, the larger the physical collecting area of the dish, the greater the sensitivity and the fainter the radio signals the radio telescope can detect. Although the problem of sensitivity to weak sources had been somewhat eased by large aperture instruments, only small gains in spatial resolution were achieved by these big telescopes.

Early Radio Astronomy at Cambridge and the Development of Radio Interferometry and Aperture Synthesis

An answer to the problem of angular resolution was provided by Martin Ryle at the Cavendish Laboratory of Cambridge University, the other major British group that gathered after World War II. A major development there was the development of radio frequency interferometry, a method of combining two or more radio telescopes together electronically to effectively produce the diameter of one large telescope. When the radio signals from two antennas separated by some baseline distance are combined, the wavefronts interfere, either constructively or destructively. The combined information produces interference fringe patterns (refer to Figure 1.21). The data collected by the pair inherently contains a time or phase delay due to the separational distance between the two antennas. A wavefront incident upon the antennas arrives at one before the other and it is this time delay ΔT, that represents the basic measurement. This delay is a function of the orientation of the baseline relative to

Figure 1.20 The proposed 400-foot (122-meter) Mark V telescope at Jodrell Bank. Large aperture telescopes such as this were planned in the mid-1960s before it was realized that the construction of apertures in excess of 100 meters is impractical. *Courtesy of Oxford University Press and Mott MacDonald.*

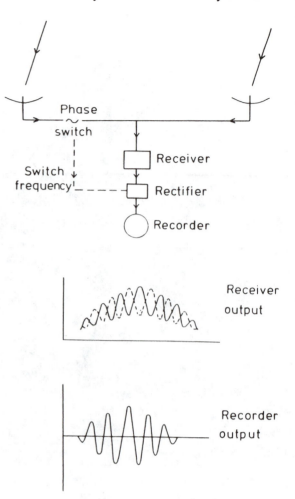

Figure 1.21 Basic schematic diagram for the phase-switched interferometer. The technique combines the signals from two telescopes to simulate a very large telescope. When the signals from two radio telescopes at a baseline distance are combined, the wavefronts interfere, producing interference fringe patterns.

the source as well as the length of the baseline. Because the precision of measurement of the delay is independent of the baseline length, the longer the baseline length the greater the precision of the determination of the orientation of the baseline relative to the source. The resolution of an interferometer is, therefore, set by the baseline distance between the antennas. In observing a source, an electronic delay is placed in the connection to one antenna to assure that the source is simultaneously observed by the two antennas. The concept is essentially an application of the principle of the Michelson optical interferometer to radio frequency observations. The technique provided a powerful tool for the observation of discrete radio sources. The phase-switched interferometer developed by Martin Ryle took the technique of interferometry a step further by phase-switching between elements and modulating the signal such that large fluctuations are removed, allowing for more accurate measurements of intensities, positions, and polarization of discrete sources.[44]

A variation of interferometry, aperture synthesis, was developed by Ryle in which the aperture of a very large dish is synthesized by a multiple dish interferometer which utilizes the rotation of the earth. In earth-rotation aperture synthesis, dishes placed on an east-west axis trace out ellip-

ses on the sky using the antenna beam as the earth rotates. Moving the elements traces out concentric ellipses, simulating a large aperture. Images synthesized by aperture synthesis techniques can simulate that of a telescope with a diameter equal to that of the distance between telescopes and provide tremendous resolution. Such an instrument is ideal for surveys of radio sources that were previously unresolved and for exploring detailed structure within galactic sources. Martin Ryle and Anthony Hewish were jointly awarded the 1974 Nobel Prize for Physics for their long-term contribution to the science of radio astronomy including new techniques of observation and the discovery of pulsars.

From the beginning, the group at Cambridge was interested in cosmic radio sources, the sun, "radio stars," and galactic objects. Martin Ryle began to study the solar radio radiation in 1945 with surplus German equipment. During the early 1950s, work on discrete sources of radio emission had replaced solar physics as the major direction in research. The techniques to investigate these sources were developed

Figure 1.22 Cambridge interferometer used in the 1958 4C survey at the Cavendish Laboratory, Cambridge, England. The interferometer is composed of four tiltable parabolic cylinder stretched wire reflectors. Dipoles are suspended by a wire stretched above the length of the cylinder. *Photo by B. Malphrus.*

Figure 1.23 Cambridge One-Mile radio telescope at the Cavendish Laboratory, Cambridge, England (one antenna of the three element interferometer is visible). Three 60-foot (18.3-meter) steerable parabolic reflectors were mounted on movable bases along an E-W line. The maximum spacing between the two farthest elements is one mile. *Photo by B. Malphrus.*

which provided increasingly higher angular resolution. Surveys of these sources of small angular extent were undertaken. The preliminary survey 1C (1st Cambridge) Survey of 1950 resulted in the detection of about 50 radio sources. In 1955 a larger interferometer had been constructed using four parabolic cylinder stretched-wire reflectors. The 2C and 3C surveys were produced with this instrument. By 1958 an aperture synthesis interferometer (consisting of two sets of parabolic cylinder reflectors, one fixed seen in Figure 1.22 and one movable) was deployed in the 4C survey, which produced over 5,000 sources.[45]

In 1962 Ryle's next telescope design was constructed. It was designed to provide detailed surveys of particular regions of sky with high resolution and sensitivity, map the brightness distribution of radio sources, and survey very weak sources. Three 60-foot (18.3-meter) steerable parabolic reflectors were mounted on movable bases along an E-W track.[46] The maximum spacing between the farthest elements is one mile (1.61 kilometers) which provides a resolution of 23 arcseconds at 21 centimeters. The system is therefore known as the Cambridge One-Mile radio telescope (Figure 1.23). The One-Mile telescope provided unprecedented detail in the brightness distribution of radio sources (Figure 1.24) particularly those of Cygnus A and Cassiopeia A,[47] which were the first real maps of discrete sources. The effect of increasing angular resolution by increasing the baseline of an aperture synthesis telescope is demonstrated using the One-Mile telescope in Figure 1.25 Other innovative instruments such as one with eight parabolic reflectors were constructed at Cambridge with carefully planned programs in mind.

The research at Cambridge initiated an inevitable marriage of radio astronomy with optical astronomy. The interferometers developed at Cambridge provided the best resolution yet achieved which allowed the most precise measurement of the positions of sources. One of the first identifications of a discrete radio source with its optical counterpart was achieved with these measurements. F. Graham Smith used an interferometer constructed of two W W II Wurzburg reflectors to obtain roughly one minute of arc accuracy for the position of the two strongest radio sources in the sky. The new precision of the measurement of these positions allowed W. Baade and R. Minkowski of Mount Palomar Observatory to make the first optical identification of radio sources in 1951.[48] They discovered that Cass A, which corresponded to a supernova remnant, was visible as a network of filaments to the 200-inch optical telescope.

The counterpart to Cygnus A, however, would provide a revolutionary implication. Cygnus A, the second strongest radio source in the sky, was discovered to correspond to an inconspicuous 17th magnitude galaxy over 100 million light-years away.[49] This distant galaxy represented a new class of celestial objects. It was later proved that radio

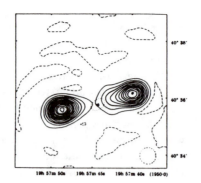
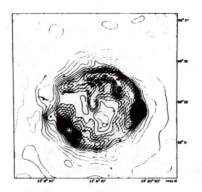

Figure 1.24 Radio brightness distribution maps of Cygnus A (left) and Cassiopeia A (right), taken with the Cambridge One-Mile telescope (Ryle et al., 1965) at 1.4 GHz. The unprecedented 23 arcsecond resolution showed great structural detail in the two sources that provided insight into the nature of the cosmic emitters. *Reprinted by permission of* Nature, © 1965, MacMillan Magazines, Limited.

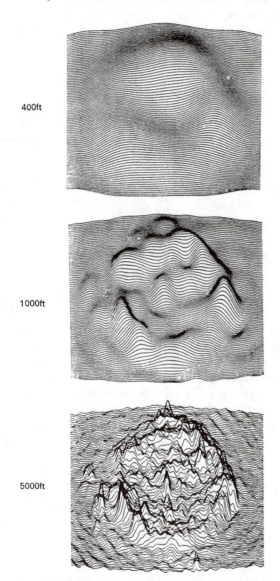

Figure 1.25 The effect of increasing angular resolution is seen in observations of Cassiopeia A, a supernova remnant, made with the Cambridge One-Mile telescope at 6 centimeters wavelength. Cass A is shown with synthesized apertures of 400 feet (122 meters), 1,000 feet (305 meters), and 5,000 feet (1,525 meters). Synthesized apertures correspond to the baseline spacing between farthest elements.

sources represented objects and regions of the universe that were invisible to even the largest optical telescopes. The result of this identification implied that radio astronomy had provided a tool for probing the most distant reaches of the universe.

Postwar Radio Astronomy in Australia: The CSIRO and the University of Sydney

Along with the two radio astronomy groups that formed in Britain after the war, a third major group formed in Australia and made significant early contributions. Since the Australian group had developed radar technology during WWII, the technology and hardware already existed to be applied to radio astronomy. After the war's end, scientists gathered at the Radiophysics Division of what became the Commonwealth Science and Industrial Research Organization (CSIRO). The division had been created in 1939 to develop radar for the Australian military. J. S. Pawsey began the first experiments in radio astronomy, measuring the temperature of the sun and of the ionosphere. He observed two distinct types of solar radiation that were associated with the quiet and disturbed sun. Pawsey observed the temperature of the undisturbed sun to be a million degrees Kelvin, far higher than expected, and published this discovery in 1946 in a classic paper.[50] Pawsey noticed that the signal from the quiet sun increased several times in intensity when sunspots were present. He later showed that these intense signals came from the areas of sunspots by using an ingenuous instrument, the sea interferometer. The design used the ocean as one reflector in the system. Water was found to reflect radio waves at the wavelength at which this telescope operated. The ocean, therefore, could be used as one of the elements in the interferometer system (Figure 1.26). This Lloyd's mirror design increased the telescope's baseline sufficiently to provide excellent resolution—resolution high enough to measure brightness distribution across the surface of the sun, proving the association between disturbed sun radiation and sunspots.[51]

The first cosmic radio astronomy research in Australia was begun with the sea interferometer. J. G. Bolton and G. J.

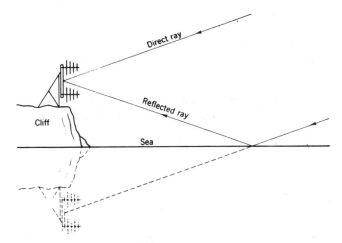

Figure 1.26 The sea interferometer, an ingenuous Lloyd's mirror design used by Pawsey et al. in Australia. The design used the ocean as one reflector in the interferometer system to increase angular resolution.

Figure 1.27 New Mills Cross at Molonglo, Australia. The enormous new Mills Cross was constructed in 1967 by the University of Sydney. The arms were constructed of mile-long wire mesh cylindrical paraboloids measuring 40 feet wide. *Photograph courtesy Sydney University.*

Stanley used it to detect many of the first "radio stars." They later showed that a strong discrete radio source in Taurus corresponded in position with the Crab Nebula, which went supernova in A.D. 1054.[52] This was the first optical identification of any discrete radio source other than the sun. They went on to discover several other radio sources with this system that were identified with optical counterparts. Centaurus A and Virgo A are notable examples.

The Australians, like the British, eventually became dissatisfied with the performance of these World War II surplus dishes and went on to pioneer several radical designs of radio telescopes. The sea interferometer was an ingenuous example. Another was the Mills Cross. In 1952 B. Y. Mills designed an aerial system utilizing two long arrays lying N-S and E-W, forming a cross to produce a very narrow beam. The two arrays producing intersecting fan beams were alternately connected in and out of phase. The switched fan

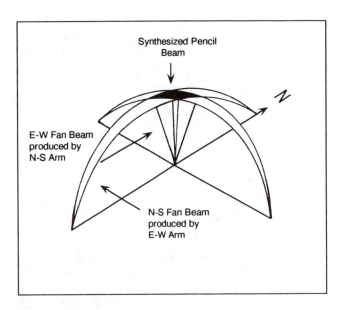

Figure 1.28 Principle of the Mills Cross radio telescope. A synthesized beam is produced by combining the east-west fan beam produced by the north-south array with the north-south fan beam produced by the east-west array. A very narrow beam and high spatial resolution are achieved in the process.

beams therefore alternately added and canceled within the pencil beam formed at their intersection. By using a switch-frequency rectifier at the receiver output to respond only to signals at the switching frequency, the system effectively simulated a narrow beam. After several relatively small prototypes had been successful produced, the first large Mills Cross was constructed in 1954 by Mills at Sydney Univeristy. The two arrays consisted of rows of cylindrical parabolic wire mesh reflectors and 250 dipoles tuned to 3.5 meters wavelength. Each arm of the array was 1,500 feet long. The telescope had an effective beamwidth of 48 arcseconds and could be steered in declination by introducing different lengths of cable between the dipole elements along the N-S line. A highly successful transit instrument with a very narrow beamwidth was produced. The design was so successful that a series of cross-type telescopes were subsequently constructed in Australia and around the world. A new Mills Cross (Figure 1.27) was completed at Molonglo in 1967. The arms were constructed of enormous mile-long, 40-foot-wide (12.2 meter) wire mesh cylindrical parabolas. Lines of dipoles were placed at the focus operating at 408 MHz producing a beamwidth of 2.8 arcminutes using the same principle as the original Mills Cross (Figure 1.28).[53] The project was partially funded by the U.S. National Science Foundation (NSF). The NSF had demonstrated serious interest in financially promoting the science of radio astronomy by partially underwriting a major instrument in Australia.

By the 1960s Australian effort began to diversify and receive major funding from the CSIRO. Two research groups had emerged with distinct philosophies, one at the University of Sydney and one at the CSIRO. A large solar observatory was constructed by the Radiophysics Division of

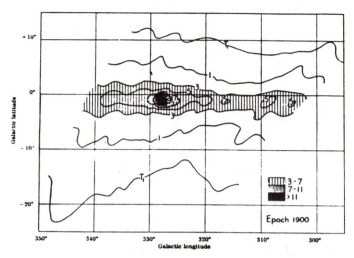

Figure 1.30 Sagittarius A, the galactic nucleus. The brightness distribution of 400 MHz radiation from the galactic center was observed with the CSIRO 80-foot (24.4-meter) reflector in 1953. The observation represented the discovery of the object at the center of the Milky Way, Sagittarius A. *Photograph courtesy* Sky and Telescope—*Chuck Baker, photo editor.*

Figure 1.29 CSIRO 210-foot (64-meter) Parkes radio telescope, Parkes, Australia, constructed in 1959. The 210-foot wire mesh reflector was alt-azimuth mounted on a supporting control tower. *Photograph courtesy J. Masterson, CSIRO Division of Radiophysics.*

CSIRO at Culgoora, NSW, in 1967. The radio heliograph, as it is called, consists of ninety-six 13.7 meter dish-shaped aerials spaced in a circle with a diameter of 3 kilometers. Each dish is automated to follow the sun across the sky, acting as a multiple-dish interferometer. Solar radio astronomy is important because the sun is the only close example of a typical main sequence star. The radio heliograph was (and is currently) used to "study the sun's atmosphere, the solar wind, and monitor flares and other active events."[54]

The Australians made many major contributions to the early history of radio astronomy. The trend toward mammoth instruments in the late 1950s found the Australian CSIRO group ready. Only the British Jodrell Bank telescope predates the Australian Parkes 64-meter in the evolution of mammoth telescopes. U.S. scientists were considering building a large aperture radio telescope in the United States on the order of 300 feet (91.5 meters) in diameter. For reasons not well documented in the literature, the telescope was built in Australia with backing from United States corporations, the Carnegie Corporation, the Rockefeller Foundation, and matching funds from the Australian government. A firsthand account of the events that led to the construction of the Parkes 210-foot telescope, which was downscaled due to costs, is provided by E. G. Brown in "The Origins of Radio Astronomy in Australia."[55] Essentially, a decision was made by Vannevar Bush, then presi-

dent of the Carnegie Corporation, and Alfred Loomis, a trustee of both the Carnegie Corporation and the Rockefeller Foundation, that the telescope could just as well be built in Australia. For these rather obscure reasons, the large aperture telescope was built at Parks, Australia, 200 miles (322 kilometers) west of Sydney. The telescope's initial design by Bowen was handed over to Freeman, Fox and Partners of London to prepare the detailed designs. The choice of Freeman and Fox, who built the Sydney harbour bridge, represents a trend of well-known bridge builders engineering the mammoth radio telescope projects around the world.

The construction contract was awarded to M.A.N., a West German firm. Construction was completed in 1959 (Figure 1.29). The 210-foot wire mesh reflector was built with surface deviations less than 9 millimeters at any orientation. The dish was alt-azimuth mounted on a supporting control tower which allowed a minimum pointing elevation of 30°. The resulting all-purpose instrument operated at wavelengths down to 10 centimeters; later an inner portion of the dish was resurfaced to operate at lower wavelengths.[56] The direction of solar research that essentially dominated the first decade of Australian radio astronomy changed as a result of the new large aperture telescope. With the Parkes telescope and the interferometers, the CSIRO effort continued to be on the cutting edge of research in radio astronomy through the 1970s and remains there even today. Among the scientific credits of the CSIRO are the discovery of 550 quasars, more than half the number cataloged by 1978,[57] and observations that established the object Sagittarius A as the galactic nucleus and provided the first detailed views (Figure 1.30).[58] The realization that the group was producing cutting edge research is vividly illustrated as early as 1952 in a statement by J. P. Wild:

Figure 1.31 25-meter radio telescope at Dwingeloo in the Netherlands. The 25-meter paraboloid constructed in 1956 was, for a time, the largest radio telescope in the world. *Photograph courtesy Sky and Telescope—Chuck Baker, photo editor.*

I can well remember the very special atmosphere that existed in the Lab at that time—perhaps, though on a humbler scale, it was something like the atmosphere at the Cavendish after Rutherford's discovery. New results were popping up all the time. Almost everything you did led to a discovery. About once every fortnight the research staff would meet together with Pawsey in the chair. Each person and team would report their progress, and Pawsey would express his approval or surprise or criticism of each in turn.[59]

The ingenious Australian telescopes, particularly the Cross installations, reflected the ideas of Pawsey on how research in radio astronomy should proceed. Both these great installations closely conformed to Pawsey's ideas of the most desirable developments in instrumentation in radio astronomy. He encouraged Mills at Sydney as well as being directly concerned in supporting Wild's design at CSIRO. By 1962, Pawsey felt that his ambition to establish front-rank radio astronomy research in Australia had been achieved.[60]

In 1962, J. L. Pawsey accepted an invitation to become director of the newly established National Radio Astronomy

Observatory at Green Bank, West Virginia.[61] Before he could assume the directorship of the NRAO, he became terminally ill. Pawsey's death was aptly described by F. J. Kerr: "His untimely death in 1962 at the age of 54 was a great blow to the development of radio astronomy, not only in Australia, but throughout the world."[62]

The histories of the early postwar research centers in Australia and England are intimately entwined and inevitably connected with the history of the NRAO. A diffusion of the great minds involved in research in radio astronomy was manifested in a redistribution of the leading researchers in radio astronomy that occurred after the war. Radio astronomers from CSIRO and Cambridge accepted positions at observatories being established all over the world. The redistribution of radio astronomers was more of a diffusion rather than a "braindrain," as the early research centers typically maintained scientific staffs. Prominent but fairly representative examples include Pawsey's trek from Cambridge to CSIRO and eventually his acceptance of the directorship of NRAO, Green Bank in 1962; Findlay's departure from Cambridge to DTM and eventually to NRAO, Green Bank; and J. G. Bolton's trek from Cambridge to CSIRO and then to Caltech in 1955, where he accepted the position of the first director of the Owens Valley Radio Observatory.[63] Ideas, techniques, and technical innovations that evolved at these early centers were diffused around the world with the movement of radio astronomers to these newly established observatories.

Early Years of Radio Astronomy in the Netherlands

The British and Australian wartime and immediate postwar accomplishments were not alone. The Americans, the Dutch, and eventually the Canadians would provide technical, theoretical, and observational discoveries that helped to revolutionize radio astronomy. Astronomy in the Netherlands has had a long and fertile history. Optical astronomers at the Leiden Observatory were among the first to appreciate the significance of radio frequency research, particularly the early research of Grote Reber. Radio astronomy began as theoretical research at the Leiden Observatory with H. C. van de Hulst's wartime prediction of the 21-centimeter radiation. Although the 21-cm line was not detected until 1951 at Harvard, the search for the 21-cm line represents the beginning of radio astronomy research in the Netherlands. The earliest observations began, as in England and Australia, with salvaged wartime radar antennas. Resolution, of course, became a problem and in 1956 a 25-meter radio telescope was constructed at Dwingeloo, in northern Holland (Figure 1.31). It was the largest telescope in the world until the construction of the Jodrell Bank Mark I. Much important work on galactic structure was performed with this telescope, primarily by G. W. Rougoor, Hugo van Woerden, and Jan Oort.[64] Research on galactic rotation util-

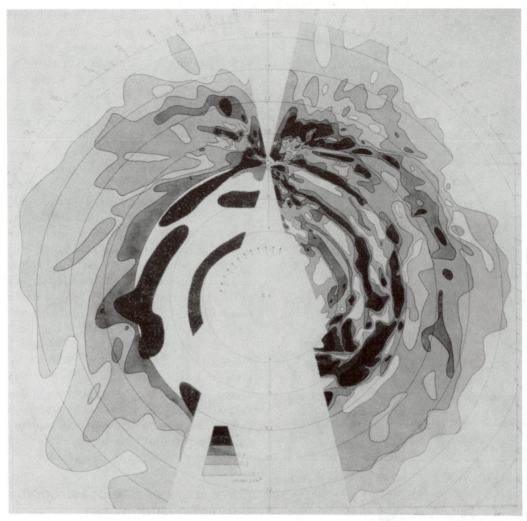

Figure 1.32 The Leiden-Sydney map of the spatial distribution of neutral hydrogen in the Milky Way. The spiral arm structure is highly detailed in the 1959 compilation of data from Leiden (Muller, Westerhout, and Schmidt) and CSIRO (Kerr, Hindman, and Gum). *Photograph courtesy* Sky and Telescope—*Chuck Baker, photo editor.*

izing the Doppler shifts of interstellar gas clouds allowed mapping the distribution of interstellar hydrogen. New insight into the structure of the Milky Way was provided. The spiral arm structure was clearly demonstrated in the Leiden survey data which was eventually combined with CSIRO southern hemisphere data (Kerr, Hindman, and Gum). The combined Leiden-Sydney map of the spatial distribution of neutral hydrogen which was produced in 1959 (Figure 1.32)[65] dramatically illustrates the spiral structure. An unexpected discovery was the anomalous "two-kiloparsec" spiral arm that is moving away from the galactic center at a tremendous rate of speed. The cause of this spiral arm's anomalous motion away from the galactic center is still not understood.[66] The instrument was recently used to make an important discovery—the detection of a previously unknown galaxy in the Local Cluster, our local galactic neighborhood. The successes at Dwingeloo motivated the planning and construction of an impressive interferometer, the Westerbork Synthesis Array. This array of 14

identical 25-meter paraboloids arranged along a 3 kilometer E-W line was completed in 1970.[67] The Westerbork Synthesis Array is operated by the Netherlands Foundation for Radio Astronomy (NFRA) and has been used to study extragalactic sources, for example, the discovery of spiral arms at radio frequencies in the Andromeda galaxy (M51) and the discovery of the "head-tail" system NGC 1265.[68]

World War II Era and Postwar Radio Astronomy in the United States

The American postwar effort initially lagged behind that of the British and Australians. There was not an evident cohesive national effort until the late 1950s with the establishment of the National Radio Astronomy Observatory. In retrospect, it seems that most of the American physicists involved in the war effort were attracted to research in nuclear physics after the war because of the high status of the field imposed by a wartime mentality. Several important

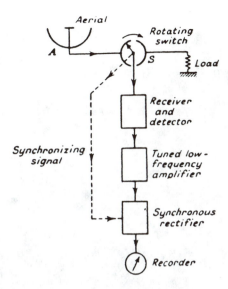

Figure 1.33 Schematic diagram of the Dicke radiometer. Receiver sensitivity is increased as the signal is continually calibrated. The ingenuous switching system greatly improved receiver stability, in terms of gain drifts.

early research centers for radio astronomy, however, were established in the United States. A few diverse efforts by independent university groups at Harvard, Ohio State, Stanford, and Caltech, and by the Naval Research Laboratory (NRL), the National Bureau of Standards (NBS), the Department of Terrestrial Magnetism (DTM), and Bell Labs proved fruitful.

An important wartime development that was a product of the development of radar in the United States, the Dicke radiometer, became critical to the evolution of instrumentation in radio astronomy. R. H. Dicke of the Radiation Laboratory at the Massachusetts Institute of Technology published a landmark paper in 1946 in which he described his work on microwave measurements and the radiometer.[69] Receiver noise, radio emission generated from the electrons in electrical currents within the receiver, is a tremendous problem at microwave frequencies. The noise generated by the receiver is considerably more intense than the signal from the astronomical source. Receiver stability, in terms of time variations of receiver gain, is a problem in the attempt to detect these weak signals. Dicke invented the radiometer as an ingenious solution to the problem. The radiometer includes a switch which effectively modulates the desired signal. Long-period gain variations are subtracted out as the signal is switched from the aerial to a calibration source, and back again (Figure 1.33). Continuous calibration greatly reduces drifts in gain. A tremendous improvement in sensitivity was provided by this form of "Dicke switching." The technique evolved to include position, frequency, polarization, and phase switching. Ryle's phase-switched interferometer is essentially based on Dicke switching between two sets of antenna lobes.[70] The development of

Dicke switching represented a critical step in the evolution of instrumentation.

Postwar research in radio astronomy in the United States began as in other nations as scientists who had worked on the wartime development of radar, primarily at the MIT Radiation Laboratory with associated work at Harvard, began to pursue scientific research in radio astronomy. The Naval Research Laboratory, the laboratory at NBS, and DTM were all assembled essentially from scientists who had been involved in the development of radar. The NBS at Washington began observations of galactic radio emission in March 1948 and published a widely cited paper. Research at NBS, however, never developed.[71]

A team had meanwhile formed at the DTM of the Carnegie Institution in Washington. Merle Tuve was appointed head of the DTM after the war. Under Tuve's direction some early important research was produced. This research culminated in the first discovery of radio waves from a planet, Jupiter. Burney Burke and K. L. Franklin serendipitously discovered the radio bursts associated with Jupiter in 1955.[72] They were making observations with a cross-type array and noticed spurious signals. After thorough analysis of the signals they identified the burst signals with the planet Jupiter. The burst type radiation was completely unexpected. Thermal radiation from the planet's surface was, of course, the expected emission. The discovery was serendipitous in that the telescope beam was set at the proper declination to observe Jupiter at a frequency of 20 MHz, which happens to be the center of a narrow band of radio emission from the planet.[73] The emission associated with Jupiter was initially confounding. The radiation does not come from the planet itself, but from the Jovian satellites as they pass through the intense magnetosphere of the planet. Some interesting physics is implied in the process. A search for emissions of other planets at radio frequencies was initiated. Despite the significant work in planetary radio emission, the DTM group dispersed in the late 1950s.[74]

Research at radio frequencies greatly enhanced our knowledge of the planets in the 1950s, particularly our knowledge of Jupiter and Venus. The Burke and Franklin discovery of the Jupiter bursts was published in 1955. The next year an NRL team led by Cornell H. Meyer announced the detection of Venus at 3 centimeters. Surprisingly, the data implied an extremely hot surface temperature of 600 K, the first evidence of a greenhouse effect.[75] Scientists at the NRL pioneered observations at very short wavelengths, centimeter and millimeter waves (at the time described as infrared radiation), using the NRL 50-foot (15.25-meter) telescope which had a very high surface accuracy and a sensitive radiometer (Figure 1.34). The group also made the first use of maser techniques (the maser had been developed by Townes and was ideal for use in low noise amplifiers).[76] By 1957 the group at NRL had estab-

Figure 1.34 The U.S. Naval Research Laboratory 50-foot (15.25-meter) radio telescope. *Photo courtesy of the Naval Research Laboratory by permission of the Defense Pentagon, and National Archives and Records Administration.*

Figure 1.35 The Ewen and Purcell fixed horn antenna used at the Lyman Physics Laboratory of Harvard University in 1951 to detect the galactic 21-centimeter radiation from neutral hydrogen. *Photo courtesy of Harvard University.*

lished surface temperatures for Venus, Mars, and Jupiter.[77] By the mid-1950s the group at NRL had established themselves as the leading U.S. research group in radio astronomy. The group at NRL would later become involved in planetary and lunar radar astronomy.

The beginnings of spectroscopy at radio frequencies and the subsequent development of molecular spectroscopy began experimentally at Harvard with the 1951 detection of the 21-centimeter hydrogen line in the Milky Way. Although the 21-cm had been predicted by van de Hulst during the war and searched for by several groups, it had not been detected by 1951. H. I. Ewen and E. M. Purcell used an unusual horn-shaped design to block out radiation from unwanted areas of the sky. (The original telescope is exhibited at NRAO, Figures 1.35 and 1.36). In September 1951, Ewen and Purcell published positive results of the detection of the 21-cm line radiation.[78] They later shared the Nobel Prize in Physics for this work. The discovery was unusual in that the theoretical prediction initiated a worldwide search by radio astronomy groups. The Australians also detected the line shortly thereafter. The 21-cm line is significant because it corresponds to the hyperfine transition of atomic hydrogen. This radio emission of atomic hydrogen allows astronomers to map the spiral arms of the Milky Way and other galaxies as the spiral arms are primarily composed of atomic hydrogen. The structure of the Milky Way can therefore be deter-

Figure 1.36 Another view of the Ewen and Purcell fixed horn antenna. The original horn antenna is on display at NRAO, Green Bank, West Virginia. *Photo by B. Malphrus.*

Figure 1.37 The Kraus two-reflector radio telescope of Ohio State University. The large reflectors, one parabolic 360 × 70 feet (109.8 × 21.35 meters) and one flat 340 × 100 feet (103.7 × 91.5 meters), were constructed of wires stretched on a metal frame. The instrument, constructed in 1962, was a brilliant design for the construction of a large and inexpensive steerable radio telescope. *Photo courtesy of the Ohio State University Photo Archives and Professor John D. Kraus.*

mined by its radio emission. The structure of our own galaxy is a very difficult thing to determine because the spiral arms are invisible at optical frequencies and the area of the nucleus and disk contains so much material. It is difficult to see the trees for the forest. Further research in mapping atomic hydrogen, primarily by the Dutch and Australians, revolutionized the understanding of galactic structure.

The effort in radio astronomy at Harvard was sporadic until the late 1950s. Bart Bok, a noted optical astronomer, started to use radio observations to penetrate the obscuring dust clouds of the Milky Way. In 1954 a 25-foot (7.6-meter) dish was in operation. In 1959, a 60-foot (18.3-meter) telescope came on-line. The first three Ph.D.s trained on the 25-foot telescope were Dave Heeschen, T. K. Menon, and A. E. Lilley. Heeschen led the group until 1957, when he left to head the Astronomy Department and eventually become director of NRAO, Green Bank. Frank Drake, an early member of the Green Bank scientific staff, was also among the Harvard group during the 1950s.[79]

By the late 1950s radio astronomy groups had assembled at other universities: Ohio State, the University of Maryland, Caltech, Stanford, Cornell, and the University of Michigan. The group at the Department of Electrical Engineering at Ohio State formed as early as 1952. J. D. Kraus pioneered a long partially steerable two-reflector telescope for source work. The Kraus design, a large flat reflector that

reflects radio signals to a portion of the parabolic reflector and then to a focus, represents an ingenious way to inexpensively construct a large steerable telescope. In 1962 the parabolic reflector was constructed of wires stretched on a metal frame that measured 360 feet long by 70 feet high (109.8 × 21.35 meters). The flat reflector (Figure 1.37) measured 340 feet by 100 feet (103.7 × 91.5 meters). The ground between the two reflectors is covered with aluminum sheeting to serve the dual purpose of simplifying the horn design and preventing the reception of thermal radiation from the ground.[80] A larger telescope was later constructed at the observatory at Nancay, France, based on the Kraus design.

Groups at Stanford, University of Michigan, and University of Maryland began research programs in radio astronomy prior to 1960. The group at Stanford assembled initially to do radar work on meteors which was similar to the early initiatives at Jodrell Bank. A 2-mile (3.2-kilometer) array for low frequency solar and extragalactic work was constructed by the group at University of Maryland at Clark Lake, California. Westerhout eventually moved from Leiden to Maryland. The group at Maryland and particularly Westerhout would later make extensive use of the national facilities at NRAO. The dependence of radio astronomy groups on national facilities, such as the National Radio Astronomy Observatory, would become evident in the 1960s. The group at the University of Maryland was the first major group to develop with no instrument program of

Figure 1.38 The Owens Valley Caltech interferometer. Two movable 90-foot (27.45-meter) equatorially mounted reflectors mounted on an E-W baseline comprise the Owens Valley interferometer, completed in 1957. Other reflectors were later added to form a N-S baseline. *Photo courtesy of the California Institute of Technology.*

Figure 1.39 The Penzias-Wilson giant horn reflector at Bell Telephone Laboratories used to discover the 3 K radiation in 1964. The discovery of the 3 K radiation provided direct observational evidence to support the big bang theory of cosmology. *Courtesy of AT&T Archives.*

its own. It was built successfully on the concept of using national observatories.[81]

The radio astronomy group at the California Institute of Technology Owens Valley Radio Observatory became the first research center in the United States to pursue interferometry techniques. The group at Caltech, under Bolton, constructed an interferometer of two equatorially mounted 90-foot (27.45-meter) reflectors which could be placed at various positions along a 1,600-foot (488-meter) E-W baseline (Figure 1.38). The interferometer went into operation in 1957. A N-S baseline was eventually added as well as a 130-foot (40-meter) telescope in the 1960s.[82] During the 1960s the group at Caltech evolved into an important research center.

Early research efforts at Cornell University culminated in the construction of the 1,000-foot (305-meter) Arecibo dish in Puerto Rico operated by the National Ionospheric and Astronomy Council housed at the university. The stationary transit dish was built in a natural depression in the hillside and came on-line in 1963. The topography of the landscape provided relatively easy construction of the enormous reflecting surface. The stationary nature of the dish reduced the need for an extensive superstructure and elaborate drive and control subassemblies. The instrument operates in transit mode, observing a swath of the sky as the earth rotates. The surface of the dish covers more than 20 acres, a larger area than the combined area of all the

other radio telescopes in the world. The enormous reflecting surface produces an exceptional spatial resolution of 3 arcminutes at 21 cm. Although the Arecibo Radio Telescope has been tremendously successful, it is highly unlikely that, due to current engineering constraints, a larger instrument will be constructed on earth in the near future. The telescope has retained the title of largest radio telescope in the world for some 33 years and seems unlikely to relinquish this title any time soon.

American postwar period contributions to the science of radio astronomy include a landmark discovery, the Penzias-Wilson discovery of the 3 K background radiation in 1964 at Bell Laboratories. The Penzias-Wilson discovery provided conclusive evidence for the big bang theory of cosmology. The background is radiation "left over" from the early expansion of the universe. The resulting photons had been predicted by several research groups to be redshifted to an energy of around 3 kelvin. Arno Penzias and Robert Wilson detected the background radiation at 2.7 K using a giant horn reflector, initially designed as a low-noise receiving station for signals bounced from echo satellites (Figure 1.39).[83] Penzias and Wilson had discovered the isotropic background radiation left over from the big bang and were awarded the Nobel Prize in Physics in 1978 for their discovery. Radio astronomers were providing direct insight into the history and large-scale structure of the universe.

Despite the efforts at Ohio State, Cornell, Stanford, NBS, and DTM there were still no single dishes in the United States bigger than 60 feet in the mid-1950s.[84] The 90-foot reflectors of Caltech's Owens Valley Radio Observatory two-element interferometer came on-line in 1957. Huge aperture telescopes in Britain, Australia, and the Netherlands were producing significant research and U.S. scientists realized they were falling behind. A large aperture telescope was critically needed to compete with cutting edge research centers in radio astronomy. Since such an instrument was well beyond the means of a single university or research center to construct, American radio astronomers began to organize the search for a solution. The stage was set for the establishment of a national observatory for research in radio astronomy.

Endnotes

1. D. S. L. Cardwell, *Technology, Science and History*, London: Heinemann Educational Books Ltd., 1972, p. 183.
2. *Ibid.*, p. 186.
3. W. T. Sullivan, "Early Radio Astronomy," published in *Astrophysics and Twentieth-Century Astronomy to 1950: 4A*, Owen O. Gingerich. Cambridge: Cambridge University Press, 1984, p. 192.
4. John Kraus, "The First 50 Years of Radio Astronomy, Part 1: Karl Jansky and His Discovery of Radio Waves from Our Galaxy," *Cosmic Search*, Fall 1981, p. 12.
5. K. Kellerman (ed.), *Serendipitous Discoveries in Radio Astronomy*, Proceedings of a Workshop held at the National Radio Astronomy Observatory Green Bank, West Virginia, on May 4, 5, 6, 1983, p. 36. Includes John Kraus, "Karl Guthe Jansky's Serendipity, Its Impact on Astronomy and Its Lessons for the Future."
6. W. T. Sullivan, "Early Radio Astronomy," 1984, p. 191.
7. W. T. Sullivan, (ed.), *The Early Years of Radio Astronomy: Reflections Fifty Years After Jansky's Discovery*, 1984, p. 218.
8. John Kraus, "The First 50 Years of Radio Astronomy, Part 1: Karl Jansky and His Discovery of Radio Waves from Our Galaxy," *Cosmic Search*, Fall 1981.
9. Karl Guthe Jansky, "Directional Studies of Atmospherics at High Frequencies," *Proceedings of the Institute of Radio Engineers*, vol. 20, no. 13, December 1932.
10. K. Kellerman (ed.), *Serendipitous Discoveries in Radio Astronomy*, 1983, p. 49.
11. Grote Reber, "The Early Years of Radio Astronomy." Lecture given at NRAO, Green Bank, West Virginia, on January 25, 1988.
12. J. L. Greenstein, "Optical and Radio Astronomy in the Early Years," published in K. Kellerman (ed.), *Serendipitous Discoveries in Radio Astronomy*, Proceedings of a Workshop held at the National Radio Astronomy Observatory Green Bank, West Virginia, on May 4, 5, 6, 1983, pp. 79–88.
13. David Edge and Michael Mulkay, *Astronomy Transformed: the Emergence of Radio Astronomy in Britain*, John Wiley and Sons: New York, 1976, p. 14.
14. K. Kellerman (ed.), *Serendipitous Discoveries in Radio Astronomy*, 1983, p. 63. John Kraus, "Karl Guthe Jansky's Serendipity, Its Impact on Astronomy and Its Lessons for the Future."
15. Grote Reber, "The Early Years of Radio Astronomy." Lecture, January 25, 1988.

16. *Ibid.*, 1988.
17. Joseph L. Spradley, "The First True Radio Telescope," *Sky and Telescope*, July 1988, p. 28.
18. John Kraus, "The First 50 Years of Radio Astronomy, Part 1: Karl Jansky and His Discovery of Radio Waves from Our Galaxy," *Cosmic Search*, Fall 1981.
19. Joseph L. Spradley, "The First True Radio Telescope," *Sky and Telescope*, July 1988, p. 28.
20. Grote Reber, "Cosmic Static," *Proceedings of the Institute of Radio Engineers*, 1940, pp. 68–70.
21. *Ibid.*, pp. 68–70.
22. Grote Reber, "Cosmic Static," *Astrophysical Journal*, vol. 91, 1940. pp. 621–624.
23. Vladimir Kourganoff, "Otto Struve: Scientist and Humanist," *Sky and Telescope*, April 1988, pp. 379–38.
24. Grote Reber, "Cosmic Static," *Astrophysical Journal*, 1941.
25. Grote Reber, "Radio Emissions from the Milky Way," *Astrophysical Journal*, vol. 100, 1944, pp. 279–287.
26. *Ibid.*, pp. 279–287.
27. John Kraus, "The First 50 Years of Radio Astronomy, Part 3: Postwar Radio Astronomy," *Cosmic Search* 13, 1983.
28. *Ibid.*, pp. 5–6.
29. D. S. L. Cardwell, *Technology, Science and History*, London: Heinemann Educational Books Ltd., 1972.
30. David Edge and Michael Mulkay, *Astronomy Transformed: The Emergence of Radio Astronomy in Britain*, p. 14.
31. *Ibid.*, p. 14.
32. *Ibid.*, p. 14.
33. W. T. Sullivan, (ed.), *The Early Years of Radio Astronomy: Reflections Fifty Years After Jansky's Discovery*, Cambridge University Press, Cambridge, England, 1984, p. 193. Contains a paper by Sir Bernard Lovell, "The Origins and Early History of Jodrell Bank."
34. *Ibid.*, p. 213.
35. *Ibid.*, p. 218.
36. *Ibid.*, p. 213.
37. *Ibid.*, p. 219.
38. The transit data of Cygnus A in Figure 1.18 was taken by Brown and Hazard with the 218-foot paraboloid at Jodrell Bank in 1949 at a wavelength of 1.89 meters. The transit data of Cygnus A in Figure 1.19 was taken by Malphrus with the NRAO 40-foot telescope in 1989 at a wavelength of 21 centimeters.
39. A. C. B. Lovell, "The Jodrell Bank Telescope," *Nature*, vol. 180, no. 4576, July 13, 1957, p. 60.
40. J. W. Findlay, "Large Radio Telescopes—1950 to 1989," NRAO Internal Publication, August 20, 1989.
41. R. M. Emberson, "The Telescope Program for the National Radio Astronomy Observatory at Green Bank, West Virginia," *Proceedings of the Institute of Radio Engineers*, vol. 46, no. 1, 1958, pp. 24–35.
42. NRAO, *Annual Report*, July 1, 1959, pp. 6–9.
43. Bernard Lovell, *The Jodrell Bank Telescopes*, Oxford University Press, Oxford, England, 1985.
44. W. T. Sullivan, *Classics in Radio Astronomy*, D. Reidel Publishing Co. Dordrecht, Holland, 1982, 115, Sir Martin Ryle, "A New Radio Interferometer and Its Application to the Observation of Weak Radio Stars," *Proceedings of the Royal Society*, A211, 1952, pp. 351–375.
45. J. S. Hey, *The Evolution of Radio Astronomy*, Elk Science, London, England, 1973, pp. 60–67.
46. Martin Ryle, "The New Cambridge Radio Telescope," *Nature*, vol. 194, no. 4828, May 12, 1962.
47. Martin Ryle, B. Elsmore, and Ann C. Neville, "High-Resolution Observations of the Radio Sources in Cygnus and Cassiopeia," *Nature*, vol. 205, no. 4978, March 27, 1965, p. 1259.
48. W. Baade and R. Minkowski, "Identification of the Radio Sources in Cassiopeia, Cygnus A and Puppis A," *Astrophysical Journal*, vol. 119, 1954, p. 206.
49. W. T. Sullivan, "Early Radio Astronomy," published in *Astrophysics and Twentieth-Century Astronomy to 1950: 4A*, Owen O. Gingerich, 1984, p. 191.
50. J. L. Pawsey, "Observation of Million Degree Thermal Radiation From the Sun at a Wavelength of 1.5 Meters," *Astrophysical Journal*, 1946.
51. W. N. Christiansen, "The First Decade of Radio Astronomy in Australia," published in W. T. Sullivan, *The Early Years of Radio Astronomy: Reflections Fifty Years After Jansky's Discovery*, 1984, pp. 113–131.
52. J. L. Greenstein, "Optical and Radio Astronomy in the Early Years,"

published in W. T. Sullivan, (ed.), *The Early Years of Radio Astronomy: Reflections Fifty Years After Jansky's Discovery,* 1984, pp. 67–84.

53. J. S. Hey, *The Evolution of Radio Astronomy,* 1973, pp. 70–82, in W. T. Sullivan, (ed.), *The Early Years of Radio Astronomy: Reflections Fifty Years After Jansky's Discovery,* 1984, pp. 85–111.

54. "CSIRO and Astronomy," CSIRO Information Service. Leaflet No. 2, January 1978, Central Information, Library and Editorial Section, East Melbourne, Victoria, Australia, p. 4.

55. E. G. Brown in "The Origins if Radio Astronomy in Australia" included in W. T. Sullivan, (ed.), *The Early Years of Radio Astronomy: Reflections Fifty Years After Jansky's Discovery,* 1984, pp. 85–111.

56. J. S. Hey, *The Evolution of Radio Astronomy,* 1973, pp. 70–82, in W. T. Sullivan, (ed.), *The Early Years of Radio Astronomy: Reflections Fifty Years After Jansky's Discovery,* 1984, pp. 85–111.

57. "CSIRO and Astronomy," CSIRO Information Service. Leaflet No. 2, January 1978, Central Information, Library and Editorial Section East Melbourne, Victoria, Australia, pp. 1–5.

58. Frank J. Kerr, "Serendipity in the Galaxy: The Galactic Warp and the Galactic Nucleus," published in K. Kellerman, (ed.), *Serendipitous Discoveries in Radio Astronomy,* Workshop Proceedings, NRAO Green Bank, West Virginia, May 4, 5, 6, 1983.

59. J. P. Wild is quoted in J. S. Hey's *The Evolution of Radio Astronomy,* 1973, pp. 73–74.

60. J. S. Hey, *The Evolution of Radio Astronomy,* 1973, p. 82.

61. *The Observer,* National Radio Astronomy Observatory Recreation Association (NRAORA)/News, vol. 1, no. 1 December 29, 1961, p. 1.

62. F. J. Kerr, "Early Days in Radio and Radar Astronomy in Australia," published in W. T. Sullivan, (ed.), *The Early Years of Radio Astronomy: Reflections Fifty Years After Jansky's Discovery,* 1984, pp. 132–145.

63. J. S. Hey, *The Evolution of Radio Astronomy,* 1973, p. 82.

64. Willem Bijleveld and W. Butler Burton, "Leiden Observatory: 350 Years of Astronomy," *Sky and Telescope,* February 1985, 119–122.

65. J. S. Hey, *The Evolution of Radio Astronomy,* 1973, p. 82.

66. Willem Bijleveld and W. Butler Burton, "Leiden Observatory: 350 Years of Astronomy," pp. 119–122.

67. Jacob W. M. Baars, et al., "The Synthesis Radio Telescope at Wester-

bork," *Proceedings of the IEEE,* vol. 61, no. 9, September, 1973, pp. 1258–1266.

68. Willem Bijleveld and W. Butler Burton, "Leiden Observatory: 350 Years of Astronomy," pp. 119–122.

69. R. H. Dicke, "The Measurement of Thermal Radiation at Microwave Frequencies," first published in "The Review of Scientific Instruments," 17, 1946, pp. 268–275.

70. M. Ryle, "A New Radio Interferometer and its Application to the Observation of Weak Radio Stars," *Proceedings of the Royal Society,* A211, 1952, pp. 351–375.

71. David Edge and Michael Mulkay, *Astronomy Transformed: The Emergence of Radio Astronomy in Britain,* 1976, p. 35.

72. B. Burke and K. L. Franklin, "Observations of a Variable Radio Source Associated with the Planet Jupiter," *Journal of Geophysical Research,* vol. 60, 1955, pp. 213–217.

73. F. Graham Smith, *Radio Astronomy,* Penguin Books, Harmondsworth, Middlesex, England, 1960, p. 203.

74. David Edge and Michael Mulkay, *Astronomy Transformed: The Emergence of Radio Astronomy in Britain,* 1976, p. 35.

75. W. T. Sullivan, "Radio Astronomy's Golden Anniversary," *Sky and Telescope,* December 1982, pp. 544–555.

76. David Edge and Michael Mulkay, *Astronomy Transformed: The Emergence of Radio Astronomy in Britain,* 1976, p. 35.

77. F. Graham Smith, *Radio Astronomy,* 1960, p. 203.

78. H. I. Ewen and E. M. Purcell, "Observation of a Line in the Galactic Radio Spectrum," *Nature,* vol. 168, no. 4270, September 1, 1951, pp. 356–358.

79. J. S. Hey, *The Evolution of Radio Astronomy,* 1973, p. 95.

80. David Edge and Michael Mulkay, *Astronomy Transformed: The Emergence of Radio Astronomy in Britain,* 1976, p. 35.

81. *Ibid.,* p. 35.

82. J. S. Hey, *The Evolution of Radio Astronomy,* 1973, p. 95.

83. Robert T. Wilson, "Discovery of the Cosmic Microwave Background," published in K. Kellerman, (ed.), *Serendipitous Discoveries in Radio Astronomy,* 1983, pp. 185–195.

84. J. W. Findlay, "Large Radio Telescopes—1950 to 1989," NRAO Internal Publication, August 20, 1989.

Chapter 2

Establishment of the NRAO: Astropolitics

The nature of scientific research changed during World War II. Prewar research could primarily be conducted by one or a few researchers using tabletop-sized instruments. Research in the physical sciences after World War II changed in scale. Physical science research evolved into "big" science in terms of the scale and complexity of instrumentation as well as the amount of money and researchers involved in single investigations. Organizing and funding large-scale research began to evolve into a complex problem.

By 1954 there was a pervasive feeling that the United States lagged behind other countries in astronomy. The American mindset toward science in the mid-1950s was dominated by the lasting effects of World War II, the Korean war which had just ended, and the ensuing cold war. The incredible achievements in applied science made by the United States during the war and shortly after relied heavily on the importation of basic science knowledge developed in Europe. The postwar era, however, represented an intellectual drain in the basic science research performed on the continent. Scientists in the United States abruptly came to the realization that they could not longer depend on basic science research performed abroad to produce the knowledge needed for the technological developments of applied science. A social philosophy of scientific research began to emerge that promoted federal government support of large venture projects in basic science research to supply the knowledge base required by a developing technological society. The defense, health, welfare, and progression of this technological society were perceived to be directly related to advancements in science. This new social philosophy of science was physically evident in the recent establishment of the National Science Foundation (NSF) in 1950 and its first major endeavor to support large-scale scientific research—the National Radio Astronomy Observatory.

An immediate problem that the newly created NSF had to deal with was support for observational research in astronomy. U.S. astronomers, particularly Otto Struve, had begun as early as 1940 to envision an observing center for optical astronomy far greater in scope than a private foundation could support. Larger scale telescopes and new technology such as photoelectric techniques were needed to contribute to basic research in astronomy and astrophysics. Simultaneously, the observational science of radio astronomy began to require large aperture telescopes to compete at the cutting edge of technology. This was the atmosphere prevalent at the time of the creation of the NSF.

The origins of the National Science Foundation, the National Radio Astronomy Observatory, and the National Optical Astronomy Observatory were intricately linked together and driven by the new social philosophy of science. The simultaneous development of plans for optical and radio astronomy observatories as well as the call for NSF support of particle accelerators, nuclear reactors, and computer systems shaped the early policy of the NSF. The National Science Board, the controlling board of trustees of the NSF, adopted a national policy predicated upon three basic postulates:

1. The federal government should support large-scale basic science facilities when the need is clear and it is in the national interest, when the merit is endorsed by panels of experts, and when funding is not readily available from other sources.

2. A national astronomical observatory, a major radio astronomy facility, and university research installations of computers, accelerators, and reactors are examples of such desirable activities for NSF.

3. Funds for such large-scale research should be handled under special budgets.

Basic research in astronomy was a prime candidate for the first big NSF venture. A need of national significance was evident, one that was consistent with the new social philosophy and could not be met by single universities or institutions. Although the developmental paths would later diverge, the National Radio Astronomy Observatory and the National Optical Astronomy Observatory had a common origin, and in a practical way shared this origin with the NSF.

The United States, where the science of radio astronomy had begun and many early contributions made, had very evidently fallen from the forefront of cutting edge research essentially because the scale of research in radio astronomy had evolved far beyond the scope of any single university or institution. American radio astronomers began to organize to consider possible courses of action. The first large-scale meeting to assess the state of research in radio astronomy was jointly sponsored by the NSF, the California Institute of Technology, and the Department of Terrestrial Magnetism of the Carnegie Institution. The conference was held on January 4–6, 1954, in Washington, D.C.[1] Participants in the conference included some 75 astronomers, physicists, and electronics engineers from the United States as well as many distinguished astronomers from around the world. The conference represented an impressive turnout of the world's prominent researchers. The organizing committee consisted of J. L. Greenstein, B. J. Bok, J. B. Weisner, M. A. Tuve, and J. Hagen. International participants included R. Hanbury-Brown, C. G. Little, F. Graham Smith, and F. Hoyle from Great Britain; B. Y. Mills and E. G. Bowen from Australia; A. E. Covington from Canada; and H. C. van de Hulst from the Netherlands.[2]

International participants gave presentations on the nature and status of research at their respective centers. B. Y. Mills reported on the Australian research on the shapes, sizes, and spectra of radio sources as well as the work on galactic structure. Mills also discussed the cross-type arrays in use in Australia and the pencil beam obtainable with low effective aperture using these innovative instruments. E. G. Bowen presented recent results of solar radio research at CSIRO using the cross-type arrays. R. Hanbury-Brown discussed the impressive British surveys made with the 218-foot paraboloid at Jodrell Bank as well as the measurement of angular diameters of sources and work on extragalactic radiation. C. G. Little discussed the Jodrell Bank program on meteors, auroras, and ionosphere and the effects of the ionosphere cosmic sources. F. G. Smith presented the Cambridge work on radio star scintillation and the program on discrete sources and solar activity. Smith also discussed the new Cavendish radio interferometer which was designed to be used in a survey of discrete radio sources (the highly successful 1C or First Cambridge Survey). R. Hanbury-Brown presented a description of the enormous 250-foot paraboloid under construction at Jodrell Bank where partial operation was expected by 1955. Great Britain would clearly have the world's foremost research facility with this enormous, all-purpose radio telescope. Hanbury-Brown's presentation must certainly have clinched the notion that the United States had fallen behind in the new science.

The American research effort was weak in comparison. Although several groups were making significant contributions, there was no cohesive large-scale effort. F. T. Haddock and J. Hagen reported on the status of research at the Naval Research Laboratories. Haddock presented work on 10-centimeter radiation from discrete sources and Hagen reported results of solar research with radio astronomy techniques. J. D. Kraus presented results of the Ohio State radio as-

tronomy project, which was just beginning and was limited by minimal funding. C. R. Burrows and W. E. Gordon described the Cornell University project which was also in the formative stages as indicated by the 204-inch (518-centimeter) telescope used at the facility, which was minuscule even by 1954 standards.

Several theoretical and technical reports were also presented. A. E. Covington reported on the Canadian solar research at 10 centimeters. H. C. van de Hulst reported on the Dutch work on galactic structure at 21 centimeters and the theoretical implications. C. H. Townes (inventor of the maser) presented a paper on microwave spectra of astrophysical interest. R. H. Dicke, E. M. Purcell, and J. Wittke gave a presentation on the excitation of the 21-centimeter line. Bart Bok presented astronomical problems connected with the 21-centimeter line. R. Minkowski gave a progress report on astronomical observations of radio sources. The conclusion at that stage in the research was that not many radio phenomena could be associated with optical objects. J. L. Greenstein presented a theoretical paper on thermal emission from gases and energy considerations in nonthermal sources. F. Hoyle presented a theoretical mechanism for radio noise generation. Hoyle favored the synchrotron process although many theoreticians at the time thought the process an implausible explanation for the generation of the intense cosmic radio waves observed.[3] The theoretical and technical reports reflect the infantile status of radio astronomy.

These presentations made it clear to the U.S. participants that the lead in radio astronomy had been taken by other countries. It was also clear that U.S. radio astronomy would fall further and further behind if immediate steps were not taken to provide observing facilities and instrumentation comparable to those existing and being planned in other countries. The science of radio astronomy had advanced to the point that large aperture instruments were now essential to be competitive in radio astronomy research. The conclusion was reached that such instruments were far beyond the scope of single institutions. R. M. Emberson and N. L. Ashton provided a summary of the conclusions reached by U.S. astronomers as a result of the conference in the January 1958 *Proceedings of the IRE.* It was noted that most U.S. radio astronomers were working with antennas smaller and less effective than those of foreign scientists, who, soon after World War II, had gained support of their respective governments for the construction of large and elaborate radio astronomy equipment. The smaller size, or gain of an antenna, can, in part, be compensated for through better electronics. A receiver with a low noise figure and extreme stability to permit integration of the very weak celestial signals over long periods of time produces a very sensitive instrument. For angular resolution which reveals details of an object's structure, which is important to all astronomers, there is no substitute for aperture size. The suggestion was made that several institutions might join together in the construction of a large antenna, larger than any single institution could support.[4] The conference represented a landmark in the history toward the establishment of a national observatory for radio astronomy. The problem had been identified and a critical need for competitive instruments recognized. How to solve it was another problem.

The January 1954 conference sparked a series of discussions on the future of radio astronomy in the United States. The NSF established a subpanel chaired by Merle Tuve to investigate the role of the agency in the future of radio astronomy.[5] Independently of the NSF subpanel discussions, Donald H. Menzel, director of the Harvard College Observatory, began discussions with Cambridge astronomers in early 1954 on the future U.S. involvement in radio astronomy. Lloyd Berkner, president of Associated Universities Incorporated (AUI), had become interested in having AUI operate a national observatory for research in radio astronomy. Berkner expressed his interest in managing and operating the facility to Donald Menzel in early 1954. In reaction, Menzel worked with Julius Stratton and Jerome Weisner of Massachusetts Institute of Technology (MIT) and Bart Bok, Fred Whipple, and Cecilia Gaposchkin of Harvard to produce a preliminary report, "Survey of the Potentialities of Cooperative Research in Radio Astronomy." The report advocated treating radio astronomy in a similar manner as nuclear physics had been treated after the war. The interdisciplinary nature of the science was emphasized implying the involvement of astronomy, physics, and electrical and mechanical engineering. The survey also pointed out that U.S. efforts would continue to fall behind other countries if a major facility were not constructed. Perhaps the most catalytic product of the survey was the parallel drawn to nuclear physics and the overt suggestion that AUI would therefore be the preferred operating institution. Menzel pointed out in the survey that the necessary facilities, because of their enormous expense, were beyond the means of any one institution. AUI had effectively solved a similar problem in the establishment and operation of Brookhaven National Laboratory as a user institute for research in nuclear physics. The nature of the work in radio astronomy was well adapted to a similar type of organization. Indeed, because of the many evident parallel relationships, the suggestion that AUI should itself support the proposal was very logical.[6]

At this point the events leading to the establishment of the National Radio Astronomy Observatory intimately involved two somewhat divergent groups AUI and Associated Universities for Research in Astronomy (AURA) led by two prominent American scientists, Lloyd Berkner and Merle Tuve respectively, and the newly created NSF headed by Alan Waterman. The story of Berkner, Tuve, and Waterman is a classic representation of two divergent ways of viewing modern science. Waterman in many ways served as an arbitrator between the two and ultimately influenced science policy. AUI and AURA represented by Berkner and Tuve became two major opposing powers in the struggle to establish an NSF-funded center for research in radio astronomy. Alan Needell has described their relationship, "Their emerging struggle to influence the character of the observatory provides an opportunity not only to look at the institutional developments, but also to contrast two important perspectives on postwar science and science policy."[7] The outcome provides insight into changing patterns of federal support of science in postwar America. The struggle between Berkner and Tuve would inevitably influence the ultimate character of the observatory and the nature of the scientific research pursued there. The events leading to the establishment of the NRAO and the decision to contract AUI for the construction and operation centered around Berkner, Tuve, and Waterman, and their battle to politically determine the institutional nature of the observatory as well as its scope and scale. Their story is an intriguing chapter in the astropolitics that would ultimately influence the way research in radio astronomy is conducted in the United States. The entire story is told in insightful detail by Needell; a brief synopsis of the events relevant to the protohistory of the NRAO is provided here.

Associated Universities for Research in Astronomy

One of the two major groups involved in the protohistory of the NRAO, was Associated Universities for Research in Astronomy (AURA), a modern consortium of universities established by American astronomers to promote large-scale research in astronomy and astrophysics. The first version of AURA represented a group of astronomers predominately from universities in the Southeast, and was initially represented in its bid to develop the NRAO by Oak Ridge Institute for Nuclear Study (ORINS). AURA was initiated by William Pollard, executive director of ORINS, and suggested by Paul Gross, dean at Duke University, as early as June 1956 as an organization to develop and operate the NRAO. AURA, however, was largely based on the control and authority of one man, Merle Tuve. Tuve (Figure 2.1) was born in Canton, South Dakota, in 1901. He received a bachelor's degree from the University of Minnesota in 1922 and a Ph.D. in physics from The Johns Hopkins University in 1926. After graduation, Tuve accepted a position at the Department of Terrestrial Magnetism (DTM) where he had a long and productive career. Tuve worked on the development of accelerators of nuclear particles for experimental work in atomic and nuclear physics and eventually led the group at DTM which performed leading research in Van de Graaff accelerator studies. During WWII Tuve served as the director of Section T of the Office of Scientific Research and Development of the Johns Hopkins Applied Physics Laboratory. After the war, Tuve returned to the DTM and began a modest research program in radio astronomy.[8]

Merle Tuve was a champion of pure science and basic

Figure 2.1 Merle A. Tuve. *Photo courtesy of the Department of Terrestrial Magnetism, Carnegie Institution, Washington, D.C.*

scientific research. He believed in the power of the individual scientist and "the concentrated effort of a scholar to perceive and verify fresh aspects of the truth in his own specialty."[9] Tuve once remarked that it was when nuclear physics began to change from science to big business that he started to lose interest. He openly criticized the postwar trends in science toward ever larger research facilities and highly specific, goal-oriented projects. He objected to the trend to confuse applied and basic scientific research and feared that the character of government-supported scientific research was becoming dominated by research in military technology.

The radio astronomy program established by Tuve at DTM was among the first centers for research in the field to appear in the United States. As an advocate of such research, he began informal conversations with NSF officials regarding the future of research in radio astronomy as early as 1953. The newly created NSF was still defining the "nature and extent of its role in supporting American astronomical research." In December 1953, Tuve delivered an address describing the DTM work to the Carnegie Institution of Washington, parent organization to the DTM. This address was a catalyst in the planning of the landmark January 1954 conference jointly sponsored by the Carnegie Institution, the NSF, and the California Institution of Technology. The January conference prompted NSF to establish a subpanel to assess what the NSF might do to support research efforts in the new science. Tuve was chosen as head of the NSF subpanel on radio astronomy. This appointment put Tuve in a prime position to influence the establishment and subsequent nature of the NRAO.

Associated Universities Incorporated

Associated Universities Incorporated (AUI) was formally incorporated in July 1946 and began by developing the Brookhaven Graphite Research Reactor, the first major nuclear reactor devoted to research, in 1950. The establishment of the AUI represented a landmark event in the history of science in the United States, the emergence of consortia of universities organized to support large-scale scientific research. These consortia and the research they would make possible represented a new scale of science.

AUI reorganized to promote growth in 1950 and named a permanent director, Lloyd V. Berkner. Berkner shaped the future of AUI by expanding the sphere of AUI research into other areas of basic science. Berkner was responsible for the pursuit of a national observatory, available to all U.S. astronomers and operated by AUI. Berkner (Figure 2.2) was born in Milwaukee, Wisconsin, in 1905. He received a bachelor of science in electrical engineering from the University of Minnesota in 1927. He enrolled in a graduate program in physics after graduation, but was never to complete the advanced degree. Instead, he accepted a position with the Bureau of Lighthouses where he was assigned the task

Figure 2.2 Lloyd V. Berkner. *Photograph courtesy of Brookhaven National Laboratory, Associated Universities Incorporated.*

of installing radio ranging systems for air navigation. In 1927 Berkner joined the Naval reserves and served as an aviator. In 1928, he joined the National Bureau of Standards (NBS) where he helped Amelia Earhart plan her first transatlantic flight. Also while at NBS, he engineered the shortwave communications and navigation systems for Richard Byrd's first Antarctic expedition. He joined Byrd on the first Antarctic expedition and served as the radio engineer. In 1930, Berkner left NBS to join the ionospheric radio research team at DTM. During the war Berkner joined the Applied Physics Laboratory at Johns Hopkins to work on the Proximity Fuse Project. There he served under Merle Tuve in the applied physics department organized for wartime research by Tuve and Vannevar Bush, who would later be considered the founding father of the National Science Foundation. As U.S. involvement in the war became imminent, Berkner was called to active duty in the U.S. Navy's Bureau of Aeronautics. He was assigned the task of organizing and heading the Radar Section, and later appointed director of the entire Electronic Material Division. In this capacity he presented the Radiation Laboratory at MIT with the task of developing a radar system to protect U.S. ships from the raids of Japanese kamikaze pilots.

After the war, Berkner assumed various planning and organizing positions within the military's research and development and science establishments. Among these positions were the position of executive secretary of the military's Joint Research and Development Board under Vannevar Bush as well as positions working with the Department of Defense evaluating weapons systems and the Department of State planning military assistance to wartime allies and assessing the importance of science in foreign policy.[10] Berkner's history of research made him an ideal candidate to be a key player in the national effort to reestablish the United States as a leader in radio astronomy.

Associated Universities Incorporated was formally incorporated in July 1946. Research began four years later with the Brookhaven Graphite Research Reactor. In 1952, the Cosmotron was constructed and a productive program for research in high-energy physics began. For years the Cosmotron was the world's most productive atom smasher. With these instruments, Brookhaven became a leading research center for nuclear physics and AUI consequently became a powerful organization in the pursuit of American scientific research. In the early 1950s AUI trustees began to seek opportunities to establish other major research facilities for American scientists that were beyond the scope of single institutions.

In the absence of projects other than Brookhaven Laboratories, it was feared that AUI officials would involve themselves exclusively with operations at Brookhaven. The fear was shared by Leland Haworth, the director of Brookhaven, as well as several trustees of AUI. To avoid the realization of these fears, AUI reorganized. The institutional restructuring of AUI called for abandoning the process of electing a president from the existing members of the board of trustees in favor of hiring a permanent president for the board who would actively seek projects for AUI to manage. The selection process for a president of AUI led to one candidate, Lloyd Berkner, whose accomplishments were familiar to AUI officials and to virtually anyone involved in postwar science.

Berkner's attitude toward postwar science was the exact antithesis to that of Merle Tuve. Berkner advocated large-scale goal-oriented research. He favored pushing technology to the physical limits to pursue scientific research problems. Needell lends insight into Berkner's philosophical views of science, pointing out that Berkner was apparently fascinated by what he perceived to be the transformation of science that had occurred during his lifetime. He was frequently quoted as saying that science had become an essential political, economic, and military force. He saw research as an endeavor that required resources, planning, and leadership far beyond what universities and the university-based science community could provide. It is likely that Berkner saw himself, with his military and leadership experience, as "called to such science statesmanship."[11] The presidency of AUI was a timely opportunity to that end. In 1950, Berkner was also being considered to become the first director of the National Science Foundation. After Berkner accepted the AUI presidency, the NSF directorship was offered to Alan Waterman.[12]

Berkner accepted the presidency of AUI conditionally. One condition required the board of trustees to formally acknowledge that AUI would diversify and expand the scope of its research activities. Other conditions included moving the AUI headquarters from Long Island to New York City and that Berkner be allowed to hire a full-time executive assistant. He chose Richard M. Emberson, a physicist and radio engineer with whom he had worked at the MIT Radiation Laboratory. Emberson, also formerly of the Harvard College Observatory and the NRL, "foreshadowed the direction in which AUI would eventually expand."[13]

Berkner perceived radio astronomy as the exact sort of science that AUI wanted to expand into. He realized that the techniques involved would have significant military and national security applications as well as important implications for science, engineering, and national prestige. Berkner believed that the recruitment of the nation's best scientists was essential to the success of the project. His thinking may have been influenced by the success of the Manhattan Project and the development of the first atomic weapons. He also believed that to attract such scientists, the administrative structure would have to ensure that the standards and traditions of science were accounted for by the observatory's designers. Needell points out that preserving those traditions required the active intervention of a powerful research organization, one that could represent the research community. Radio astronomy was clearly just the sort of endeavor Berkner and AUI had been looking for.[14]

This was essentially the state of affairs at AUI when the January 1954 conference was called. American scientists were in critical need of a major radio astronomy observatory to again become competitive in the field. The parallel between nuclear physics and radio astronomy had been made in the Cambridge-Harvard-MIT "Survey of the Potentialities of Cooperative Research in Radio Astronomy." Lloyd Berkner saw radio astronomy as an ideal opportunity for the corporate expansion of AUI. The timing was perfect.

AUI and the Establishment of the NRAO

In April 1954, the AUI board of trustees met to discuss corporate expansion. Berkner distributed copies of Menzel's "Survey of the Potentialities of Cooperative Research in Radio Astronomy." Berkner's report to the trustees updated the survey. Although Menzel's group at Cambridge had originally envisioned a joint Harvard-MIT facility, they had come to prefer a facility operated by an independent corporation with a wider scope, one that would serve the entire research community. This step was important in the evolution of the conception of a national facility. The trustees agreed that radio astronomy was a highly appropriate field for corporate expansion and granted Berkner authorization to establish an AUI committee to further investigate the possibility.[15]

Berkner then proceeded in a manner similar to the steps that led to the AUI establishment of the Brookhaven National Laboratories. He invited 27 prominent astronomers to participate in an "ad hoc group on cooperative radio astronomy." Astronomers invited to participate in this planning conference included all of the authors of the survey; John Hagen of the NRL; McMath and Tuve, the NSF subpanel chairmen; Harold Ewen, Edward Purcell, John Van Vleck, and Norman Ramsey of Harvard; Robert Dicke of Princeton; Charles Townes of Columbia University; and Lee DuBridge and Jesse Greenstein of the California Institute of Technology. Berkner had the insight to invite Alan Waterman, director of the NSF; Raymond Seeger, also of the NSF; and Emmanuel R. Piore of the Office of Naval Research.[16]

The composition of the group was insightful almost to the point of being visionary. Berkner had consulted with Menzel about this in a series of meetings. The West Coast astronomers were invited to preclude the notion of any geographic bias among users of the institution, since AUI was composed of powerful Northeastern universities. The addition of the NSF observers was intended to involve NSF officials in early aspects of the planning. In fact, Berkner and J. R. Zacharias of AUI had met with Menzel, Seeger, and Piore prior to the ad hoc conference to deliberate the

government support of a national research facility for radio astronomy. Quoting Needell: "At those meetings Seeger had indicated that the NSF was greatly interested in the project described by Menzel and would make every effort to support whatever could be agreed upon by Berkner and Tuve."[17] The divergent viewpoints and imminent conflict between the two major driving forces behind the establishment had been foreseen.

A document outlining the essence of the ad hoc conference was circulated to the proposed committee members prior to the meeting. The document asserted that AUI was acting in response to the initiatives of the January conference and at the request of the Harvard-MIT group. The lagging status of research in radio astronomy in the United States was attributed to the high cost of competitive equipment, the "relative scarcity of scientists with sufficient training and background to include broadly all aspects of the subject," and "the absence of an agency to act in their behalf and to provide the opportunity for the necessary studies." Berkner implied in the document that the scientists who had initiated the proposal "already agreed that at present the greatest need was for a permanent facility where the large and basic equipment, such as large antennas, steerable dishes, and pulse generators, 'can be erected and made available to any of the various groups who are qualified and have need to use them.' "[18]

The ad hoc group on cooperative astronomy met on May 20, 1954. The outcome of the conference was a recommendation that AUI request funds to conduct a feasibility study for the planning and construction of a national observatory for radio astronomy. The principal objectives of the study were to conduct:[19]

1. A survey of opinions among scientists actively participating in research in radio astronomy or interested in pursuing such research in order to set up a program of research objectives

2. An examination of the suggestions made regarding the major instruments to gain some understanding of the technical problems of design and construction

3. An examination of proposed sites and their comparative desirability

4. An examination of any other expenditures essential to the establishment of the observatory including access roads, power lines, and support buildings

5. Preliminary estimates of the costs involved in the construction phase

6. Preliminary estimates of the organization and staff necessary to operate the completed facility and methods of proposed cooperation among interested institutions

AUI also agreed to the preparation of detailed plans with a view to submitting a request to the NSF for financial support. The request to fund the feasibility study was to be sub-

Figure 2.3 Alan Waterman. *Photo courtesy of U.S. National Archives.*

mitted to the NSF upon completion of a draft proposal by the end of June 1954.

In the proposal, the complex task of planning the facility was organized into a three phase process. Phase I was the feasibility study. Phase II was the production of a planning document including detailed designs of the instrumentation, facility infrastructure, and institutional organization. Phase III was the actual construction. This process would allow NSF to decide whether or not to proceed with the project at the end of Phase I. The selection of an operating institution to continue with the later phases would also be made at the conclusion of Phase I.[20]

A steering committee was set up to guide the study. The committee members were picked during discussions in July 1954 by Tuve, Emberson, and members of the Cambridge group. The steering committee was composed of J. P. Hagen of the NRL (chairman); B. J. Bok, Harvard College Observatory; A. J. Deutsch, Mt. Wilson and Palomar Observatories; H. Ewen, Harvard; L. Goldberg, University of Michigan; J. D. Kraus, Ohio State; A. B. Meinel, University of Chicago; J. B. Weisner, MIT; and H. E. Tatel and M. A. Tuve. J. P. Hagen served as chairman until he was selected to head the Vanguard project (the American International Geophysical Year satellite program). B. J. Bok then served as

chairman until he left to assume the directorship of the Australian National Observatory.[21]

The inclusion of NSF subpanel members, including its chairman, Merle Tuve, was only allowed after consultation with NSF lawyers. The lawyers reasoned that this inclusion was allowable since so few experts in the field existed. Overlapping was therefore considered necessary.

The steering committee approved a proposal for a grant of $105,000. The proposal was submitted to the NSF in July. Alan Waterman (Figure 2.3) met with the members of the National Science Board (NSB), the governing body of the NSF, to consider the proposal. Copies of the proposal were sent to Tuve and the radio astronomy subpanel.[22] Berkner seemed confident of receiving NSF approval and left for Europe to tour radio astronomy observatories with a view to gaining insight into the construction and operation of major radio astronomy facilities. The conflicts of interest that were to emerge were not yet anticipated.[23]

Conflicting Ideologies: Little Science v. Big Science, Regionalism v. Nationalism, and the Division of Astronomy

Despite the overlapping of committee members and Berkner's confidence in the proposal, several objections were raised by members of the astronomical community. There were four major objections to the proposed NRAO governance by AUI, principally voiced by Merle Tuve:

1. Management of the observatory by nuclear physicists with little connection to radio astronomy
2. Regionalist advocacy of the managerial body who were from powerful Northeastern universities (Harvard, MIT, Cornell, Princeton, Yale, Columbia, Pennsylvania, Johns Hopkins, and Rochester)
3. The perceived division of optical and radio astronomy
4. The proposed scale of the first instrument

The first important objection raised was that astronomers and not nuclear physicists should be the operators of the facility. Tuve claimed that, in the absence of radio astronomers on the AUI board of trustees, AUI could not provide continuity in using the proposed instruments and would have to rely on Harvard astronomers for that purpose. This argument led to a perceived North v. South regionalist advocacy of the operation of the observatory. Tuve was objecting to the perceived regionalism exhibited by AUI in the conspicuous absence of astronomers from non-Northeastern universities on the AUI board. Objections raised from three southern members of the National Science Board, Paul Gross, Joseph Morris, and William Houston, echoed this sentiment.[24] It was in this atmosphere that the AURA was created by astronomers from predominantly Southeastern universities. Resolutions to these objections would lead to the development of a new social philosophy of science that was perhaps first evidenced in the creation of the NRAO, an observatory with a truly national character.

The third major objection to an AUI-operated observatory was the division of optical and radio astronomy. Tuve firmly believed that all of the newly developed instrumentation techniques in observational astronomy, including photoelectric photometry and other newly developed optical techniques, and the new techniques in radio astronomy were different aspects of the same investigations and should therefore be controlled by one national laboratory for astronomy. This unification, however, was not to materialize. The development of the NRAO and NOAO would ultimately, in a managerial sense, represent a division of astronomy into subdivisions of optical and nonoptical research technologies. The NRAO and the NOAO had parallel origins in the proposal by some version of AURA to develop large-scale observatories funded by NSF and managed by a national consortium of universities with an interest in astronomy. The astropolitics of creation, however, ultimately led to divergent paths for the NRAO and the NOAO.

The primary objection voiced by Tuve, was, however, against what he considered preconceived notions about the scale of the planned instruments. Tuve believed that AUI members were planning to construct an enormous paraboloid, larger even than the world's largest existing telescope, the 250-foot paraboloid at Jodrell Bank. Tuve feared that such an expenditure, as would be required for the gargantuan instrument, would seriously jeopardize other astronomy programs in the United States that were government funded. Tuve also questioned whether the scientific need for such an instrument was indeed proven.[25] He suggested a scaling down of the feasibility grant and further investigation of the types of instrumentation considered. He suggested a stepwise approach to a far more modest facility. Tuve's suspicions that the AUI planners were thinking in gargantuan terms was well justified.

The Age of Gargantuan Telescopes That Never Was

Scientists involved in the planning and feasibility study had envisioned the eventual construction of enormous paraboloids ranging from 600 feet in diameter to a titan paraboloid of nearly one-half mile in diameter. These visions were entertained well into the early 1960s before it was realized that such instruments were impractical and physically impossible to construct with existing technologies. The fantastic vision of American astronomers was not unique. Scientists all over the world were envisioning radio telescopes of gargantuan proportions, for example, the proposed Jodrell Bank 400-ft Mark V telescope, the NRAO's proposed 600-ft telescope, and the U.S. Navy's 600-ft Sugar Grove instrument.

A limited understanding of the existing construction and

materials technologies led to these fantastic proposals. In a 1958 document, "The Telescope Program for the National Radio Astronomy Observatory at Green Bank, West Virginia," E. M. Emberson and N. L. Ashton (who was later the primary designer of the 140-foot (42.7-meter) telescope) in reference to the AUI feasibility study suggested that the diameter of the (proposed) paraboloid must be carefully considered. It was estimated that the cost of a radio telescope varies as a power of 2.7 of the diameter. The predetermined surface specification and the structural problems of the telescope were important factors that could greatly affect the cost. Early in the feasibility study, structural engineers were asked to estimate at what size a conventional steerable paraboloid, such as the Jodrell Bank telescope, would be limited by the strength of the materials available for its construction. According to Emberson and Ashton, "If an extreme design were adopted for a telescope, a truly enormous aperture might be obtained"; but if more or less conventional designs are visualized, "apertures of two to three thousand feet should be possible before the physical characteristics of the structural materials would be limiting."[26]

Another example of the scope of these proposed instruments is found in the planning as late as 1960. Emberson reflects the scale on which the planners were thinking in a paper presented to the New York Academy of Sciences: "Plans at Green Bank are looking ahead to the construction of a telescope with an aperture of about one-half mile (0.8 kilometer), this gain being attained by sacrificing sky coverage and restricting it to a small zone a few degrees each side of the zenith." The 1,000-ft (305-meter) telescope was being planned by Cornell University under a Department of Defense contract at the time.[27] It is clear that Emberson was familiar with the proposed design of the 1,000-ft telescope and envisioned the Green Bank instrument as being similar in design. Yet another example of planning of a giant telescope by the NRAO scientists recurrent in the historical literature is a 65-meter telescope for millimeter wave observing first proposed in 1972. This size aperture is enormous for very short millimeter waves because the surface accuracy required over the entire dish for such high frequency observations requires an elaborate supporting superstructure and positioning system. Although the giant millimeter wave telescope has appeared periodically in various planning documents and proposals, the instrument was never constructed.

Early in the study, it was recognized that the planners placed a premium on paraboloids of enormous aperture. There was a general consensus among the engineers consulted during the feasibility study that, from purely structural considerations, a paraboloid of several thousand feet in diameter could be mounted to be fully steerable.[28] There was a unanimous agreement that the cost of such a telescope would be beyond the budget available for the initial establishment of the observatory. The steering committee, therefore, decided that a structural design study should be undertaken for construction of a paraboloid of the more "modest" size of 600 feet.

The design study for the 600-foot telescope was undertaken by Jacob Feld. The steering committee provided guideline specifications to incorporate as minimum goals. Feld was asked to investigate the feasibility of constructing a parabolic surface true to within 1 inch over the entire 600-foot aperture and true to within 5/8 inch over the inner 300 feet. A surface tolerance of 1/4 inch over the entire paraboloid was set. The reflector was to be mounted in such a manner that the pointing accuracy would be within 7 seconds of arc at any given declination. The maximum permissible angular rate of motion, or slew rate, was to be 30° per minute. The Feld study was completed in July 1955. The conclusions showed that it was technically feasible to construct a 600-foot steerable paraboloid. The study indicated, however, that the extreme tolerances imposed difficult structural problems. Feld concluded that the most practical design would be one that would incorporate various types of servo mechanisms to keep the components of the telescope structure in the desired shape and adjustments.[29]

By the summer of 1956, it became evident that the Navy would embark upon a similar project to build a 600-foot telescope. Merle Tuve's arguments against the feasibility and utility of an enormous paraboloid were highly controversial to officials at AUI, who were well aware of the perceived importance of the secretive Sugar Grove project. J. P. Hagen, the initial chairman of the steering committee and NRL staff member, had often argued in favor of building the largest possible telescope at the earliest possible time. Hagen's position represented the vast majority of committee members. "In all likelihood, Tuve did not know that the push for a super-large dish went far beyond Hagen (and the entire committee), his own personal interests and questions of scientific utility."[30] The perceived critical need for an enormous telescope had in fact evolved far beyond the realm of radio astronomers.

The Sugar Grove Project and Its Relationship to the Establishment of the NRAO

The NRL and the Office of Naval Research (ONR) by 1956 had embarked upon a fantastically ambitious, highly classified "scientific" project. NRL and ONR planned to construct a fully steerable 600-foot paraboloid in the remote Allegheny mountains of Sugar Grove, West Virginia. The Sugar Grove site is less than 30 miles away from the site chosen for the NRAO. The project was to be the largest intelligence intercept station ever conceived. The basic idea was that if a large and sensitive enough telescope could be constructed, Soviet radio signals reflected from the surface of the moon could be intercepted. "The world's biggest bug was planned for the backwoods of Sugar Grove."[31]

AUI and NSF decided to treat the NRL-ONR Sugar Grove project independently from the NRAO project. AUI had already solicited the Feld preliminary design study for a 600-foot telescope and it was recognized that "for practical reasons the two efforts must be kept in touch."[32] In reality, however, only a few members associated with the NSF project were aware of the Sugar Grove project in the planning stages. Allan Needell expounds on the subtlety of the relationships between members of both projects: "One consequence of the relationship between the NSF, AUI and the Navy project was that Edward F. McClain (at the request of Homer Newell, leader of the Atmosphere and Astrophysics Division at the NRL) and Warren Ferris, who oversaw the design contracts for the 600-foot radio telescope were asked to attend the July conference." Needell further points that McClain, who later wrote a popular, unclassified account of the planned Navy device, had become the leader of the NRL radio astronomy group and had been on the AUI radio astronomy steering committee. A major practical result of the cooperation was that details of the Navy program were made available to Emberson, who was given the necessary clearance in May 1956. They were also made available to Waterman, Seeger, and Frank Edmondson, the NSF program director for astronomy.[33] Although the two projects were not related per se, the awareness of the Sugar Grove project on the part of AUI members was beneficial. The familiarity of the AUI scientists with the Navy project was, in all likelihood, considered an important advantage by the NSF officials[34] and figured in the decision to contract AUI for the NRAO project.

The Sugar Grove project was supported by the highly secretive National Security Agency (NSA), which had by many accounts in the late 1950s become the force in power behind the CIA. The NSA, engulfed in the mass paranoia of the cold war, had constructed elaborate intercept stations around the world for the purpose of intelligence gathering through the Signals Intelligence System (SIGINT). The Sugar Grove project was the "ultimate in both ambition and failure" of all intercept stations built during the 1950s. The project was also the ultimate intercept station in terms of cost. In the wake of the reaction to Sputnik 1, the NSA (insightfully referred to as the Puzzle Palace by James Bamford) and the Office of Naval Research were awarded a hefty $60 million for the project in 1969. The figure is reflective of the scale of the thinking. One year later, with only the foundation laid, the NSA and ONR requested authorization of an additional $18 million.[35]

By January 1960, the construction of the Sugar Grove telescope had begun and the "supposed" nature of the project declassified. The instrument was ostensibly being constructed to conduct research in ionospheric physics, navigation and "communications." The Navy did, however, promise to make half of each day's observing time available to the nation's astronomers. The 600-foot instrument (Figure 2.4) was to have a reflecting surface of more than 7 acres and

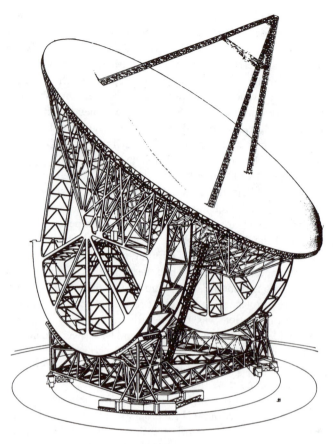

Figure 2.4 The U.S. Navy Sugar Grove 600-foot (183-meter) telescope. The proposed instrument would have stood 665 feet high
and was planned to incorporate 20,000 tons of steel and 600 tons
of aluminum. The scale is suggested by the human figures at the
base. The instrument would have been the largest movable structure ever constructed. It was never completed. *Illustration by Irving
Geis, Copyright © 1960 Scientific American, Inc. All rights reserved.*

tower to the height of a 66-story building. The design of the
superstructure would require 20,000 tons of steel. The mesh
surface would require 600 tons of aluminum and 14,000 cubic yards of concrete would be required to construct the circular base to provide motion in azimuth.[36] The Sugar Grove
instrument would have indeed been the largest movable
structure ever built.

The engineering problems turned out to be as enormous
as the instrument itself. By the summer of 1962, Sugar
Grove had clearly become a "solid steel albatross." After 2
years of labor only the twin erection towers, each 420 feet
high with derricks on top boasting 200-foot booms, and the
enormous circular base had been completed (Figure 2.5).
The real engineering problems, however, lay in the compu-
tations required for materials and stress points. The mathematical calculations required for the project were, in the
words of one engineer, "almost beyond comprehension." As
many as thirteen components had to be joined together at
one point, which demanded up to 92 separate formulas to
be worked out simultaneously, a feat that would have taxed
the capability of even the largest commercial computer

Figure 2.5 Twin erection towers used in the construction of the
U.S. Navy Sugar Grove 600-foot telescope. Each tower was 90 feet
square (27.45 meters square) and 420 feet (128.1 meters) high.
Construction booms to be placed on top were 200 feet (61 meters)
long. *Photograph by David Linton, Copyright © 1960 Scientific
American, Inc. All rights reserved, reproduced by permission of Ann
H. Linton.*

then available. Although a U.S. Navy IBM 704 computer
had worked on design specifications for over six months,
construction had advanced no farther than the rotating
tracks and pintle bearings.[37]

During the summer of 1962, the decision was finally

made to abandon the Sugar Grove project. Jerome B. Weisner, science adviser to President Kennedy, indicated in a memorandum to the president that he and CIA Director John McCone believed that "from the intelligence point of view it is a very marginal project." Weisner also indicated in the memo that he expected Secretary of Defense McNamara to cancel the project.[38] Cancellation of the project was at this point imminent. The Sugar Grove project was abandoned despite an expenditure of some $78 million.

The Navy continues to operate a receiving facility at the Sugar Grove site. Instruments at the site include a huge set of Wullenweber antennas, a 60-foot paraboloid constructed in 1957 as a prototype for the 600-foot, and an array of microwave receiving antennas ranging in size from 30 feet to 150 feet in diameter. The 150-foot antenna was constructed in part from materials left over from the 600-foot project. The suspected function of these antennas is to intercept incoming signals destined for the Communications Satellite Corporation (COMSAT) in Etam, West Virginia. More than half of the commercial, international satellites entering and leaving the United States pass through COMSAT.[39] Sugar Grove, after all, has proven to be an essential SIGINT intercept station.

The failure of the Sugar Grove project was not completely in vain. There were implications beneficial to the NRAO. The design studies for the 600-foot and its consequent failure would eventually preclude the attempt to construct the proposed 600-foot NRAO telescope. The Sugar Grove project was also influential in establishing a radio quiet zone by the West Virginia state government. The two groups worked closely together to have a radio quiet zone established that provides a sanctuary from electromagnetic interference at radio frequencies.

At the time that lands were being acquired for the NRAO site, steps were taken to establish the radio quiet zone. Contact was made with William Marland, governor of West Virginia, through the offices of Arthur D. Little, Inc. on behalf of both groups. Marland and members of his staff were briefed on the needs of the observatory and the Sugar Grove project to be protected from the encroachment of sources of radio interference. The legislative leaders were favorably disposed toward the plan and drafted a zoning act. A special session of the assembly and the senate was convened and the Radio Astronomy Zoning Act was enacted on August 9, 1956, by the West Virginia state legislature. The bill restricts operation within 10 miles of the observatory of unlicensed electrical equipment which generates interfering field strengths greater than a specified limit.[40] The zoning act was scientifically significant in that it represented the first legislation in the world to protect research in radio astronomy.

The West Virginia Zoning Act is directed toward unlicensed, local sources of radio communications. The Federal Communications Commission has jurisdiction over licensed

transmissions. The radio noise problem of the observatory and Sugar Grove was ultimately taken to Washington. After much debate and several hearings, special rules were created to establish a National Radio Quiet Zone to protect both Green Bank and Sugar Grove. The radio quiet zone is 100 miles across in the east-west direction and 120 miles across from north to south. The enactment requires that applications for transmitters within the zone be brought to the attention of the NRAO director who, in turn, is responsible for bringing the application to the attention of the Navy officials at Sugar Grove. A coordinated response is then submitted to the Federal Communications Comission.[41] The arrangement has effectively provided a radio interference sanctuary for both projects.

In the early planning stages, the relevance of Sugar Grove to the NRAO project was the perceived critical need and planning of an enormous 600-foot telescope. The ambitions of the Sugar Grove project were paralleled in the NRAO feasibility study. AUI members familiar with the Sugar Grove project commissioned the study which resulted in the proposal for a 600-foot NRAO instrument. Interaction with the Naval officials on the part of AUI members would inevitably effect the proposed telescope program at the NRAO as well as NSF's selection of AUI to operate the national radio astronomy facility. The conclusions of the AUI feasibility study implicitly paralleled the visions of the planners of the Sugar Grove project down to the exact size and type of the proposed instruments.

Conclusions of the Feasibility Study

In November 1954, the NSF radio astronomy subpanel reached several conclusions regarding the proposed feasibility study. Despite Tuve's objections to the size and scope of the proposed instruments and the institutional organization, the subpanel recommended approval of the AUI proposal. The recommendation, however, was conditional. The subpanel included the stipulations that the feasibility study be limited to instruments of modest size and that the search for a site for the observatory be limited to the region within a 300-mile (483 kilometer) radius of Washington, D.C. The size stipulation was included to appease the objections of Merle Tuve. The AUI steering committee guiding the feasibility study interpreted the "modest size" stipulation to delineate an aperture not in excess of 600 feet. The site stipulation was made, primarily, as an attempt to geographically balance the nation's major astronomy facilities since all of the major optical observatories were located in the West. The subcommittee also reasoned that many of the nation's active radio astronomy groups, such as Naval Research Laboratories, Department of Terrestrial Magnetism, National Bureau of Standards, and the group at the University of Maryland were located in or near Washington.[42]

The 300-mile stipulation seemed reasonable to the AUI steering committee, which of course was more than willing to comply with the stipulation to locate the observatory in the Southeast. The instrumentation size stipulation was sufficiently ambiguous to also be acceptable. Both stipulations were therefore acceptable to Berkner and the AUI. NSF quickly approved the feasibility study. On February 18, 1955, the NSF granted $85,000 to AUI to support the feasibility study.[43] The establishment of a national radio astronomy observatory seemed close at hand.

With the feasibility study approved, the steering committee focused its efforts on defining what sort of instrumentation was most appropriate for the observatory. The feasibility study for the telescope program of the observatory ended in late 1955 with a proposal for a series of six fully steerable parabolic radio telescopes:[44]

1. A 600-foot telescope, the largest completely steerable telescope at the observatory and, indeed, in the world

2. A telescope of 250- to 300-foot aperture

3. A 140-foot telescope of the highest attainable precision

4. Two telescopes of the "standard" 60-foot or 84-foot size, both to be devoted to observational research programs requiring interferometric measurements

5. A relatively small (28 feet in diameter) telescope to be devoted to pattern measurements and to the testing of receivers or other components under development at the observatory

The AUI steering committee members reached a consensus by April that the first instrument to be constructed would be the 140-foot high precision fully steerable paraboloid. The members of the committee agreed that further studies and experience would be needed before any attempt to construct the 600-foot telescope.[45]

NSF, NSB, AURA, and AUI: Management of the NRAO and a Social Philosophy of Science

By April 1955 Alan Waterman had decided that it was time that the NSF formally address the issue of whether or not to commit to supporting the creation and operation of large user-type facilities. Waterman requested that AUI produce a general planning document for an NSF-sponsored facility to be presented to the NSB Committee on the Mathematical, Physical, and Engineering Sciences (MPE).[46] AUI provided the requested documents, including an extensive planning document containing design proposals solicited for the 140-foot instrument. Berkner and Emberson provided the "standard arguments" for the creation of an NSF-supported facility. They also presented an argument that had been previously only implied—the military implications of such a facility. In a preliminary draft of the argument to be presented to the MPE, Berkner and Emberson explained that most radio astronomy signals are incredibly weak by normal communication engineering standards. The

demands of the astronomers have forced advances in electronic techniques which have had many practical benefits. A few examples include tremendous improvements in the noise factors of receivers, in the precision of constructing and controlling various types of antennas, in discrimination and integration techniques, in broad-banding radio frequency components, and in data display devices.[47] The value of such advances to military security was obvious.

The military references incited intense emotional reactions. Tuve was apparently outraged. Berkner and Tuve were both permitted to make verbal and written presentations to the NSB Committee on MPE Sciences prior to its consideration of the question of support for large facilities.[48] In his written summary, Tuve presented his standard arguments. He expressed the fear among optical astronomers that large expenditures for radio astronomy would decrease support for research in other areas of astronomy. He questioned whether AUI was an appropriate agency for the implementation of government-funded research in radio astronomy. A postscript ended the written presentation that supposedly represented the feelings of the NSF subpanel. In it he stated that the subpanel regarded radio astronomy as part of astronomy and did not believe it should compete with the proposed national observatory and new photoelectric work. The subpanel regarded these as complementary developments and took the position that it was more logical to do both at modest rates than to choose one and eliminate the other. Quoting Tuve, "Radio astronomy is a study of the heavens, not just glorified electronics."[49] Tuve did, however, inform the subpanel that there was a general consensus that radio astronomy was a field that should be supported by the NSF and that the first logical step was the planning of an intermediate telescope on the order of 150 feet in diameter.[50]

The NSB met on May 18, 19, and 20, 1955, to formally discuss the NSF role in the support of large-scale scientific research. On May 20, the full NSB passed a resolution "committing the NSF to support large-scale basic scientific facilities . . . The NSB cited 'a National Astronomical Observatory, a major radio astronomy facility, and the university installation of computers, accelerators and reactors' as examples of the kind of facilities appropriate for NSF support." In addition, the NSF concluded that support for these facilities should be provided for by "special budgets" to preclude encroachment on the regular established programs of the NSF.[51] The May 20 resolution marked a major turning point in the history of radio astronomy and large-scale scientific research in general in the United States; a trend toward big science was beginning to be realized.

Encouraged by the May 20 NSF resolution, the AUI steering committee met in New York on May 31, 1955, to discuss details of the planning work. The committee "attempted to navigate an intermediate course between the purist attitude of Tuve and the grandiose plans of Berkner."[52] Detailed planning work was undertaken, including the inspection and evaluation of potential sites, the design of an intermediate dish set at 140 feet in diameter at this conference, and examination of the preliminary studies of the 600-foot dish. Tuve and the NSF radio astronomy subpanel approved the immediate plans and AUI was granted an additional $140,500 to continue its preparatory activities.[53] The detailed planning document requested by the NSF was, at this point, the predominant task of the preparatory activities to be supported by the additional funding.

Although the planning of instrumentation was well underway and an intermediate course between Tuve and Berkner had apparently been sought, several problems still existed that would prove to be a source of confrontation. The scope and type of instruments were still in question although the sizes had been determined. The choice of site and the institutional nature of the facility were still in question. The big question, according to Tuve and several other subpanel members, was the choice of the appropriate contractor to construct and operate the facility.

Tuve, partly because of the military implications brought out by Berkner, reacted against the possible choice of AUI as the management institution. The primary result of Tuve's reaction was an attempt to organize the formation of a new consortium to manage the radio astronomy facility. Tuve had, in fact, envisioned a consortium that would manage the newly created national optical observatory as well as the radio astronomy facility. To assuage Tuve and his allies on the NSB, alternatives to AUI were considered by the NSF. AURA was created to this end. AURA was, in this manner, inspired by Tuve. This alternative to AUI management was initiated in June of 1965 and was ultimately proposed by William Pollard, executive director of the Oak Ridge Institute of Nuclear Studies (ORINS). Paul Gross, dean at Duke University, had been contacted by the MPE division of NSF to suggest possible alternatives to AUI management. Gross suggested that ORINS could be a vehicle through which a new consortium could be created. Gross worked with William J. Pollard who assumed the responsibility to develop a proposal to present to the NSF.

Twelve potential incorporators were identified including Leo Goldberg, Jesse Greenstein, John Hagen, Irvin Stewart, Donald Menzel, and Otto Struve. Tuve had simultaneously contacted Jesse Beams, president of the University of Virginia, and Irvin Stewart, president of West Virginia University, about the possibility of creating a consortium of universities with the help of ORINS. The proposed consortium, initially called the Association of Universities for Radio Astronomy (AURA), ultimately evolved into a more broadly representative proposed organization, the Association of Universities for Research in Astronomy (AURA).[54] The perceived advantage of the proposed consortium of universities over AUI was the reliance on astronomers as opposed to nuclear physicists as managers of the observatory. The inclusion of institutional members was invisioned to be very flexible. All universities with serious research interests in astronomy would eventually be allowed to join the organization. Geographic preferences would in this way be precluded.

The timing was critical, the NSF's MPE committee had called a conference to discuss the AUI feasibility study and the management of the radio astronomy observatory. The Conference on Radio Astronomy Facility was held in Washington on July 11, 1956. The central focus of the conference was a session concerning the possible organization and management to ensure the national character of the radio astronomy facility. Two proposals were presented, recommending AUI and AURA. The potential of AURA as a viable leadership alternative of the radio astronomy observatory was established at this conference. Inspired by this recognition, Pollard and Tuve began to organize. The initial meeting of the AURA consortium was planned for August 21, 1956. At this meeting the incorporators would adopt the bylaws and accept the institutional members. The respective institutions would thereafter appoint trustees to govern the corporation.[55]

The AURA consortium, as originally envisioned however, was never to materialize. Menzel reacted negatively to his AURA invitation on behalf of the Harvard Observatory and refused to join. Leo Goldberg also declined on behalf of the University of Michigan.[56] Waterman requested that the members postpone the incorporation until the NSB could make a decision on the management institution. He perhaps suspected that incorporation of the consortium would be unnecessary after the NSB decision. This version of AURA never came into existence; only the name and acronym survived. A social philosophy of basic science began to emerge in these events. The attempt to establish AURA reflected this emerging social philosophy, one that was based in a reaction to the need of big science research ventures. Much basic research had grown too big to be supported by single institutions. Consortia of universities began to emerge as the organizational structure needed to manage big science changed. The organizational structure of these big science ventures would begin to resemble corporate structure—a quasi-independent governing body supported by external (or generated) funding. Although this version of AURA would have possessed an organizational structure consistent with this emerging social philosophy, an organization based on this corporate structure with experience in the management of big science ventures already existed—the AUI.

The NSB by late July had reduced the management choices to three. Waterman identified the three choices as: (1) management of the national radio astronomy facility by a single university, (2) management by AUI, (3) and management by a new organization which almost certainly indicated the Tuve-Pollard AURA consortium.[57] The NSF internal deliberations, as uncovered by Needell, are recorded in NSF memoranda, correspondence, and personal memoranda of Alan Waterman and several members of the NSB. According to these documents, Waterman reportedly never favored a Tuve-Pollard alternative to AUI. His first choice was management by West Virginia University. His personal admiration for Irvin Stewart certainly must have influenced this proclivity toward governance by a single institution.[58]

During July, Waterman engaged in several discussions with Irvin Stewart regarding the management and operations of the national radio astronomy facility. Stewart determined that the "peculiarities of West Virginia law would have made the financial arrangements unduly complex, and suggested that another university be found."[59] The complete absence of an astronomy group at West Virginia University must also have figured in the decision to decline the invitation as a management choice for the observatory.

The disinterest on behalf of West Virginia University and the disintegration of the AURA effort in radio astronomy, even before its incorporation, weighed heavily in the favor of AUI as the management choice. The experience of AUI as well as its established contacts, particularly the contacts involved with the Navy's "parallel" Sugar Grove project, began to seem an undeniable advantage. Berkner agreed to have astronomers appointed to the board of trustees as vacancies occurred, which appeased the argument against management by nuclear physicists to a certain degree. Berkner also indicated that the AUI board would "consider modifying its bylaws to permit the addition of up to three 'trustees-at-large,' and that these would be from outside the Northeast."[60] The advantages of AUI as the management institution began to add up. The alternatives seemed weak, at best, by comparison.

On August 24, 1956, the NSB awarded the project to AUI. By November 1956, an establishment contract between AUI and NSF had been completed. The NSF was still, at this point in its history, an infantile institution in the postwar scientific research community. The commitment to construct a national facility for research in radio astronomy represented the "entrance of the NSF into the realm of big science."[61]

It is clear that many of the NSF members considered Tuve's attitude anachronistic. There was a general belief that the nature of an AURA consortia-operated facility would reflect Tuve's ideals toward postwar research. Needell explains: "It was also clear that Tuve's notions about science and the importance of keeping it small and centered on the individual researcher were not shared by many astronomers, nor were they shared by the National Science Foundation."[62] Most scientists and science administrators also apparently believed that Tuve's ideas were out of step with the world of 1956. Needell lends insight into the prevailing attitude toward the scale of the new science and its federal sup-

port, stating that they rationalized that the development of scientific tools of unprecedented power and therefore science itself was enormously spurred by close association with massive government-sponsored development projects. AUI was an organization that was formed to manage such projects. In fact, AUI was the first modern university consortium and the only alternative with experience in the type of large-scale scientific research that Tuve so strongly opposed.

Although the NSB indicated that serious consideration would be given to the possibility of a common management for the national radio and optical astronomy observatories, this was not meant to be. A version of AURA did survive and ultimately became responsible for the management of the National Optical Astronomy Observatory (NOAO). The site survey was undertaken early in 1955, at nearly the same time as the initial grant was awarded to AUI. In fact, the search for the optical site was still in progress when the contract for the NRAO was signed with AUI. AURA was not incorporated until October 28, 1957. Kitt Peak was selected as the NOAO site on March 1, 1958; a Southern Hemisphere site at Cerro Tololo was selected in 1962. Other instruments were subsequently added and combined to form the National Optical Astronomy Observatories.[63] The NRAO and NOAO thus had a common origin but ultimately led divergent paths that resulted in two independent organizations. In a recent sequel to this story, the management of the Hubble Space Telescope, the first large aperture optical space telescope, was awarded to AURA in 1982. The chief competing contractor for the Space Telescope, interestingly, was AUI.

In retrospect, it must be stated that although much of Tuve's purist and simplistic approach to science was apparently out of place in the postwar science of 1956, many of his ideals for the national radio astronomy facility were visionary. Tuve had foreseen the enormous engineering problems associated with construction of a 600-foot paraboloid. His objections to attempting the construction of large aperture instruments were again echoed in the emerging troubles that plagued the construction of the NRAO 140-ft. telescope in the early 1960s. Ironically, in his objection to any connection with the Sugar Grove project, he had also foreseen the eventual attitude of the NRAO toward the military implications of the research performed there.

The members of the NRAO staff have historically taken great pride in the fact that the research in radio astronomy performed at the observatory has no malevolent application. Research in postwar physical sciences has unleashed tremendous powers. These powers have often been used for destruction, the primary example being the phenomenal destructive power unveiled by research in nuclear physics. Radio astronomy is seen as a benevolent science by many researchers, particularly staff members at the NRAO. Richard Fleming, in reference to the access to research performed at the NRAO and the military implications, points out that instrumentation developed at the NRAO is shared openly with other observatories and is freely printed and distributed. Military and independent scientists, and private industries have the same access to NRAO developments. Anyone interested in building better receivers or improving receiver technology, radio frequency amplifiers, or antennas is allowed access to the technological developments of the NRAO.[64] This openness and freedom of information are reflective of the observatory's user institution administrative structure.

The relationships of the NRAO to Sugar Grove and the NRL reflects this openness. NRL has historically done much of the work for Sugar Grove and has been intimately related to NRAO. For example, the NRAO has done time keeping for NRL with the interferometer system. Since Sugar Grove is a military installation, most of its work is done by contractors or by NRL. These organizations have the same access to NRAO research and therefore share technology.[65] This attitude pervades the history of research at the observatory. The primary goal of the National Radio Astronomy Observatory is and has been the advancement of the science of radio astronomy.

Planning Document

With the NSB decision to contract AUI as the NRAO management institution and the conclusion of Phase I, the feasibility study, the AUI steering committee could devote its efforts to Phase II, the planning document. Emberson was selected as acting project director of the planning document. The size, type, and nature of the observatory's instrumentation had been defined by the conclusions of the feasibility study. The purpose of Phase II was to provide more detailed plans of the primary instruments as well as define the institutional structure of the facility, its infrastructure, and the selection of the site.

The planning document provided a brief history of radio astronomy in the United States as an introduction and outlined the potential usefulness of the planned instruments. Their greatest value lay in the unknown and in their ability to extend still further the frontiers of science. Hagen indicates, "The availability of such instruments will enable astronomers to solve many of the vexing problems, raised by the limitations of present equipment, facing us today, but more important will bring to light many things that are today unknown and unpredictable."[66] Other topics treated in the planning document included a summary of the feasibility study and the proposed instruments defined within the study; research objectives and potential telescope programs; site specifications, selection, and development; preliminary specifications for the 140-foot and 600-foot; three potential designs to be considered for the 140-foot; staff structure, personnel policies, and procedures; and the construction budget.[67]

Selection of Green Bank as the NRAO Site

The choice of site had been carefully deliberated by the AUI planners. In the production of the planning document the steering committee adopted eight specifications to guide the selection of the site:

1. *Radio Noise.* The level of noise or interference at wavelengths below 10 meters must be extraordinarily low. Several conditions were necessary to meet this criterion. The number of nearby inhabitants must be extremely low. Similarly, the site should be at least 50 miles away from any city or concentration of people or industries. The site should not be near high-tension power lines, which generate radio noise through corona discharges. The site should not be near commercial air routes. The site should be in a valley surrounded by as many ranges of high mountains in as many directions as possible to attenuate direct propagation from radio stations and to reduce diffraction of tropospheric propagation into the valley.

2. *Location South.* The location should be as far south as possible to permit observation of the center of the Milky Way and other objects having southern declinations.

3. *Location North.* The site should be in northern latitudes to permit research involving auroras, ionospheric scintillation, and polar blackouts.

4. *Ice and Snow.* The site should not be in an area of excessive ice and snow that would create snow loads on the telescopes and cause excessive down-times.

5. *Winds.* The site should not be in an area subject to violent winds and tornadoes. High winds may cause dangerous vibrations in large structural units.

6. *Humidity.* The climate should be reasonably mild to avoid excessive physical deterioration of materials.

7. *Size.* The size should be large enough to allow adequate separation among the installations of many types and sizes of telescopes and arrays. A total area of 5,000 to 10,000 acres is desirable.

8. *General Surroundings.* Within the limits set by the basic requirements, the site should provide as many of the attributes of a university campus as possible and be accessible by plane, rail, or automobile.

A ninth specification, that the search area be limited to within a 300-mile radius of Washington, D.C., was added by the NSF advisory panel.[68]

An ad hoc panel was created to search for an appropriate site. Emberson organized the panel largely of persons having direct knowledge of the search area as well as an interest in performing research at the proposed observatory. The search area was greatly reduced initially by the high winds criterion. Hurricane and tornado information furnished by

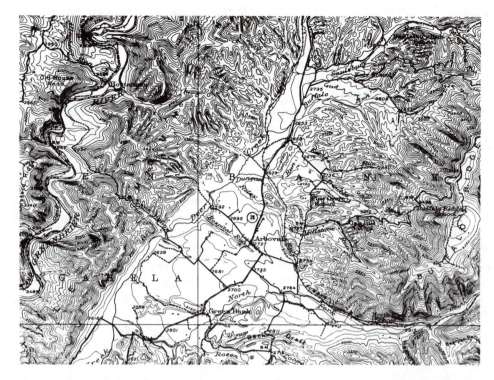

Figure 2.6   U.S. Geological Survey map of the NRAO Green Bank Site. The Deer Creek Valley is surrounded by mountain ranges on all sides providing ideal protection from radio interference.

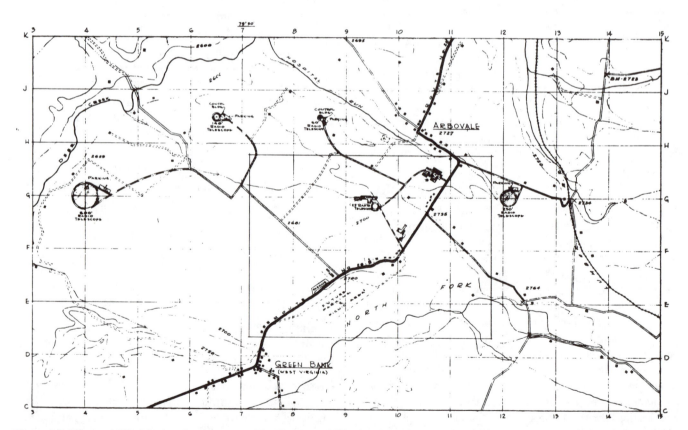

Figure 2.7   General NRAO site plan as prepared by Eggers and Higgins, an architectural firm in New York City. Sites for the proposed 123-foot (37.5-meter), 60-foot (18.3-meter), 40-foot (12.2-meter), 250–300-foot (76.25–91.5 meter), and 600-foot (183-meter) telescopes are shown. Courtesy NRAO/AUI.

1946	Incorporation of AUI. Est. of Brookhaven National Laboratory
1950	Establishment of NSF. Government Support of Science Research
January 1954	State of Astronomy Conference. NSF, DTM, and California Institute of Technology. Concludes U.S. lags behind other countries in radio astronomy; need is recognized for large telescopes and improved observing equipment.
	Harvard College Observatory and MIT discussions lead to "Survey of the Potentialities of Cooperative Research in Radio Astronomy." AUI expresses interest in managing a national radio astronomy facility.
April 1954	AUI Board of Trustees meet. Concept of a national facility to serve the interests of the university research community discussed.
May 1954	AUI "ad hoc group on cooperative astronomy" includes astronomers from Harvard, MIT, Columbia, CIT, NRL and NSF. Request of funds for feasibility study from NSF to produce a planning document with detailed instrumentation designs.
June 1954	Proposal for Feasibility Study submitted. Three phase process suggested.
July 1954	NSF creates advisory Committee for radio astronomy. Tuve serves as chair. Approves AUI request for 105K for Feasibility Study.
November 1954	NSF Subpanel recommends funding AUI proposal with site stipulation.
February 1955	NSF approves Feasibility Study, 85K awarded to AUI.
April 1955	AUI Steering Committee plans instruments; 120–150 feet diameter set for the 1st telescope.
May 1955	NSB passes resolution committing to "large-scale basic science facilities" citing the establishment of a National Observatory" for radio astronomy.
	AUI Steering Committee meets. First telescope size fixed at 140-feet and preliminary studies begun for a 600-foot telescope.
July 1955	Feld Study for Sugar Grove Begun, AUI officials become aware of the U.S. Navy project
December 1955	AUI Steering Committee meets in Washington. Green Bank, West Virginia chosen as site.
June 1956	Tuve/Pollard organization attempted: ORINS/AURA proposal developed
July 1956	Conference on Radio Astronomy-2 proposals submitted: AUI and ORINA/AURA
August 1956	NSB reduces management choices to : single university, AUI, or a new organization (i.e. AURA).
November 1956	NSB decides in favor of AUI. West Virginia Radio Astronomy Zoning Act Passed. Contract establishing the NRAO signed by NSF and AUI. Berkner appointed interim director.
October 1957	Ground breaking ceremonies at Green Bank on October 17, 1957.

Figure 2.8 Timeline of significant events leading to the establishment of the National Radio Astronomy Observatory.

the U.S. Weather Bureau eliminated many regions in the search area. Regions in the Northeast were eliminated by the ice and snow criterion. The initial stage of the search ended with a list of two dozen possible sites compiled by the search committee. Of these, many were eliminated upon visible inspection because of existing urban and industrial centers.

Arrangements were made with the Naval Research Laboratory to provide instruments for the examination of the remaining sites for radio interference. Engineers of Jansky and Bailey, Inc. were contracted for the purpose of making radio noise measurements with the NRL equipment. The data from these measurements permitted the assignment of an index for interference levels at the five remaining sites. On the basis of these measurements, one site stood out. The radio interference measurements indicated that Green Bank, West Virginia, was in a class by itself. The

Green Bank site was also first on the basis of population studies, first in the nearby urban areas criterion, and first in the aviation activities criterion.[69] The site also proved ideal on the basis of protection from radio interference by surrounding mountain ranges.

The Green Bank site consists of a portion of the Deer Creek Valley in Pocahontas County. The average elevation is 2,700 feet above sea level. A U.S. Geological Survey map (Figure 2.6) of the Deer Creek Valley included in the planning document shows that the site is completely surrounded by mountain ranges, many of which rise to nearly 4,000 feet above sea level. The mountains provide ideal protection from spurious radio interference.

In the December 11–13, 1955, AUI steering committee meetings, Green Bank was designated as the preferred site choice. The committee urged that nearly all of the Deer

Creek Valley be acquired to ensure local protection against interference. The NSF adopted the steering committee's recommendation and authorized AUI to obtain purchase options. After completion of the establishment contract between AUI and the NSF in November 1956, site procurement was an immediate priority. The U.S. Army Corps of Engineers had vast experiences with the process of land procurement and was therefore engaged to acquire the land on behalf of the NSF. The site consisting of 2,700 acres (4.2 square kilometers) was acquired for $550,000.[70]

Initial site development plans were produced by the New York firm of Eggers and Higgins. The steering committee provided specifications for the initial development. Instrumentation defined by the feasibility study was incorporated into the site as well as a central laboratory and administrative building, a site maintenance building and garage, a telescope maintenance building, control buildings for each major instrument, a dormitory and apartment building with a cafeteria, and several on-site residences for staff members and visiting scientists. The general plan for the initial site development as prepared by Eggers and Higgins is seen in Figure 2.7. For the actual construction, it seemed more logical to engage a local firm which would be familiar with the regional building situation. AUI accordingly contracted Irving Bowman and Associates of Charleston, West Virginia, to perform the necessary architectural and engineering work for the infrastructure at Green Bank.[71] The design, engineering, and construction of each instrument would be contracted out to firms to be decided at later planning stages. Construction of the site infrastructure began in May 1957.

Establishment of the NRAO

On November 17, 1956, a contract for the establishment and operation of the National Radio Astronomy Observatory was signed by Alan T. Waterman and Lloyd V. Berkner. As discussed in this chapter, the NRAO came into existence through a series of complex scientific, technical, and political events. The history of the observatory's creation provides important insight into the establishment and operation of postwar scientific research facilities. By way of summary a timetable of significant events is provided (Figure 2.8). Lloyd V. Berkner was appointed interim director of the NRAO. The establishment of the first field office and the groundbreaking ceremonies took place on October 17, 1957, a significant event in the evolution of radio astronomy in the United States.

Endnotes

1. "Proceedings of Washington Conference on Radio Astronomy, January 4–6, 1954," *Journal of Geophysical Research*, vol. 59, p. 140, March, 1954, "Radio Astronomy Conference," *Science*, vol. 119, p. 588, April 30, 1954.

2. John Hagen, "Radio Astronomy Conference," *Science*, vol. 119, p. 588, April 30, 1954.

3. "Proceedings of Washington Conference on Radio Astronomy, January 4–6, 1954," *Journal of Geophysical Research*, vol. 59, p. 140, March 1954.

4. R. M. Emberson and N. L. Ashton, "The Telescope Program for the National Radio Astronomy Observatory at Green Bank, West Virginia," *Proceedings of the IRE*, vol. 46, no. 1, January, 1958, p. 23–35.

5. Merle A. Tuve Papers, Untitled draft of a speech delivered to the trustees of the Carnegie Institution of Washington on December 1, 1953, Box 326, Manuscript Division, Library of Congress.

6. "Survey of the Potentialities of Cooperative Research in Radio Astronomy," included in a personal memo from Menzel to Berkner, April 13, 1954, AUI RA file, BNL, "Lloyd Berkner, Merle Tuve, and the Federal Role in Radio Astronomy," *OSIRIS*, 2nd series, 1987, 3, 261–288.

7. Allan A. Needell, "Lloyd Berkner, Merle Tuve, and the Federal Role in Radio Astronomy," *OSIRIS*, 2nd series, 1987, 3, 261–288.

8. *Ibid.*, pp. 261–288.

9. *American Men and Women of Science*, 13th ed., New York: Bowker, 1976, vol. II, p. 1186.

10. Allan A. Needell, "Lloyd Berkner, Merle Tuve, and the Federal Role in Radio Astronomy," *OSIRIS*, 2nd series, 1987, 3, 261–288. Needell summarizes Lloyd Berkner's career based on *American Men and Women of Science*, 13th ed., New York: Bowker, 1976, vol. II, p. 21.

11. *Ibid.*, pp. 10–17.

12. J. Merton England, *A Patron for Pure Science: The National Science Foundation's Formative Years*, Washington, D.C.: NSF, 1982, p. 123–127.

13. Allan A. Needell, "Lloyd Berkner, Merle Tuve, and the Federal Role in Radio Astronomy," *OSIRIS*, 2nd series, 1987, 3, 261–288.

14. *Ibid.*, pp. 261–288.

15. AUI board of trustees minutes, April 16, 1954, AUI records. Also described in Needell's, "Lloyd Berkner, Merle Tuve, and the Federal Role in Radio Astronomy," *OSIRIS*, 2nd series, 1987, 3, 270.

16. Berkner to Paul Hammond, May 11, 1954, personal memo with attachments, AUI RA file, BNL, "Lloyd Berkner, Merle Tuve, and the Federal Role in Radio Astronomy," *OSIRIS*, 2nd series, 1987, 3, 270.

17. AUI executive committee minutes, May 21, 1954, AUI records, "Lloyd Berkner, Merle Tuve, and the Federal Role in Radio Astronomy," *OSIRIS*, 2nd series, 1987, 3, 270.

18. Allan A. Needell, "Lloyd Berkner, Merle Tuve, and the Federal Role in Radio Astronomy," *OSIRIS*, 2nd series, 1987, 3, 271, Berkner to Hammond, May 11, 1954, AUI RA file, BNL.

19. R. M. Emberson, "National Radio Astronomy Observatory," *Science*, November 13, 1956, vol. 130, no. 3385, pp. 1307–1318.

20. Richard Emberson, Lloyd Berkner, and Charles Dunbar, "Draft Research Proposal to the National Science Foundation for a Grant in Support of a Feasibility Study of a National Radio Astronomy Facility," June 29, 1954, AUI RA file, BNL.

21. R. M. Emberson and N. L. Ashton, "The Telescope Program for the National Radio Astronomy Observatory at Green Bank, West Virginia," *Proceedings of the IRE*, vol. 46, no. 1, January 1958, p. 24.

22. AUI board of trustees minutes, July 14, 1954, and AUI executive committee minutes, July 26, 1954, AUI RA file, BNL.

23. Allan A. Needell, "Lloyd Berkner, Merle Tuve, and the Federal Role in Radio Astronomy," *OSIRIS*, 2nd series, 1987, 3, 271.

24. J. Merton England, *A Patron for Pure Science: The National Science Foundation's Formative Years*, Washington, D.C.: NSF, 1982, p. 282.

25. Allan A. Needell, "Lloyd Berkner, Merle Tuve, and the Federal Role in Radio Astronomy," *OSIRIS*, 2nd series, 1987, 3, 272.

26. R. M. Emberson and N. L. Ashton, "The Telescope Program for the National Radio Astronomy Observatory at Green Bank, West Virginia," *Proceedings of the IRE*, vol. 46, no. 1, January 1958, p. 25.

27. R. M. Emberson, "Radio Astronomy and the New National Observatory," *Annals of the New York Academy of Sciences*, presented at a meeting of the Division on March 18, 1960.

28. R. M. Emberson, "National Radio Astronomy Observatory," *Science*, November 13, 1956, vol. 130, no. 3385, pp. 1307–1318.

29. Jacob Feld, "Design Study for the Construction of a 600-foot Radio Telescope," *Annals of the New York Academy of Science*, vol. 70, 1955, p. 153.

30. Allan A. Needell, "Lloyd Berkner, Merle Tuve, and the Federal Role in Radio Astronomy," *OSIRIS*, 2nd series, 1987, pp. 3, 281.

31. James Bamford, *The Puzzle Palace: A Report on America's Most Secret Agency*, New York: Penguin, 1983, pp. 218–221.

32. R. M. Emberson, memorandum, "Relationship of the National Radio

Astronomy Facility to Military Projects," June 18, 1956, AUI RA file, BNL.

33. Allan A. Needell, "Lloyd Berkner, Merle Tuve, and the Federal Role in Radio Astronomy," *OSIRIS*, 2nd series, 1987, 3, 281, "The 600-foot Radio Telescope," *Scientific American*, January, 1960, 202(1):45–51; and Emberson memorandum, "Relationship of the National Radio Astronomy Facility to Military Projects," June 18, 1956, AUI RA file, BNL.

34. *Ibid.*, p. 281.

35. James Bamford, *The Puzzle Palace: A Report on America's Most Secret Agency*, pp. 218–221.

36. Edward F. McClain, Jr., "The 600-Foot Radio Telescope," *Scientific American*, January 1960, 202(1):45–51.

37. James Bamford, *The Puzzle Palace: A Report on America's Most Secret Agency*, pp. 218–221.

38. *Ibid.*, pp. 218–221.

39. *Ibid.*, pp. 218–221.

40. J. W. Findlay, "Noise levels at the National Radio Astronomy Observatory," *Proceedings of the IRE*, vol. 46, no. 1, January, 1958, pp. 35–38.

41. R. M. Emberson, "National Radio Astronomy Observatory," *Science*, November 13, 1956, vol. 130, no. 3385, pp. 1307–1318.

42. Greenstein to Tuve, personal memorandum, November 17, 1954 and McMath to Tuve, personal memorandum, October 15, 1954, Tuve Papers, Box 326; and Berkner memorandum, "Conference with NSF Panel on Radio Astronomy," November 18, 1954, AUI RA file, BNL, "Lloyd Berkner, Merle Tuve, and the Federal Role in Radio Astronomy," *OSIRIS*, 2nd series, 1987, 3, 273.

43. R. M. Emberson, "National Radio Astronomy Observatory," *Science*, November 13, 1956, vol. 130, no. 3385, pp. 1307–1318.

44. *Ibid.*, pp. 1307–1318.

45. AUI executive committee minutes, March 18, 1955, AUI Records.

46. Allan A. Needell, "Lloyd Berkner, Merle Tuve, and the Federal Role in Radio Astronomy," *OSIRIS*, 2nd series, 1987, 3, 274.

47. *Ibid.*, p. 274.

48. *Ibid.*, p. 274.

49. *Ibid.*, p. 274.

50. *Ibid.*, p. 274.

51. *Ibid.*, p. 275.

52. *Ibid.*, p. 275.

53. Allan A. Needell, "Lloyd Berkner, Merle Tuve, and the Federal Role in Radio Astronomy," *OSIRIS*, 2nd series, 1987, 3, 276, Bok to Tuve, May 31, 1955 and Bok to Deutsch July 22, 1955, AUI RA file, BNL; as well as Tuve to Seeger, September 29, 1955, and Waterman to Berkner October 11, 1955, Tuve Papers, Box 327; and AUI executive committee minutes, June 17, 1955, July 14, 1955, October 18, 1955, AUI Records.

54. Allan A. Needell, "Lloyd Berkner, Merle Tuve, and the Federal Role in Radio Astronomy," *OSIRIS*, 2nd series, 1987, 3, 284, Draft, Agreement of Incorporation, June 22, 1956, Tuve Papers, Box 328; and circular letter William Pollard to Distribution, July 16, 1956, NSF History File.

55. Allan A. Needell, "Lloyd Berkner, Merle Tuve, and the Federal Role in Radio Astronomy," *OSIRIS*, 2nd series, 1987, 3, 285.

56. *Ibid.*, p. 285.

57. *Ibid.*, p. 285.

58. *Ibid.*, p. 285.

59. *Ibid.*, p. 285.

60. Allan A. Needell, "Lloyd Berkner, Merle Tuve, and the Federal Role in Radio Astronomy," *OSIRIS*, 2nd series, 1987, 3, 285, AUI executive committee minutes, July 20, 1956, AUI Records.

61. Allan A. Needell, "Lloyd Berkner, Merle Tuve, and the Federal Role in Radio Astronomy," *OSIRIS*, 2nd series, 1987, 3, 283.

62. *Ibid.*, p. 283.

63. Frank K. Edmondson, *AUI, Kitt Peak, and Cerro Tololo—The Early Years*, unpublished manuscript, March 28, 1992.

64. J. Richard Fleming, Interview with Benjamin K. Malphrus, National Radio Astronomy Observatory, Green Bank, West Virginia, March 31, 1989.

65. J. Richard Fleming, Interview with Benjamin K. Malphrus, March 31, 1989.

66. "Planning Document for the Establishment and Operation of a Radio Astronomy Observatory," prepared for the National Science Foundation by Associated Universities, Inc., New York, New York, 1956.

67. *Ibid.*

68. R. M. Emberson, "National Radio Astronomy Observatory," *Science*, November 13, 1956, vol. 130, no. 3385, pp. 1307–1318.

69. *Ibid.*, pp. 1307–1318.

70. *Ibid.*, pp. 1307–1318.

71. "Planning Document for the Establishment and Operation of a Radio Astronomy Observatory," 1956.

Chapter 3

Early Days of the NRAO

The first NRAO staff was appointed in June 1957. A small scientific, engineering, and technical staff was assembled to lay the groundwork for the facility. The primary duties of the scientific staff were directed toward organization and program planning, supervision of equipment and instrumentation development and contracting, and the initiation of research projects. The engineering staff was concerned with the design and eventual construction of the major instruments. The technical staff was concerned with the development of electronics equipment to be used with the telescopes. R. M. Emberson was appointed chairman of the Major Design and Construction Department. John Findlay was appointed chairman of the Research Equipment Development Department. Two astronomers, D. S. Heeschen and F. D. Drake, were appointed to the permanent staff. The appointment of the first director to replace Lloyd Berkner, the interim director, was carefully considered. Otto Struve, a highly productive and respected optical astronomer, was selected for the position. The selection of Struve reflects the ambitions of the observatory's organizers to create a cutting edge research facility. Struve was appointed first director of the NRAO on July 1, 1959.[1] The intentions of the planners is recalled by James F. Crews, director of Telescope Services and original member of the Green Bank staff since 1958: "One of the observatory's biggest goals was to develop as quickly as possible an international respect... because the United States just did not have that in radio astronomy at that time." Crews reflects upon the ambitions of the initial staff members: "One of the things that we did was to work very hard to do everything in a very professional way to bring Green Bank to the forefront in a very few years."[2]

The first observing program had actually begun before the groundbreaking ceremonies and site development. A simple interferometer was set up in the fall of 1956 with the assistance of the radio astronomy branch of the Naval Research Laboratory. The interferometer, a simple total power design, was used to observe bright radio sources and measure the noise on the site at a frequency of 30 MHz.[3] The instrument consisted of two corner reflectors 38 × 38 × 50 feet (11.6 × 11.6 × 15.25 meters), set up on a 2000-foot (610-meter) east-west baseline. It was a temporary instrument, primarily used for gaining experience with noise levels at the site. The operating frequency was raised to 40 MHz in 1957 to observe an unnatural phenomenon—transits of the first Russian Sputnik satellite in October 1957.[4]

The first major instrument originally slated for construction by the steering committee was the 140-foot equatorially mounted paraboloid. The specifications set for the 140-foot and its basic concept dictated that a radical design be created. The 140-foot was not an off-the-shelf design and time would be required for the detailed design, fabrication of the components, and erection. Soon after Dave Heeschen was appointed to the permanent staff, he reviewed the situation and concluded that a smaller telescope that could be readily constructed should be acquired to begin research operations immediately.[5] The NSF staff approved of the idea to expedite research at the new observatory.

Performance specifications were prepared by the staff, calling for a paraboloid of 60 feet in diameter or greater. Several research institutions in the United States were in the process of planning or acquiring instruments of similar aperture. The Harvard College Observatory had recently acquired such an instrument; the Naval Research Laboratory was placing an order for an 84-foot (25.62-meter) instrument; the California Institute of Technology was well along in planning the construction of the two 90-foot (27.45-meter) telescopes of its interferometer; the National Bureau of Standards, the University of Michigan, and several other institutions were at various stages of the planning and construction of radio telescopes of similar aperture.[6] In view of these activities, it was apparent that several companies might submit proposals for the immediate construction of a telescope without a long design and development period.

The Tatel 85-Foot Telescope

The final specifications set by the NRAO staff were similar to those of the planned University of Michigan telescope. In the early days of the NRAO, there was a close al-

Figure 3.1 The NRAO Tatel 85-foot (25.9-meter) telescope. The 85-foot was designed and constructed by the Blaw-Knox Company and erected in 1959. The solid surface parabolic high precision reflector was equatorially mounted to allow tracking of radio sources. *Courtesy NRAO/AUI.*

The telescope design specifications called for a solid surface paraboloid on an equatorial mount (Figure 3.1). The instrument superstructure is supported on an equatorial mount consisting of a series of nearly equilateral triangles forming a structure attached to three concrete bases, two on the north side and one on the south. The polar shaft is located at the axis of a right-circular cone, whose apex coincides with the south end of the polar shaft and whose base forms the truss of the polar drive gear. The polar gear is 48 feet in diameter. The sides and base of the cone are composed of triangular elements of structural steel. Motion in declination is provided through a declination gear that is 20 feet (6.1 meters) in diameter. The circular declination gear and truss structure are visible in Figure 3.1. A triangular structural frame extends from the north end of the polar drive shaft to the ends of the declination shaft. Many such examples of triangular systems are incorporated into the mounting structure. The triangle is, in fact, the basic structural element of the mounting design.

The reflector is a solid surface paraboloid made up of panels covered with aluminum sheeting. The paraboloid f/d (focal length/diameter) ratio is 0.43, the same as that of the planned 140-foot telescope. The coincident ratio is desirable so that the feeds are interchangeable. The panels are adjusted to fit an overall paraboloid accurate to ±1/4 inch over its entire surface. Each panel is adjusted to the bolting surface to within 1/8 inch tolerance. The tolerances were specified such that the telescope can be effectively used at wavelengths as short as 3 centimeters.[10]

The equatorial mount provides sky coverage similar to that of optical telescopes. Sky coverage ranges in declination from the southern horizon to the pole (−52° to 90°) and in hour angle from the eastern to the western horizon (6^h to 6^h). The drive and control system provides slew and scanning motions about the polar axis and slew, scan, and tracking motions about the declination axis. The electronic motor drives are accurate to two minutes of arc.[11]

Howard Tatel developed a unique gear system to provide this accuracy inexpensively. The telescopes available at the time used high precision expensive gearing systems and accompanying mounts. Tatel developed the concept of the 85-foot mount with large gearing and used higher gear ratios in cooperation with Blaw-Knox who did the actual mechanical design.[12] Tatel had found a unique and effective solution based on a basic physical principle. Tatel and the Blaw-Knox Company produced the telescope with 60- and 85-foot diameters.[13]

The electronic equipment used initially with the 85-foot was essentially composed of off-the-shelf components, in keeping with the general telescope concept. A dual feed, operating at 3.75 and 21 centimeters was acquired from Jasik Laboratories. The dual feed horn consisted of two concentric horns with identical phase centers. Both horns were tested and showed good primary patterns and a satisfactory

liance between the two groups and the similarities in design and specifications were a direct consequence of this relationship.[7] Of the four proposals received, two were most consistent with the specifications and low in cost. These two proposals were from the D. S. Kennedy Company and the Blaw-Knox Company. The decision to award the contract to the Blaw-Knox Company of Pittsburgh, Pennsylvania, was based on judgments of the relative research value of telescope characteristics that were not identical in the two proposals. The University of Michigan also selected Blaw-Knox to construct its proposed telescope. The two telescopes would therefore be identical except for minor changes due to the difference in latitude between the two observatories.[8]

The basic design concept for the Blaw-Knox 85-foot (25.9-meter) telescope came from the radio astronomy group working with Merle Tuve at the Department of Terrestrial Magnetism. Bob Hall was then head of the Blaw-Knox antenna division and oversaw the design of the 85-foot telescope. Howard E. Tatel of the DTM was one of the principal contributors to the design. Tatel died while on a field trip in 1957 before the telescope was constructed; the NRAO 85-foot was officially named the Howard E. Tatel Telescope in his memory.[9]

Figure 3.2 The NRAO 85-foot telescope control room, circa 1959. Position indicators, data storage and processing equipment, and back-end receiver components are located in the control room. Courtesy NRAO/AUI.

impedance match over their working frequency ranges. Two conventional total power receivers were obtained, a 3.75-centimeter (also referred to as 440 MHz) receiver built by the Ewen-Knight Corporation and a 21-centimeter (1,420 MHz) receiver built by the Airborne Instruments Laboratory. Early during the observational work with the telescope, it was found that the two receivers could be mounted and operated simultaneously with little cross interference. A 75-centimeter feed and receiver, both designed and built by the NRAO research equipment development staff, were later added.[14] Simultaneous observations at three wavelengths were then possible. Observations at multiple frequencies are desirable to give some idea of the spectra of cosmic phenomena. The three sensitive front-end receivers and the capability of simultaneous observation at three frequencies made the Tatel 85-foot a versatile instrument.

The control room (Figure 3.2) contains the position indicators that give the telescope's position in right ascension and declination, the data storage and processing equipment that controls its motion and data collection, as well as the back end of the total power receiver. The original back end consisted of a 30 MHz intermediate frequency amplifier followed by a detector and integrating and display circuits. Such a back end can be used with various combinations of local oscillators, mixers, and preamplifiers to give a total power receiver capable of operating at the various frequencies required.

A calibration signal is introduced at each front-end receiver input through a direction coupler. The calibration signal is obtained from a standard noise generator utilizing an argon gas-discharge tube. In addition to the usual analog outputs, a digital output system was initially used with the receivers. A schematic diagram of the calibration signal and digital output systems used with the receivers is shown in Figure 3.3. The DC output of the final detector is converted to a frequency proportional to the instantaneous amplitude of the detector output voltage. This varying frequency is averaged for some specified time in a frequency counter. The average frequency over the specified time is then proportional to the integrated value of receiver output.[15] Varying integration times for observations are thereby possible. This capability is desirable because long integration times are required for weak sources. The output of the frequency counter was originally printed and punched on paper tape in a suitable form for entry into a digital computer. The data reduction, enhancements, and corrections were performed with the observatory's first main computer, an IBM 360.

A completely digitized indicator system to measure and record the antenna position and observation times was

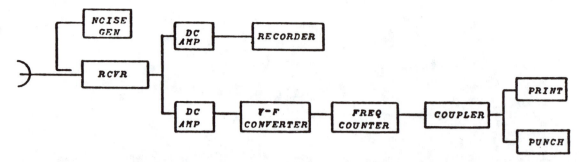

Figure 3.3 Block diagram of the calibration signal and digital output systems used with the receivers in the 85-foot telescope system. The first digital data collection system in radio astronomy was developed by the NRAO for the 85-foot telescope.

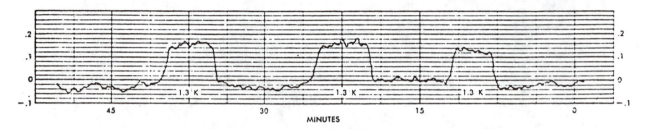

Figure 3.4 Typical record from the 1,400 MHz total power receiver, using an integration time of 20 seconds. The time scale increases to the left. Perceived intensity increases vertically. The square wave functions are calibration signals, corresponding to a temperature of 1.3 kelvin.

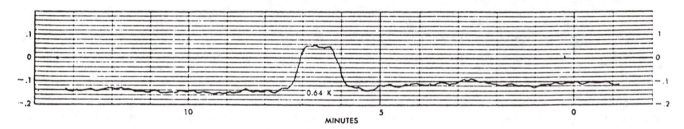

Figure 3.5 Typical record from the 440 MHz total power receiver, using an integration time of 6 seconds. The time scale increases to the left. Perceived intensity increases vertically. The square wave functions are calibration signals, corresponding to a temperature of 0.64 kelvin.

built by the Control Equipment Corporation. This Digital Antenna Position and Time Indicating System called DAPTIS could read and store position and time information automatically at 10 or 60 second intervals or at random, if desired. The shaft positions of the polar and declination axes of the telescope were measured by the use of 12-inch (30.5 centimeters) Farrand inductosyns. The inductosyn resolvers measure the angle of rotation of the axis shafts. The shaft positions were converted to digital outputs to be read instantaneously. The indicator system allowed reading and printing the sidereal time and frequency of the observation and the receiver output simultaneously. The indicator system allowed precise and automated data collection.

The overall gain of the total power receiver of the 85-foot was relatively stable. The problems of achieving good sensitivity had been reasonably well overcome. A key component of the total power receiver, a crystal diode front-end switch, was developed at the NRAO. The total power receiver is essentially a switched radiometer in which a front-end switch

is incorporated to switch the receiver input at some frequency switching rate between the antenna and a comparison load for calibration. The crystal diode switch developed by the NRAO electronics staff independently makes adjustments for insertion loss and isolation. Tests with the switch in operation at 1,400 MHz gave a receiver stability of about 0.1 K over a period of several hours.[16] The relative stability of the total power receiver is evident in the baseline data and calibration signals taken with the 1,400 and 440 MHz receivers (Figures 3.4 and 3.5). Receiver stability is important for the long integration times required when observing weak signals. Stability at this level allowed the detection of very weak signals.

Erection and Performance Characteristics of the Tatel 85-Foot Telescope

A contract was signed with the Blaw-Knox Company on October 1, 1957, for the construction of the 85-foot tele-

Figure 3.6 Results of the photogrammetric survey of the 85-foot telescope surface deformations, while the telescope is pointed toward the zenith. The contour interval is 0.005 foot (1.52 mm.).

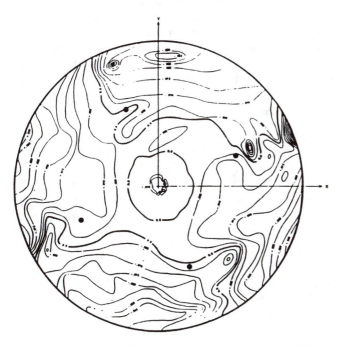

Figure 3.7 Results of the photogrammetric survey of the 85-foot telescope surface deformations, while the telescope is pointed toward the horizon. The contour interval is 0.005 foot (1.52 mm.).

scope. Erection of the telescope started at Green Bank in the summer of 1958. The erection phase including the necessary tests and adjustments was slightly underestimated. The NRAO staff, particularly members of the Astronomical Department under Dave Heeschen, worked regularly on

the instrument to expedite the erection and adjustments. The control building, designed to be a self-contained unit, was also constructed in the summer of 1958. The Howard E. Tatel Telescope was dedicated in a ceremony on October 16, 1958.

The 85-foot telescope is designed to work at relatively high frequencies, up to 8 GHz. A high precision surface is necessary to operate at such frequencies. The resolution of a single parabolic reflector, the smallest angle of sky that can be seen, is equal to the wavelength observed divided by its diameter (times a correction factor of 1.22). Large apertures are therefore required for long wavelengths. For short wavelengths the surface accuracy becomes important because small deformations may be a large fraction of the wavelength observed. At high frequencies deformations in the surface can critically affect the aperture efficiency and therefore the overall gain of the antenna. Setting the surface accurately is therefore very important to the overall performance of the telescope. The surface of the reflector must be carefully set to approximate a perfect paraboloid when pointing to the zenith. To operate at 3.75 centimeters, the panels must be adjusted to fit the specified tolerances of an overall paraboloid to ±1/4 inch (0.64 centimeter). Each panel must be adjusted to the bolting surface to within 1/8 inch (0.32 centimeter) tolerance. In 1959, the technology dictated that the surface panels be measured and set by hand by technicians on the surface of the reflector. The telescope was pointed to the zenith and measurements of the surface positions were taken with a theodolite (a high precision surveyor's instrument) and steel tape. This method was later applied to initially set the surface of the 300-foot and 140-foot telescopes.

The 85-foot panels were later more precisely measured to determine the aperture efficiency. A photogrammetric survey of the telescope surface was made in 1962. Photogrammetry is essentially the practice of obtaining surveys by means of photography. A photogrammetric survey applies photographic recordings of stereoscopic images to survey relative positions. Automated stereoplotting scanning devices are used to sense surface slope to a high precision. Photogrammetric surveys of both the 85-foot and the 300-foot telescopes were made by D. Brown Associates, Incorporated of Eau Gallie, Florida. Photogrammetric surveys were an innovation because the accuracy of the surface setting could be obtained for telescope positions other than stow. Large aperture paraboloids deform under the weight of their own gravity, and the deformations change as the telescope tilts to point in declination. Adjustments of the surface panels are therefore necessary to compensate for these deformations. Results of the photogrammetric survey of the 85-foot surface are seen in Figures 3.6 and 3.7. Measurements of the surface were taken with the telescope pointing to the zenith and toward the horizon. The contours show areas of equal deviation of the surface from a best-fit paraboloidal surface. Precise measurements of the surface defor-

Table 3.1 Results of the Photogrammetric Survey of the Surface of the 85-Foot Telescope

Postion of Reflector	RMS Departure from Best Fit Paraboloid	Focal Length of Best Fit Paraboloid
Toward Zenith	3.16 mm	10.944 ± 0.002 m
Toward Horizon	5.71 mm	10.903 ± 0.0005 m

Table 3.2 Aperture Efficiency of the 85-Foot Telescope

Weighted RMS Surface Errors	Frequency	Measured Aperture Efficiency	Calculated Aperture Efficiency
0.275 cm	1,420 MHz	56%	57%
	3,000 MHz	52%	52%
	5,000 MHz	45%	42%
	7,600 MHz	32%	28%

mations determine the surface precision and therefore the telescope's performance. Results of the surface measurements are summarized in Table 3.1.[17]

Testing the performance characteristics of the antenna is essential to effectively interpret the observations. The most critical performance characteristics of an antenna system are aperture efficiency and beamwidth. Precise measurement of the surface deformations allowed the aperture efficiency to be calculated with a high level of accuracy which therefore allowed a precise calculation of the overall gain of the parabolic reflector. Aperture efficiency is a measure of the ability of the antenna to concentrate radiation at a specific frequency and is determined by measuring the far-field radiation pattern of the antenna. The far-field radiation pattern must be measured for a precise determination of the aperture efficiency at this frequency. Beam symmetry of the radiation pattern of the antenna is necessary to reveal true source structure. Sidelobes must be kept at minimum to avoid detecting radiation from sources close to the field of view and from ground radiation when observing close to the horizon. Measurements of the aperture efficiency of the 85-foot were made by F. D. Drake, C. M. Wade, and P. G. Mezger by measuring the radiation pattern reflected from a transmitter. Aperture efficiency, in effect, determines the gain of any aperture antenna at the operating wavelength. High gain is necessary to collect and concentrate the weak cosmic radio signals. The calculation of the antenna gain of the reflector was initially based on the formula derived by Mezger in 1952:

$$G = \eta \left(\frac{\pi D}{\lambda} \right)^2 \exp\left(-\frac{4\pi}{\lambda} \epsilon \right)^2$$

where

η = the aperture efficiency
D = the diameter of the dish
λ = the operating wavelength
ϵ = the reflector rms deviation from the best-fit paraboloid

The formula was derived by statistical methods and is based on the assumptions that the deviations from the best-fit paraboloid are random and distributed in a gaussian manner and that the errors are uniformly distributed over the antenna aperture.[18] Aperture efficiency is always less than one in practice because there is a reduction from full utilization of the aperture area due to nonuniform illumination, energy loss over the edge of the reflector, and blockage of the aperture by the feed and feed supports. The calculations were later modified to take into account the variation of the primary feed illumination over the surface of the dish. This procedure can be done if the feed pattern is known by using a weighted value of the measured root mean square (rms), an average of the magnitude of surface deviations. The photogrammetric measurements provided these values with a high level of precision and allowed Drake, Wade, and Mezger to calculate the aperture efficiency of the 85-foot quite accurately. The results of these measurements are provided in Table 3.2 for comparison with the calculated values. The photogrammetric measurements showed that the error deviations are essentially gaussian. The close agreement between the calculated and measured aperture efficiency values showed that the simple theory of the effect of phase errors could accurately predict the performance of a parabolic antenna.

The performance of a radio telescope is largely based on the sensitivity of the system, which is ultimately defined by the minimum flux density of radiation that can be detected by the antenna. The minimum detectable flux density is dependent on the performance characteristics of the antenna and the radiometer systems. These performance characteristics can be characterized overall by the aperture efficiency and the system temperature. The theoretical minimum detectable flux density may be determined from these two values. The mathematical expression to determine the minimum detectable flux density is given as

$$\Delta S = \frac{2k K_S T_{SYS}}{\eta A \sqrt{\Delta v \Delta t}}$$

where

ΔS = minimum detectable flux density
T_{SYS} = system temperature
K_S = sensitivity constant (dependent on receiver type)
k = Boltzmann's constant = 1.38×10^{-23} J/K
η = aperture efficiency of antenna
A = geometric surface area of aperture

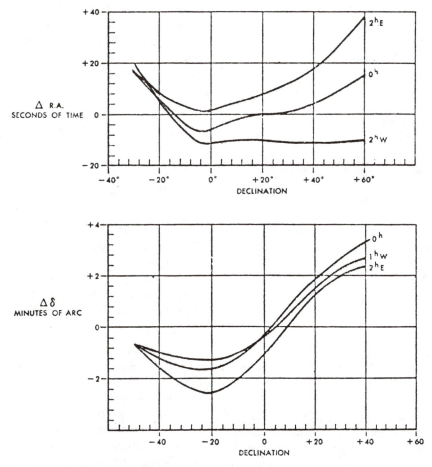

Figure 3.8 Pointing errors of the 85-foot telescope. The calibration curves are determined from observations of bright radio sources. Pointing errors in right ascension (top) and declination (bottom) are shown.

Δv = bandwidth of receiver

Δt = integration time

The system temperature (T_{SYS}) is determined from contributions made by the receiver noise temperature and other variables:

$$T_{SYS} = T_A + T_{LP}[1/\epsilon - 1] + T_R/\epsilon$$

where

T_A = antenna noise temperature, K

T_{LP} = physical temperature of transmission (waveguide or coaxial line) between antenna and receiver, K

T_R = receiver noise temperature, K

ϵ = transmission line efficiency, dimensionless $(0 \leq \epsilon \leq 1)$

T_R represents the internal noise ineherently produced by the receiver system; T_A is the effective noise temperature of the antenna radiation resistance. Typically, approximately half of the T_{SYS} is produced by the receiver and transmission and half by the antenna.

The actual signal received from a source ΔT is the difference between the temperature of the area of sky observed T_{sky} and T_{SYS}. The value of T_{sky} includes the source as well as contributions from the area of sky on which the source appears superimposed; this contribution also includes T_{3K} due to the cosmic microwave background radiation of the Big Bang. The theoretical minimum detectable flux density, measured in janskys (jansky = 10^{-26} W m^{-2} Hz^{-1}) determines the observing program in terms of defining the nature and number of observable radio sources.

Pointing errors of the 85-foot telescope were determined from observations of bright radio sources to accurately assess observations. The polar axis of the telescope was adjusted to be parallel to the earth's axis of rotation to within 10 seconds of arc. The declination axis was adjusted to be perpendicular to the polar axis to within 20 seconds of arc. Measurements of these alignment errors were made by optical observations of bright stars. Pointing errors of the telescope are due to the axial misalignments, deflections of the feed, and deflections of the dish structure itself. These errors were then calibrated by radio observations of intense sources whose positions were accurately known. Observations were made at 3.75 cm and 22 cm wavelengths. The re-

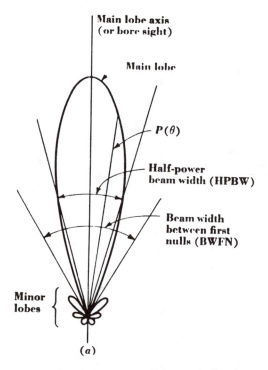

Figure 3.9 Antenna beams pattern for a parabolic telescope. The telescope beamwidth is determined from the half-power method.

sulting calibration curves, taken at 22 cm, are shown in Figure 3.8. Correcting for the calibrated pointing errors allowed positions to be set to an accuracy of one minute of arc. Most of the remaining error is inherent in the inductosyns used to measure the position of the axes. Improved synchros used to replaced the inductosyns later improved the position determination to an accuracy of nearly 20 arcseconds.[19]

Other important performance characteristics of radio telescopes are the antenna beamwidth and beam pattern. The response of an antenna as a function of direction defines the antenna pattern. The antenna pattern for parabolic dishes typically consists of a main beam accompanied by a number of minor lobes. The lobe structure is caused by radiation blockage by the feed and feed support structure and by radiation spillover. Beamwidth determines the angular resolution of the instrument at the operating frequencies. A sufficiently small beamwidth is necessary to reveal structural detail in radio sources. The beamwidth deviates from the theoretical resolving power due to the shape of the beam pattern. In practice, the beamwidth is measured by the half-power method. The beamwidth is determined by placing a transmitter as far away from the antenna as possible and measuring the resulting field intensity pattern as a function of angle and distance. If the transmitter is far enough away, the distance causes no change in the resulting pattern and a far-field pattern is obtained. The far-field pattern was obtained for the 85-foot by placing a transmitter on a distant mountain. At lesser distances the measurements change according to distance and angle, and a near-

field pattern is obtained. The resulting antenna pattern is typically plotted in polar coordinates on a linear power scale (Figure 3.9). The beamwidth is determined at the half-power intensity point, where the intensity drops to half its maximum intensity.[20] The theoretical beamwidth may be calculated by

$$HPBW = \frac{58\lambda(m)}{D(m)}$$

where

$HPBW$ = half-power beamwidth in degrees
λ = operating wavelength in meters
D = aperture diameter in meters

The measured antenna beamwidths for the 85-foot antenna initially were 110 minutes of arc at 440 MHz, 37 minutes of arc at 1,400 MHz, and 7 minutes of arc at 8,000 MHz. Near sidelobes measured at all three frequencies were down more than 20 dB from the main lobe gain.[21]

Accurate measurements of the performance characteristics of a radio telescope are essential to effectively interpret the observations. The NRAO staff historically has developed technical innovations in measurement techniques of various telescope characteristics to accompany the developments in front- and back-end receiving systems, data collection and reduction systems, and special equipment for specific observations. Telescope operations with the 85-foot marked an important step for the NRAO technical staff. The technical staff had become capable of the development of world class hardware. By 1959, the NRAO staff had developed hardware, particularly receivers, amplifiers, and feeds, as good or better than any available. This technical development of world class hardware was a critical step toward the forefront in radio astronomy research.

85-Foot Telescope Programs

The first observations with the NRAO 85-foot Tatel telescope began on Friday, February 13, 1959. The first year's observation program with the instrument had defined the NRAO as a national and international user institution. A list of the users during the first few years is quite impressive and includes: D. Heeschen, F. Drake, M. Roberts, H. Hvatum, G. Westerhout, T. K. Menon, R. Stockhausen, J. W. Findlay, G. Field, Grote Reber, D. Hogg, C. Wade, S. Weinreb, T. Orhaug, Y. Terzian, M. DeJong, I. Pauliny-Toth, and P. Mezger.[22] Research programs with the 85-foot included early important work in planetary science and research in galactic and extragalactic radio astronomy. The initial observations began with position determination programs and sky surveys. The first search for extraterrestrial intelligent life, Project OZMA, was initiated with the 85-foot telescope. Project OZMA was a landmark experiment and represented the beginning of what many scientists believe to be the most significant experiment undertaken by mankind. The possible conclusions of such an experiment could be more far-reaching than any other scientific endeavor.

Position Determination Program

One of the earliest programs undertaken with the NRAO 85-foot telescope was the accurate determination of radio source positions. The interpretation of subsequent observations would depend on the accurate knowledge of the position of radio sources. The pointing accuracy achieved by the inductosyn system and DAPTIS position and timing system allowed Frank Drake and the astronomy staff to provide precise position measurements. The inductosyn and timing readings were combined with observations of radio sources made to measure antenna deflections under gravity (prior to the photogrammetric survey) and with optical observations of stars to determine axis alignments and zero-point errors. The combination of this data made possible the establishment of a coordinate system on the sky in which the angular distance between any two points was known to an accuracy of about three arcseconds. The coordinate system was then made absolute by observations of several planets. The observations of planets were used to establish absolute positions on the coordinate system because the planets were the only radio sources for which the optical centers and radio centers were known to coincide within a few seconds of arc.

The apparent positions of radio sources were obtained for the program from drift curves taken during tracking observations. Tracking observations produce drift curves of varying radio intensity as the beam of the telescope is passed across the source. The most accurate positional information comes from the portions of the drift curve with the greatest slope. The intensity of a source was therefore observed near the half-power points with the telescope tracking. The positions of zero intensity on either side of the source response were observed in order to detect any slope in the background brightness temperature or any radiometer drift, as well as to provide relative measures of the radiometer deflection at the half-power points. The results of the observations produced two deflections at very well determined positions. The antenna beam had previously been determined to be gaussian by Drake, Wade, and Mezger. A gaussian function was fitted to the two deflections to obtain the apparent source position. The curve, however, could not be fitted with only two deflections unless either the amplitude or the dispersion was known. The assumption was avoided by observing a third point on the drift curves, the amplitude or peak. The gaussian function used was

$$f(x - x_0) = Ae^{-b(x-x_0)^2}$$

where

A = signal amplitude
b = beamwidth

The gaussian function provides a value of x_0, the apparent source position when applied to the obtained point deflections. The observatory's Bendix G-15 computer was used to apply the gaussian function to the three observed deflections for each source. An accurate value for the source position, intensity, and angular size was thereby obtained. The initial NRAO position determination program provided positions of 30 sources to within accuracies of the order of 10 arcseconds.[23] It is interesting to note that based on the results of the 85-foot position determination program, it was predicted that the 140-foot telescope would provide accurate positions for "a very large number of sources, on the order of 300."[24]

The 85-Foot and Early Surveys

With the performance characteristics of the 85-foot telescope well understood and the positions accurately known for bright radio sources, the astronomy staff at the NRAO was ready to initiate more ambitious research programs. Several early sky surveys were begun. The most significant of these early surveys was Dave Heeschen's Four Color (Frequency) Survey. Heeschen observed a series of standard, relatively well known sources, at four different frequencies. The purpose was to gain some insight into the spectra of the sources.

Heeschen took advantage of the multifeed system allowing simultaneous use of the 85-foot receivers. The receivers were used for observations at 440 MHz, 1,200 MHz, 1,400 MHz, and 8,000 MHz. To achieve a high level of internal consistency in the data, the observations were made in the form of ratios to a standard source. The intense source IAU 23N5A or Cassiopeia A (Cass A) was selected as the standard source. Cass A was chosen because of its intensity and because it had been frequently observed and was thought to be well understood, particularly in terms of the value of its flux density. It was not known at the time that the flux density of Cass A is not constant. Cass A is decreasing in intensity because it is a supernova remnant and the outer gaseous shell is dissipating as it expands, thereby reducing its flux density at radio frequencies. Although the standard source chosen for the survey did not provide an absolute flux density value, the results of the project did provide valuable information regarding the spectra of radio sources.

Heeschen used two methods to compare the source flux densities to Cass A. In the first method, the output deflections due to source and standard source were obtained within as short a time as possible. The ratio source/Cass A was formed directly from the observed deflections after corrections for atmospheric extinction and receiver nonlinearity had been applied. In this method receiver gain must remain constant over the time required for both observations. Differences in sky background and ground emission also vary for the different positions, introducing an error in the comparison. The second method added an intermediate

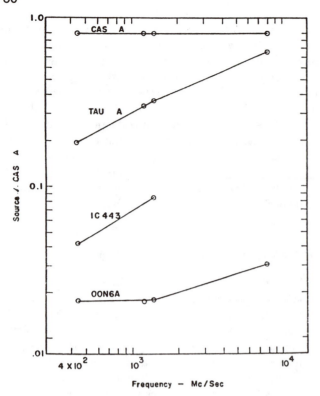

Figure 3.10 Spectra of galactic sources observed relative to Cass A, determined by the 85-foot Four Color Survey in 1959. *Courtesy of D.S. Heeschen.*

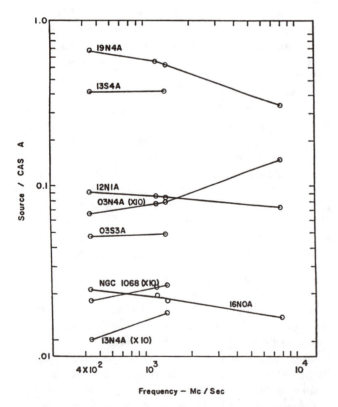

Figure 3.11 Spectra of extragalactic sources relative to Cass A, determined by the 85-foot Four Color Survey in 1959. *Courtesy of D.S. Heeschen.*

step to compensate for these problems. The deflection due to a source was compared immediately with the deflection from the internal calibration signal. The calibration signal was later removed and source/Cass A ratios obtained using the measured Cass A/calibration ratio. This comparison method largely eliminated the effects of ground and background sky emission as well as reducing the observation time and movement of the antenna, allowing for a higher receiver gain stability.

To eliminate systematic and long term effects, four sets of observations were made in March, July, and September of 1959 and in January of 1960. Long period variations in antenna gain and receiver stability were thereby eliminated. A mean value of the source/calibration ratio was calculated from all observations at a given frequency.

Results were obtained for 14 sources, 4 galactic and 10 identified with extragalactic objects. Corrections were made for source size for the sources whose extent was large compared to the telescope beamwidth. Sources of large angular extent were observed and the data was represented using contour maps. The contour features were integrated to obtain the total flux density. The total flux density was then compared to the standard. The results revealed important spectral features in the sources.[25]

Spectral differences between extragalactic sources were far more pronounced over the frequency range 440–1,400 MHz than the galactic sources. The principal spectral features of the galactic sources are seen in Figure 3.10. The spectral features of the extragalactic sources are shown in Figure 3.11. The two "normal" galaxies appeared to have flatter spectra than the peculiar extragalactic sources. The spectrum of M33 (NGC 598), a "normal" galaxy, appeared to be similar to a thermal source. This observation allowed Drake in 1959 to theorize that at high frequencies there may be a contribution from the thermal emission from ionized hydrogen distributed throughout the galaxy and from thermal emission from the galactic nucleus.[26] The initial 85-foot Four Color Survey produced important spectral information for galactic and extragalactic sources that provided new insight into the nature of the cosmic objects and the mechanisms that produced the radio frequency radiation. The Four Color Survey represented initial steps toward a trend to observe cosmic phenomena over a wide range of frequencies to broaden the understanding of the nature of these phenomena.

Planetary Science

The NRAO 85-foot telescope was brought into action at an opportune time in the history of planetary science. Prior to 1955, the planets had only been observed at optical frequencies. Research with optical telescopes had essentially come to a lull. There were many unresolved mysteries about the members of the solar system that observations with earth-based optical telescopes could not explain. The atmosphere of Venus could not be penetrated to observe the surface. In fact, surface characteristics of few planets could be determined. The surface temperature, composition, and structure in many cases were a mystery. In other cases it was thought that the physics of the planets were well understood based on optical studies. Observations at radio frequencies turned up many totally unexpected discoveries.

Jupiter was the first planet to be observed. Jupiter bursts were serendiptiously discovered by Burke and Franklin in 1955 who happened to be observing in the narrow band between 15 and 25 MHz where the bursts occur. Further investigations by Burke and Franklin at the DTM showed that the radio radiation was often strongly circularly polarized, primarily exhibiting a right-hand polarization, but occasionally left-hand polarization.[27] This suggested the presence of a magnetic field. The radio radiation from the planets was expected to be thermal, although the Jupiter bursts were mysteriously nonthermal. This discovery led other groups to try to detect the planet at various additional wavelengths. In 1958, Mayer and colleagues at NRL detected Jupiter at centimeter wavelengths with the 50-foot (15.25-meter) high precision telescope. Their measurements at 3 cm implied an effective temperature of 145 K which indicated that the radio emission at this wavelength was primarily thermal. In 1959, however, Sloanaker observed a temperature of 600 K at 10 cm. Also in 1959, Drake and Hvatum observing at 70 cm with the 85-foot Tatel telescope discovered that the temperature had risen to 50,000 K. The intense emission at decimetric wavelengths was undeniable evidence of a nonthermal component. These observations led Drake and Hvatum to suggest that the intense radio frequency radiation might be attributed to synchrotron radiation from electrons trapped in an intense Jovian magnetic field, analogous to the earth's Van Allen belts.[28] It seemed practical that a large planet with a strong magnetic field would give rise to intense Jovian Van Allen regions. Their hypothesis was verified in 1960 when Radhakrishnan and Roberts observed an equatorial radio diameter three times the size of the planet with the Caltech interferometer. Later it was shown that the bursts were strongly related to certain orbital positions of Io, the closest major satellite. Io apparently causes a tidal or magnetic disturbance in the magnetic field that stimulates synchrotron radiation from Jupiter's magnetosphere. With this theory, the NRAO astronomy staff had begun to make major contributions to theoretical astrophysics.

Another early surprise in planetary exploration at radio frequencies was strong emission from the planet Venus. Venus was first observed by the NRL group, who accomplished much of the pioneering work in radio planetary astronomy. The NRL 50-foot high precision telescope was ideally suited for planetary radio astronomy because of the short wavelengths involved. Mayer, McCullough, and Sloanaker at NRL discovered the radio brightness temperature

at wavelengths of 3 cm and 10 cm to be 600 K. This intensity of radio waves emerging through the dense Venusian atmosphere was the first direct evidence of an extremely high surface temperature. Clark and Kuzmin later showed in 1965 that the major component of the 10 cm radiation originated from the solid surface of Venus. The extreme surface temperatures have been explained by a greenhouse effect caused by the high carbon monoxide concentrations in the dense Venusian atmosphere. The original NRL observations were confirmed at other wavelengths by groups at Green Bank and at the Lebedev Institute of Moscow.[29] The biological significance of this discovery is profound. The temperatures involved are far too high to support carbon-based life forms. Biochemical reactions become destructive at such high temperatures. The conditions on Venus ruled out the possibility of any form of life.

Observations made in 1959 at the NRAO, the NRL, and the Lebedev Institute consistently showed that the apparent disk temperature rises after the planet passes inferior conjunction, when the plane moves ahead of the earth in its orbital motion around the sun. The sunrise portion of the surface was found to be cooler than the sunset portion.[30] This temperature differentiation showed that Venus rotates in a prograde direction around its axis. The NRAO observations of Venus provided evidence of the planet's rotation, showed that the microwave radiation comes from the surface, and indicated a surface temperature of 707 ± 18 K (813 ± 65 °F), very close to the now accepted value of 750 K (900 °F).[31] The NRAO Venus research program provided another major contribution to planetary science.

Other early research programs in solar system studies include Hvatum's study of the radio brightness distribution of the moon at 3.75 and 22 cm. Hvatum found that there were small but well-determined changes in the brightness distribution that corresponded with lunar phases. The brightness distribution findings provided information about the nature and conductivity of the lunar surface at the depths from which these wavelengths are emitted.[32]

Research in planetary radio astronomy with the NRAO 85-foot telescope resulted in many successes. It was built at an ideal time in the early stages of the science. Advances in theoretical work, discoveries, and numerous publications attest to the early NRAO contributions to planetary science. Several of the observatory's astronomy staff members, including Drake, Heeschen, and Hvatum, produced major publications. Burke began a long and prolific history as a user of the NRAO facilities after his discovery of the Jupiter bursts. Such research at radio frequencies in general revolutionized planetary science. The associated science of radar astronomy evolved symbiotically with planetary radio astronomy. Among the contributions of radar astronomy are the determination of extremely accurate planetary distances, radar mapping of the moon and the surface of Venus, and the measurement of the rotational period of Mercury.

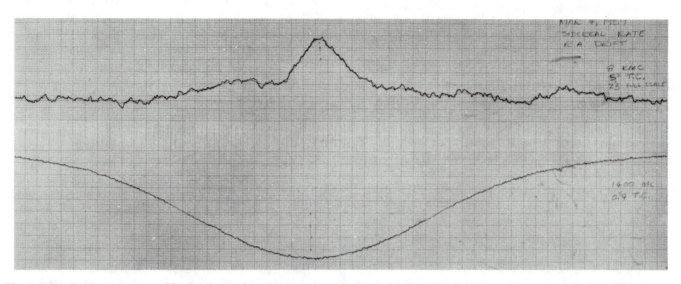

Figure 3.12   Drift scan curves of Sagittarius A, the galactic nucleus obtained with the NRAO 85-foot telescope on March 4, 1959. A comparison of these two scans reveals the higher resolving power of the smaller beamwidth: 6 arcminutes at 3.75 cm (top scan). The 22 cm scan producing a 36 arcminute beamwidth (bottom) shows one deflection at the galactic nucleus. The 3.75 cm scan shows four deflections, thereby showing the first source structure in the nucleus of the Milky Way. *Courtesy of Frank Drake.*

Thus, planetary science was revolutionized by radio observations in the 1950s and 1960s.

Galactic Astronomy

By the late 1950s it was evident that radio astronomers would also revolutionize the understanding of galactic astronomy, particularly in the area of galactic structure. The spiral arm structure of the Milky Way had been revealed by radio observations of atomic hydrogen associated with the spiral arms. An intense source was discovered at the galactic center by radio astronomers. An entire coordinate system was adopted by astronomers in 1959 based on this discovery. The galactic coordinate system uses the intense source in Sagittarius as the origin because it closely corresponds to the rotational center of the galaxy. The nucleus of the Milky Way had been a mystery because obscuration caused by gas and dust prevented optical telescopes from viewing the center. A high concentration of stars in the nucleus also obscured the view. It was hard to see the proverbial trees for the forest. Radio waves, which can penetrate the obscuring gas and dust, provided the first views of objects associated with the galactic nucleus. Radio resolution of the galactic nucleus was accomplished by the NRAO staff using the 85-foot telescope.

The radio resolution of the galactic nucleus was undertaken by Frank Drake in 1959. Drake and his colleagues resolved the nucleus into at least four components, two of which were not located at the rotational center. Drake observed the region in Sagittarius simultaneously at two wavelengths, 22 cm and 3.75 cm, using the two concentric feed horns. The antenna beamwidths were 36 and 6 minutes of

arc, respectively. Observations were made taking drift curves in both right ascension and declination across the most intense portion of the region. An original strip chart recording of the drift curves at the two wavelengths is seen in Figure 3.12. A comparison of the drift scans reveals a striking feature of the higher resolution provided by the 6 minute beamwidth (top scan). The 6 minute beam resolves the region into four distinct sources, whereas the 36 minute beam (bottom scan) perceives one signal. The advantage of the higher resolution afforded by observations at shorter wavelengths is evident in the drift scans. The complexity of the nucleus was revealed. The right ascension and declination drift scans of the region provided a great deal of structural information about the objects. Typical drift scan curves are shown in Figure 3.13 and 3.14. By reducing the two sets of scans independently and then comparing them for agreement, a high degree of accuracy and important information can be obtained. Here, the broadening of the strongest component near the center indicated an object about 20 parsecs in diameter. This size matched optical measurements of the complex regions of other galaxies. The size of the 20 parsec region agreed with Baade's calculated typical diameter of galactic nuclei based on these observations. Drake concluded that the brightest of the four objects in the 20 parsec region was actually the galactic center.[33]

Drake's 3.75 cm survey of the region revealed two very small thermal sources at the galactic center, one of which was extremely bright, and two other sources lying in the galactic plane, symmetrically oriented with the respect to the center. The two outer sources were found to be nonthermal, suggesting the production of synchrotron radiation in the structure oriented along the galactic plane. Drake mapped

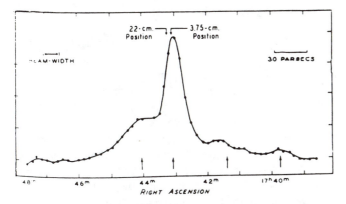

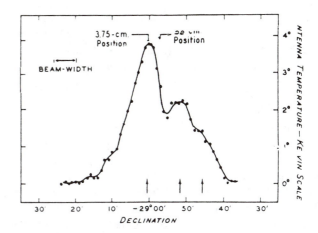

Figure 3.13 Right ascension drift scan curve of Sagittarius A at 3.75 cm taken with the NRAO 85-foot telescope. The curve was produced by the average of eight scans. The horizontal scale is right ascension and the vertical scale is the radio energy perceived by the telescope. *Courtesy of Frank Drake.*

Figure 3.14 Declination drift scan curve of Sagittarius A at 3.75 cm taken with the NRAO 85-foot telescope. The curve was produced by the average of eight scans. The horizontal scale is declination and the vertical scale is the radio energy perceived by the telescope. *Courtesy of Frank Drake.*

the region of Sagittarius A from data taken in the project showing its complex structure. A contour map of the brightness temperature (Figure 3.15) reveals the complex structure of the objects at the nucleus. Weak sources also were detected away from the plane which are ionized hydrogen (HII) regions associated with the galactic nucleus.[34]

The resolution of the objects in the nucleus was a major contribution to galactic astronomy. The resolution of Sagittarius A into four components was the first view into the nucleus of our own galaxy. Such a view was not obtainable with optical telescopes. The question of whether one of the

objects was the nucleus and the other objects subordinate to it or whether the nucleus consisted of a conglomerate of similar objects of physically similar structures was open to debate. The first clue, however, had been provided about the nature of the galactic nucleus and the mechanisms involved with the production of the intense radiation observed.

Other research programs in galactic astronomy using the 85-foot were initiated in 1959 by astronomers who investigated HI (neutral, atomic hydrogen) and HII (ionized hy-

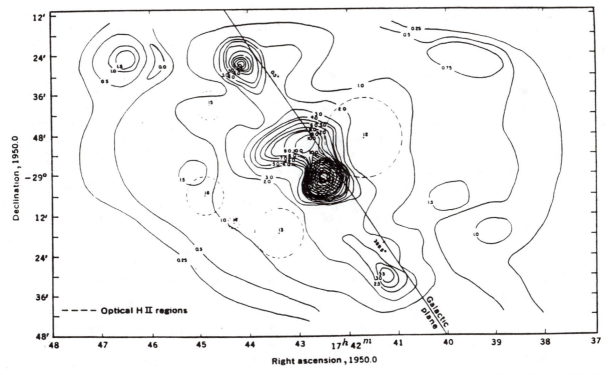

Figure 3.15 Contour map of the galactic center based on radio observations at 3.75 cm with the NRAO 85-foot telescope (Drake, 1959). The contour unit is 1 K and the resolution is 6 arcminutes. *Courtesy of Frank Drake.*

drogen) regions, supernova remnants (SNRs), globular clusters, planetary nebula, galactic magnetic fields, and secular variation of the radio intensity of galactic objects. T. K. Menon's observations of HI emissions provided insight into the structure of the Cygnus Loop and the Orion nebula. The distribution of neutral hydrogen in the vicinity of the Cygnus Loop was shown to be associated with the optical structure. His investigations of the Orion nebula at 3.75 cm provided high resolution mapping of the inner regions. The radio brightness of this complex region associated with starbirth agreed fairly well with the visible structure. Roberts attempted to detect 21 cm radiation from globular clusters. In one of the earliest programs, Roberts obtained negative results observing the bright globular clusters M3 and M13. An upper limit of 0.4 K emission of neutral hydrogen in the globular clusters was set by the observations. The intensity of neutral hydrogen emission gives an indication of the mass of the cluster (hydrogen is the principle component of the interstellar medium and of the stars within the cluster) and is therefore an important characteristic to understand. Even the negative results he obtained provided insight into the structure and distribution of mass in globular clusters.

Heeschen initiated an early program to observe supernova remnants (SNRs) in an attempt to determine their radio brightness and spectra. Early studies showed that the luminosities of these SNRs was relatively consistent but their spectra greatly differed, ranging from the flat spectrum of Taurus A to the very steep spectrum of Cass A. Osterbock and Stockhausen attempted to observed planetary nebula at 3.75 cm and 22 cm in a program also begun in 1959. Radio emissions of these objects combined with the optical observations, it was hoped, would give clues to the electron densities and temperatures of the planetary nebula.[35]

An interesting international program of Hogbom and Shakeshaft used various radio telescopes around the world for a period of 6 years to produce evidence for the secular variation of the intensity of radio emission from Cass A. Hogbom and Shakeshaft searched for a long term variation of the intensity of Cass A at 1,400 MHz. Their measurements began in 1954 using the NRL 50-foot telescope. Observations used the Leyden Observatory in 1965, the Bonn telescope in 1958, and the NRAO 85-foot telescope in 1959 and 1960. All measurements compared the flux density of Cass A to Cygnus A to produce a ratio of intensities. All of the observations indicated a 1 percent decrease in flux density of Cass A relative to Cygnus A. The observations were conducted with instruments of similar aperture and corrected for atmospheric extinction and receiver nonlinearity.[36] The observations provided important insight into the nature and structure of supernova remnants. The long term variation program represented the early stages of international efforts and cooperation in radio astronomy. The nature of the NRAO as an international user institution had become a reality.

Extragalactic Astronomy

Several early programs investigating extragalactic phenomena were conducted with the 85-foot telescope. Heeschen's 1959 Four Color Survey primarily observed extragalactic objects (10 of the 14 sources selected). The survey included galaxies of all four major types—spiral, elliptical, peculiar, and irregular. Peculiar galaxies, so named because of their abnormal appearance, were investigated. These galaxies were shown to have steeper spectra than normal galaxies. The survey gave insight into the little understood extremely violent phenomena occurring in these peculiar galaxies and into the nature of irregular galaxies, those with amorphous or assymmetrical or distorted shapes.[37]

In another early observing program Field attempted to detect the presence of intergalactic hydrogen by observing Cygnus A at a number of frequencies to look for absorption of radiation by the hydrogen. Cygnus A was observed at 1,250 MHz and 1,500 MHz to investigate the absorption. The intergalactic hydrogen was predicted to produce a broad absorption line in the spectrum of the source. The underlying premise of these investigations was to resolve the missing matter problem. Other extragalactic programs investigated the neutral hydrogen content and structure of galaxies.

The observations of clusters of distant galaxies revealed another interesting phenomena, that of radio "jets." Observations with optical telescopes had discovered jet structures emerging from the nuclei of certain galaxies. C. M. Wade and colleagues observed the Virgo cluster of galaxies with the 85-foot telescope at a frequency of 1,390 MHz which produced a half-power beamwidth of 36 arcminutes. Wade observed a faint extension of Virgo A that had been detected by B. Y. Mills at 85.5 MHz with the Mills Cross telescope near Sydney. The Mills Cross antenna, however, could not detect any source structure. The 85-foot observations provided much higher resolution and a receiver sensitive enough to detect antenna temperatures down to 0.1 K. Wade's observations revealed a separate discrete source of a very small angular size. The source intensity was roughly 0.04 that of Virgo A. According to Wade, there was "a suggestion on the records of a very faint 'bridge' connecting the new source with Virgo A." The identification of the new source with M84 was announced by Wade, based on the precise positional information obtained with the 85-foot system.[38] Radio jets are now identified as a whole new set of phenomena. Many extragalactic radio sources are now known to contain intense, well-columnated jets. Jets have been detected in the full range of radio galaxies from weak nearby galaxies to quasars. The radio jets are assumed to be radiative losses associated with energy pipelines linking the cores of active galaxies and quasars to their expanded emission. A continuous-beam source model is implied by the jet structures. Large-scale jet structures have subsequently been found in 70 to 80 percent of the sources in nearby radio galaxy surveys. Interestingly, the detection rate is far lower in powerful sources.[39]

During the course of these early programs, phenomena were identified that affected the precision of the observations. Gart Westerhout found that the receiver output fluctuated during foggy or cloudy atmospheric conditions while surveying the galaxy at 3.75 cm. This fluctuation agreed with the theory of changing atmospheric emissivity caused by variations in the water vapor and O_2 content of the atmosphere. Observations requiring long integration times at this wavelength were therefore found to be possible only during good weather conditions. Early observations at 22 cm showed a notable difference between daytime and nighttime observations.[40] Solar radiation being picked up by weak sidelobes was found to be the cause. The observational efficiency was increased during these early programs not only through technical innovations and hardware improvements, but also through observing experience.

Project OZMA: The First Search for Extraterrestrials

The first active search for extraterrestrials was initiated at the National Radio Astronomy Observatory at Green Bank with the 85-foot Tatel telescope. The search for extraterrestrial intelligence (SETI) could potentially represent the most profound discovery of the scientific endeavor. The possibility of extraterrestrial life remained apparently unanswerable before the advent of radio astronomy. During the early 85-foot telescope program in 1959, Frank Drake of the NRAO astronomy staff began to wonder if radio astronomy techniques could lead to an answer.

The observing techniques of radio astronomy once again were seen as the only possible means to pursue a new scientific investigation. An interesting view into the feeling of the times is provided by Frank Drake:

We were quickly looking for projects to do with it (the 85-foot telescope). It was a time when it was easy to make discoveries, such as the Jupiter radiation belts, structure of the galactic center; things like that just rolled out very easily.[41]

It was under these circumstances that Frank Drake one day tried to calculate how far away a telescope like the 85-foot could detect radio signals comparable to the strongest radio signals then leaving the earth. Drake worked through the math and concluded that with the new sensitive receivers and the large collecting area, the 85-foot could detect our strongest radio signals at a distance of 10 or 20 light-years.[42] There was a remarkable significance to that distance. It is the distance to the nearest stars.

Frank Drake approached Otto Struve with his calculations and prediction that the 85-foot telescope could detect reasonable signals from nearby solar-type stars. Drake was hopeful of the possibility: "We ought to look because for all we know, practically every star in the sky has a civilization

that is transmitting."[43] Struve believed that many stars had planetary systems, based on his optical work in stellar rotations. Struve approved the program, but with the stipulation that it be kept as secret as possible. He was concerned that news of a SETI program would damage the reputation that the NRAO was trying to establish as a national observatory. Drake secretively embarked upon a landmark experiment which he called Project OZMA, "named for the queen of the imaginary land of Oz—a place very far away, difficult to reach, and populated by exotic beings."[44]

Drake set upon a design for the SETI receiver system in the spring of 1959. The observatory happened to be planning a program to attempt to detect the Zeeman splitting of the 21 cm line. The receiver system required to search for the Zeeman effect had many of the characteristics that Drake saw desirable for the SETI system. Two channels were needed for both programs as well as good frequency stability and a narrow bandwidth. The frequency chosen for the OZMA search was therefore set at 21 cm. According to Drake "that is why the frequency was picked, not for any profound reason like magic frequencies or water hole or anything else."[45]

The Project OZMA receiver was designed by Drake and the NRAO staff. Kochu Menon joined the NRAO staff and became involved with the construction of the radiometer and later used it in an attempt to observe the Zeeman effect. The basic design of the receiver system is seen in Figure 3.16. The system was designed around a DC comparison radiometer. The DC comparison radiometer was an innovation designed to overcome gain instabilities and noise in the receivers. It had a broad band channel and a narrow band signal and took the difference of the two while scanning the narrow band channel. A spectrum was produced by this method. A comparison band and a signal band were tuned and the difference taken so that the narrow band signals could be picked out. The system thereby filtered out all radiation but a narrow band signal centered around 21 cm. Continuum sources, which all cosmic sources were then known to be, were filtered out in this manner. The system could filter out the radio radiation of all astronomical sources.

The system was basically a superheterodyne receiver, which converts the incoming radio signal to a lower predetermined carrier frequency, known as the intermediate frequency (IF). This was typically done by mixing the incoming signal with that from a local oscillator tuned simultaneously to the input stage of the receiver. The difference in signals produced is therefore the pre-set IF. The IF signal is amplified and later demodulated to obtain the desired signal. The incoming signal was converted from 1,420 MHz down to 30 MHz and then amplified and fed to a device, a Hammerlund Superpro, which was tuned to the desired signal frequency.[46] The device downconverted the frequency, provided the tuning, and produced a broad band of frequencies from which the desired bands could be filtered.

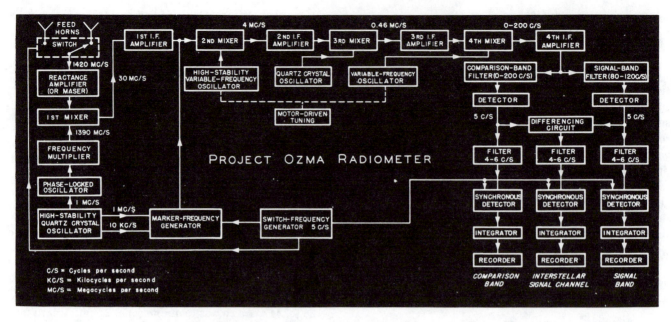

Figure 3.16 Block diagram of the Project OZMA radiometer installed in the NRAO 85-foot telescope. The system was essentially a combination Dicke switching superheterodyne receiver and DC comparison radiometer.

The receiver system utilized two feed horns and a Dicke switching system to alternate between horns. The concept was that switching between horns and then "differencing" the signal would eliminate terrestrial signals. Terrestrial signals could, however, come in one horn and not the other and pass through the system and be detected, so another solution was eventually found. The receiver system was essentially a combination Dicke switching superheterodyne receiver and DC comparison radiometer.

The receiver system constructed for Project OZMA employed an innovation that would revolutionize receiver systems in radio astronomy: the reactance or parametric amplifier. The front-end amplifier is a crucial component in the system's operation. The original Adler amplifier, which was an electron beam amplifier initially used on the receiver, never really worked. Dana Atchley of Microwave Associates had, however, developed a new highly sensitive, low noise, high frequency amplifier called a parametric amplifier, or paramp, he claimed was extremely good and stable. Stability had previously been the prohibitive problem for parametric amplifiers. He loaned the prototype to the NRAO for the project along with his chief engineer to help operate it. This innovation was the first paramp that was stable enough to be used for radio astronomy.

The parametric amplifier uses an active element whose reactance is varied over another high frequency. Amplification of weak signals is achieved through a nonlinear modulation or signal-mixing process that produces additional waves at other frequencies. This process provides negative-resistance amplification of the applied signal wave and an increase in power in the new frequencies generated. This process is advantageous because of its low level of noise generation. The paramp was therefore ideal for the first stage of input for microwave receivers in which extreme sensitivity is required. Atchley saw the newly developed paramp as ideal for use in radio astronomy.

The paramp had an isolator and a Dicke switch and a direction coupler that injected calibration signals into the system. The paramp replaced the Adler electron beam first stage amplifier. Since first stage of amplification in the front end is the most critical, the new paramp was essentially the heart of the system. Although it required delicate front-end tuning, the paramp worked beautifully. The 85-foot paramp installed for Project OZMA was the first ever to be used in radio astronomy. The new paramp reduced the system temperature of the 85-foot from 1,500 to 350 K.[47] A revolution in receiver electronics, in which paramps became state of the art, ensued as a result of the greatly improved performance of the 85-foot receiver. This development was a critical step for the NRAO electronics division. The NRAO electronics staff had begun to employ and develop receivers that were among the lowest noise, most sensitive receivers of any appreciable bandwidth in the world, an accomplishment for which they would be known from that point on. Staff members that participated are pictured in Figure 3.17. The NRAO telescopes have been highly sought after because of their research potential since the early 1960s.

The OZMA project was initiated quietly but soon received sensational media coverage. During the construction of the receiver in 1959 a paper was published in *Nature* by Philip Morrison and Giuseppe Cocconi describing the potential of attempting a search for interstellar communica-

Figure 3.17 The 85-foot NRAO telescope and members of the original Project OZMA research team. (Front row, left to right): Bob Viers, Dewey Ross, Bill Meredith, Troy Henderson, and Bob Uphoff. (Back row, left to right): George Grove, Fred Crews, Omar Boyer, Frank Drake, and Kochu Menon. *Courtesy NRAO/AUI.*

tions. The proposed experiment ironically was nearly identical to Project OZMA which was secretly underway.[48] The paper advocated a search at 21 cm which was becoming thought of as a "magic frequency" because it was emitted by the most abundant element in the universe, neutral hydrogen. Cocconi and Morrison's theoretical experiment also proposed observing the four nearest solar-type stars, two of which were observed in Project OZMA. Otto Struve decided that the credit and publicity would be attributed to Morrison and Cocconi for a program that had been initiated at Green Bank for 6 months. He therefore publicly announced Project OZMA at a lecture he presented at MIT.[49] One consequence of this presentation was the expected descent of the media upon the observatory. Another conse-

quence of the media release was that Dana Atchley of Microwave Associates became aware of the project for which he thought his newly developed paramp would be ideal.

Along with the use of parametric amplifiers in radio astronomy receivers, Project OZMA represented another revolutionary concept in radio astronomy instrumentation— digital data collection. All of the data at the observatory had been taken on strip chart recordings since operations began. Project OZMA produced thousands of feet of strip chart recordings all of which had to be analyzed for anomalous events. This process was laborious and time consuming and led Drake and the astronomy department to develop a means of recording the data digitally. "At the time it was considered a far out and rebellious way to proceed. Everyone knows that astronomers read charts," recalls Drake.[50] Drake and colleagues used a digital voltmeter which they connected to the output of the receiver by means of an RC circuit. The telescope operator would engage the voltmeter at the beginning of the observations and write down the numbers that the system produced. An upgraded version was later developed in which the output frequency, which was proportional to the voltage, was connected to a counter. The radiometer signal would then cause the frequency to change and the count of the total number of cycles over a certain period of time, the integration time, became the output. The system was augmented to produce the output on punched paper tape, which could then be manipulated and represented on the IBM 610 computer. The digital data collection system went into operation in 1960 and became the first digital system in astronomy.[51] This innovation in data collection developed by the NRAO would become standard in radio astronomy instrumentation.

In April 1960 the system went into operation and observed two targeted stars, Tau Ceti and Epsilon Eridani. These two stars were chosen because they are nearby solar types and were thought to be single star systems. Drake's calculations had shown that reasonable radio signals could be detected from planets in orbit around these two stars whose distances were both about 10 light-years from earth.

The program ran for 2 months with, of course, negative results. An exciting false alarm occurred, however, on the very first day. An intense very regular signal was obtained when Epsilon Eridani was observed. It pulsed eight times a second but disappeared after a few minutes. After 10 days of observing, the signal reappeared. By this time, Drake had set up a second small receiver with a small horn pointing out of the window. It also detected the signal, showing that it was terrestrial in origin. In retrospect, Drake remarked that it had all the earmarks of being an electronics countermeasure system, probably airborne, based on the time scale during which it was observed. For a period of about 10 days, Drake thought that he might have succeeded in discovering an extraterrestrial civilization. With the wisdom of an experienced scientist, though, he decided not to release the results until he could verify the event. The difficulty that Drake experienced of distinguishing the extraterrestrial signals from interference foreshadowed a major problem for all SETI searches. Yet, Frank Drake, for a brief time, had a glimpse of what it would be like to discover a civilization on another world.[52]

After the conclusion of Project OZMA, the first modern scientific SETI program, Frank Drake continued to investigate the problem on a more theoretical level. He undertook the task of quantitatively calculating a value for the number of extraterrestrial civilizations in our galaxy. The calculation involves many unknowns and would seem impossible

to determine. Drake identified the variables that are necessarily involved and derived the SETI-Drake equation which calculates the number of advanced technical civilizations in the Milky Way galaxy:

$$N = N_* f_p n_e f_l f_i f_c f_L$$

N_* = the number of stars in the Milky Way Galaxy
f_p = the fraction of stars with planets
n_e = the number of planets ecologically suitable for life
f_l = the fraction of those planets on which life evolves
f_i = the fraction that develops intelligent life
f_c = the fraction that communicates
f_L = the fraction of a planet's life occupied by the communicating civilization

The first variable is reasonably well known. The number of stars in the Milky Way is roughly 300 billion.[53] Each subsequent variable in the equation becomes more speculative than the one preceding. The SETI-Drake equation provides a range when worked out by different scientists. The solutions range from less than one to enormous numbers of civilizations. A reasonable median value of the solutions predicts roughly 4,000 technological civilizations in the Milky Way. The numbers are staggering considering there are billions of other galaxies in the known universe. The implications of the equation are profound. In the SETI-Drake equation Drake encompassed everything from our evolution to our possible extinction. The SETI-Drake equation demonstrates one of the most important consequence of SETI; it forces us to view our past and future from a cosmic perspective.[54]

Project OZMA, in the end, provided major contributions to the science of radio astronomy: the first successful use of paramps for receiving systems, the first digital data collection system, and the first direct use of computers for radio astronomy instrumentation. The project also resulted in the SETI-Drake equation, a quantitative means of estimating the number of extraterrestrial civilizations in the galaxy. Perhaps the major contribution was that Frank Drake and the SETI team initiated a new scientific endeavor, the search for exobiological life forms. Subsequent SETI studies were inspired by Project OZMA. SETI projects have been pursued by groups worldwide. Although more than 40 SETI searches have been carried out since Project OZMA, only a minute portion of the possible frequencies have been observed. Major programs have recently been initiated, such as the NASA SETI program which utilizes a revolutionary multichannel spectrum analyzer capable of observing over 8.25 million channels. The NASA SETI program employs a dual-mode search strategy. The SETI sky survey will scan the entire sky at a high degree of sensitivity. A SETI targeted search will conduct a much higher sensitivity search of all solar-type stars within 85 light-years of earth.[55] Although many critics have accused SETI research of searching for "the needle in the cosmic haystack," a new

Figure 3.18 The NRAO 12-foot (3.66-meter) precision paraboloid obtained in 1959. The 12-foot was used primarily as test instrument for antenna studies of the major instruments. *Courtesy NRAO/AUI.*

field of scientific investigation has nevertheless evolved. Project OZMA was a landmark experiment in that it opened a new field of scientific investigation.

Planned 85-Foot Telescope Programs

The accomplishments of the 85-foot telescope were numerous. The innovation and foresight in planning on the part of the NRAO staff provided the tools with the necessary precision and sensitivity to support cutting edge research. Improvement programs in electronics during the early days with the 85-foot set the precedent for the level of innovation and technology development that earned an international respect for the National Radio Astronomy Observatory. The 75-cm receiver and feed designed and built by the NRAO staff for the 85-foot telescope demonstrated

that the electronics division could build equipment comparable to any then available. The evolution of technology during the 85-foot program was tremendous. Innovations incorporated into the early 85-foot system contributed the first parametric amplifiers used in radio astronomy as well as the first digital data collection techniques, both installed during Project OZMA. The first use of computers in radio astronomy instrumentation was credited to the NRAO staff also during Project OZMA.[56] Innovation in instrumentation techniques and foresight in planning are evident even in the early work of the NRAO electronics and astronomy groups.

It was clear that from the beginning the astronomy group had planned to use the 85-foot as an interferometer. The off-the-shelf design was suitable for interferometry because of its reproducibility. The planning toward interferometry is evident in a description of the foundation design for the 85-foot telescope. The observatory staff constructed the 85-foot foundation according to the Blaw-Knox detailed design. A modification of the design was incorporated that would facilitate movability for interferometric measurements. The planning stages of the 85-foot foundation were described in the 1959 NRAO Annual Report. A. J. Deutsch, a member of the NRAO Advisory Committee, recommended that the base plates and foundations be modified to facilitate putting the telescope on rails so that it could be moved with respect to the 140-foot telescope as an interferometric pair.[57] The 85-foot was later incorporated into a productive interferometer system with two and later three identical elements. Remote elements were eventually added to the interferometer. Although 85-One remained the stationary element in the NRAO Green Bank interferometer, a trend toward interferometry is evident in the early thinking of the planners of the major instruments.

Other Early Telescope Programs

Other early telescope programs that were undertaken in the observatory's formative years include antenna studies and atmospheric research programs utilizing a 12-foot (3.6-meter) reflector and later a 20-foot (6.1-meter) parabolic reflector. A precision 12-foot parabolic reflector with an f/d ratio the same as the 85-foot and 140-foot was obtained from the Electronic Specialty Company of Cohasset, Massachusetts, in 1959 (Figure 3.18). The reflector was an off-the-shelf model constructed of spun aluminum. The antenna was mounted on a surplus radar turret that provided 360° of azimuth motion and limited motion in altitude. The 12-foot was used primarily by the staff for test purposes, particularly for pattern measurements in developing feeds and feed supports.

One early problem investigated with the 12-foot was aperture obscuration. The feed and feed support structure of a telescope obstruct an appreciable area of the telescope aperture. This obscuration results in decreased aperture illu-

Figure 3.19 The Little Big Horn used in the NRAO standardization program. A horn design was utilized in the daily observation of Cassiopeia A to determine the absolute flux density of the source, thereby providing a standard calibration source. *Courtesy NRAO/AUI.*

mination and therefore reduces the gain of the telescope as well as contributing side lobes to the beam pattern. The problem is extremely difficult to deal with theoretically. A study of the diffracted pattern of the aperture obstruction by feed supports of the 140-foot was made using the 12-foot as a model. The 12-foot was used to extend the theoretical predictions that the feed support design would obscure 5 percent of the aperture.[58] The NRAO staff has subsequently contributed much to the understanding and improvement of aperture illumination and efficiency through studies of aperture blockage and the development of innovative feed designs.

The 12-foot telescope was also used for atmospheric research. Short term variations in the atmosphere were investigated at the 3.75 cm wavelength by Torleiv Orhaug to see how atmospheric conditions would affect observations with the major telescopes. Atmospheric absorption was found to significantly affect observations at 3.75 cm. The effect was found to decrease as the observing frequency decreases, as was expected.[59] A 20-foot wire mesh paraboloid designed to operate at 6 cm was later acquired (in 1962) to continue the study.

Two major research projects were initiated during the early stages of the observatory in addition to the 85-foot projects: antenna studies and atmospheric research. A stan-

dardization program (1959) and an investigation of long term variation in radio sources (1961) were undertaken by the early sixties. Two innovative instruments were built for these research projects, the Little Big Horn and the NRAO 40-foot radio telescope.

The Standardization Program and Little Big Horn

The determination of absolute values for the intensity of radio sources was an early major problem faced by radio astronomers. Standardization of observations could only be achieved if the observations were measured relative to a source whose absolute intensity was well known. The feasibility study had pointed out that the observatory would undertake the establishment of standards of measurement for radio and radar astronomy. The writers of the feasibility study saw the establishment of standards of measurement as a responsibility of the national observatory. A calibration horn was constructed to conduct the standardization program. The instrument would observe one source, Cass A, to accurately determine its radio flux density for a standard of measurements. The calibration horn (Figure 3.19), designed by John Findlay and constructed by the NRAO staff, was better known as the Little Big Horn because of its design.

The Little Big Horn is 120 feet long, and the aperture at the upper end is a rectangle 13 by 17.5 feet. The aperture at the lower end is 3 by 6 inches, the size of a standard L-band waveguide. The horn was inclined to a fixed position of 58° 32', the declination of the source.[60] The transit of Cass A was observed by the instrument every day to determine an absolute flux density of the source. The design of the horn was such that the collected power could be calculated precisely from its well-known linear dimensions. The horn design allowed a less ambiguous view of the observed area of sky. All other areas of sky, except the one observed, are effectively blocked out by the horn design. The Little Big Horn could thereby provide a standard of measurement for the other instruments because of its precise determination of the absolute flux density of Cass A.

The flux of power reaching the earth from a radio source is measured in janskys, the amount of radio energy striking a given area (one square meter) in a given frequency range (one Hz); 1 jansky = 10^{-26} watt per square meter per Hz. The problem was to measure the power flux ($S(\nu)$) as accurately as possible. The method used was to collect the radiation from the source in the antenna with a well-known collecting area (A) measured in square meters. The collected power is then $S(\nu)A$ watts/Hz. This power was fed to the receiver and detector which are sensitive to radio frequency power over a frequency range centered around ν (MHz). The bandpass of the receiver is an irregular function of frequency represented by a response function $B(\nu)$, which varies with frequency. The energy detected is therefore written

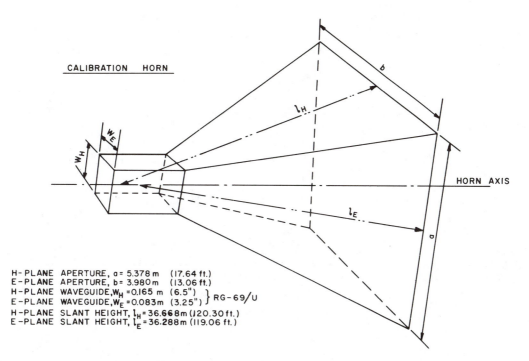

CALIBRATION HORN

H-PLANE APERTURE, a= 5.378 m   (17.64 ft.)
E-PLANE APERTURE, b= 3.980 m   (13.06 ft.)
H-PLANE WAVEGUIDE, W_H =0.165 m (6.5")
E-PLANE WAVEGUIDE, W_E =0.083m (3.25") } RG-69/U
H-PLANE SLANT HEIGHT, l_H = 36.668m (120.30 ft.)
E-PLANE SLANT HEIGHT, l_E = 36.288m (119.06 ft.)

Figure 3.20 Diagram of the internal dimensions of the calibration horn antenna. The horn design efficiently observed a given area of sky by blocking all other areas out. The linear designed also allowed for a precise calculation of the aperture surface area, a value that needed to be known precisely to determine the absolute intensity of the radio signal detected. *Courtesy H. Hvatum.*

$$\Delta P_a = \tfrac{1}{2} \int_0^\infty A S(\nu) B(\nu)\, d\nu .$$

The method of measuring $S(\nu)$ used was to measure P_a, the power received. Cass A initially gave a value of 5×10^{-14} watt when observed at 1,400 MHz with the 85-foot telescope using a receiver with a pass band 5 MHz wide. The measurement of ΔP_a employed was to make a comparison by substituting a source of noise of known intensity for the antenna as a calibration signal. A resistor at an absolute temperature ΔT as used to deliver a maximum power ΔP_{as} into the receiver where

$$\Delta P_{as} = k\Delta T \int_0^\infty B(\nu)\, d\nu .$$

The receiver has essentially the same response to the heated resistor as it does to the antenna. The temperature T_s may be adjusted so that the receiver deflection is the same for the antenna and the heated (thermal) resistor. The power is therefore defined by the relationship

$$\Delta P_a = \Delta P_{as}$$

therefore,

$$\tfrac{1}{2} \int_0^\infty A S(\nu) B(\nu)\, d\nu = k\Delta T \int_0^\infty B(\nu)\, d\nu .$$

Both A and $S(\nu)$ are very slowly varying functions of frequency compared to $B(\nu)$. The following approximation can therefore be made

$$\tfrac{1}{2} A S(\nu) \int_0^\infty B(\nu)\, d\nu = k\Delta T \int_0^\infty B(\nu)\, d\nu ,$$

which gives

$$S(\nu) = \frac{2k\Delta T}{A} .$$

The derivation of this relationship showed that to measure $S(\nu)$, the flux from the source, only the measurements of ΔT_s and A are required.[61] The Little Big Horn was therefore designed to obtain these measurements.

The dimensions of the horn were designed such that the collecting area could be easily and accurately calculated from linear dimensions. The internal dimensions are given in Figure 3.20. The effective area of a pyramidal horn antenna may be calculated from its dimensions, according to Schelkunoff in 1943. The calculation of the area of Little Big Horn from its dimensions produces an effective collecting area of 9.5 square meters. Measurements of the horn dimensions and its loss allowed calculation of A to within 2 percent.

The horn was built from 1/8 inch aluminum sheets supported by an aluminum superstructure. An RG-69 waveguide was utilized at the small end. A standard NRAO receiver, a simple radiometer, and calibration system (Orhaug and Waltman, 1962) with various front ends were used. A block diagram of the receiver is shown in Figure 3.21.

The first observations were made in October 1959 at 21 cm. The frequency spectrum from 900 to 1,400 MHz was observed to investigate the flux measurements over time. By 1964 the investigation had produced results of 2,370±26 janskys measured for the flux density of Cass A. The measured value of the flux is the best-fit-line through the data points collected over the period of 1960 to 1964 (Figure 3.22). The measurements were corrected for atmospheric absorption. The secular reduction of the flux was measured

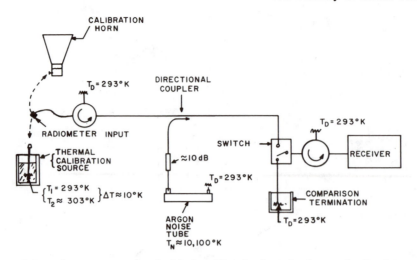

Figure 3.21 Block diagram of the radiometer system used with the calibration horn for the standardization program. A standard NRAO receiver (Orhaug and Waltman, 1962) consisting of a simple radiometer and calibration system was used. *Courtesy H. Hvatum.*

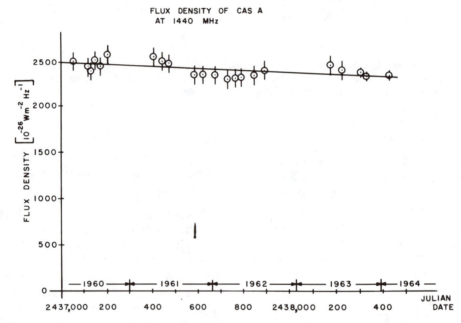

Figure 3.22 The flux density of Cassiopeia at 1,440 MHz. A best-fit-line through datapoints collected during the period 1960 to 1964 was used to provide a standard value of 2370±26 janskys for the flux density of Cass A. *Courtesy H. Hvatum.*

to be 1.75±0.52 percent per year, which improved the measurements made by Hogbom and Shakeshaft in 1961.[62] The precise measurement of the absolute flux density of Cass A provided an accurate standard to which other observations could be compared. A precise value for the standard flux density source Cass A provided a means of calibration for all other observations. The accuracy and credibility of other observations at the observatory were assured by the Little Big Horn standardization program.

The 40-Foot Radio Telescope

Another important early NRAO telescope program was an attempt to observe long term variation in radio sources.

The long term variation program was initiated by Dave Heeschen in 1961. The idea was to observe 10 or 12 sources every day for a period of years to determine if there were any long term variation in the intensity of radio sources. The visual brightness of stars and other cosmic phenomena had long been known to vary over time. A secular reduction of Cass A was in the process of being determined. Although the long term variation of radio sources was a high priority problem, it was deemed impractical to tie up the major instruments for the amounts of time required by such a study. The long term variation program at the NRAO therefore resulted in the design and construction of an automated 40-foot transit telescope to monitor several sources daily. The NRAO 40-foot (12.2-meter) shown in Figure 3.23 became

Figure 3.23 The NRAO 40-foot (12.2-meter) radio telescope. The 40-foot transit telescope was constructed in 1962 as a fully automated instrument to investigate long term variation in several radio sources by observing the same sources daily for 5 years. *Courtesy NRAO/AUI.*

the one of the first fully automated telescopes in radio astronomy.[63]

The 40-foot telescope, a wire mesh paraboloid, was designed to operate at 750 and 1,400 MHz simultaneously. A Jasik feed was used with two standard NRAO switched receivers.[64] An automated positioner was developed to point the telescope to positions corresponding to eight cosmic sources and two calibration sources, and make observations each day. The program was designed to examine these sources for secular variation over a period of 5 years. The change in intensity of the sources was expected to be very small and barely, if at all, detectable. The variation was obscured in the uncertainty of the results of previous programs. The 40-foot telescope would be exclusively dedicated to this problem for several years in an attempt to reduce the uncertainties and provide conclusive evidence for long term radio source variation.

A control system was developed to completely automate the positioning of the telescope as well as the collection and recording of data. The control system positioned the telescope in declination 52 sidereal minutes prior to the transit time of each source. The system would then instruct the telescope to record the output of the two receivers as the source moved through the telescope beam. The output was integrated for 117.981 sidereal seconds and then punched on paper tape. The sidereal time and the output signal were digitally recorded. A typical observation would have several basic automated steps. Observation of the background sky prior to the transit of the source established a zero level for the background sky. After a baseline was thus established, a calibration signal was introduced before and after the transit. The source was then observed for a set time, depending on its declination. The control system would digitally record the data, print the output on punched tape, and then move to the next source.[65]

The long term variation program went on-line in March of 1962. The program continued until 1968. No long term variation in the intensity of the eight sources observed was detected. Secular variation was detected in radio sources in the late 1960s. The 40-foot receivers at the time performed at the borderline of the sensitivity required to detect the secular variations. Had it observed the brightest 12 sources instead of the brightest 8, it might have detected the long term variation that does occur in radio sources.

The 40-foot telescope was refurbished in 1968 and used in various other programs. In 1987 the front-end receiver was completely upgraded for use in a teacher education program jointly sponsored by the NRAO and West Virginia University and funded by the National Science Foundation. Two modernized 21-cm receivers were installed and the telescope positioning system converted back to manual, with automation capabilities to be added later.[66] The sensitivity has been tremendously upgraded with the new system. The 40-foot is now capable of observing in excess of 100 sources and is used primarily as an educational tool. New uses for a historical telescope have been discovered at the NRAO which allows the 40-foot to continue to contribute to the science of radio astronomy.

Early Staff

The early days of the National Radio Astronomy Observatory were marked by a period of rapid growth and discovery. The innovations in instrumentation and observing techniques earned an international reputation for the young observatory. An astronomy and electronics staff was assembled that had begun to provide the tools necessary for the pursuit of cutting edge research.

The directorship of the early NRAO changed hands in 1961 when Otto Struve resigned to devote more time to his research and writing. Joseph L. Pawsey was appointed the new NRAO director on December 18, 1961. Pawsey was an internationally known radio astronomer and assistant chief of the Division of Radiophysics of Australia's Commonwealth Scientific Research Organization. He accepted the appointment but never actually acted as director because of health problems. D. S. Heeschen had the position of acting director, succeeding Otto Struve in December 1961.[67] Heeschen was appointed director of the NRAO by the AUI Board of Trustees on October 19, 1962, a position he would occupy for nearly 20 years.

The early NRAO staff was extremely prolific in terms of observing programs, technical innovations, discoveries, and publications. The staff also made major contributions to the planning and construction phases of the major instruments, namely the 140-foot, the 300-foot, and the interferometer. The early days of the evolution of the National Radio Astronomy Observatory came to an end when these major telescopes began to come on-line.

Endnotes

1. *National Radio Astronomy Observatory Annual Report*, July 1, 1959. NRAO, Green Bank, West Virginia, operated by Associated Universities Incorporated, under contract with the National Science Foundation.
2. James F. Crews, Interview with Benjamin K. Malphrus, National Radio Astronomy Observatory, Green Bank, West Virginia, operated by Associated Universities Incorporated, under contract with the National Science Foundation, March 3, 1989.
3. The convention in the 1950s and 1960s dictated using the term *cycle* as opposed to *Hertz* to describe frequency; Mc/s, Mcs or even Mc appear in the historical literature rather than MHz.
4. *National Radio Astronomy Observatory Annual Report*, July 1, 1959.
5. *Ibid.*, p. 4.
6. *Ibid.*, p. 4.
7. *Ibid.*, p. 6.
8. *Ibid.*, p. 6.
9. *Ibid.*, p. 4.
10. Richard M. Emberson, "National Radio Astronomy Observatory," *Science*, vol. 130, no. 3385, November 13, 1956, pp. 1307–1318.
11. *Ibid.*, pp. 1307–1318.
12. James F. Crews, "The Tatel 85-foot Telescope," *The Observer*, National Radio Astronomy Observatory, Green Bank, West Virginia, vol. 1, no. 5, March 31, 1964, p. 1.
13. *Ibid.*, p. 1.
14. Richard M. Emberson, "National Radio Astronomy Observatory," pp. 1307–1318; National Radio Astronomy Observatory Annual Report, July 1, 1959, p. 8.
15. D. S. Heeschen, "Observations of Radio Sources at Four Frequencies," *Astrophysical Journal*, vol. 133, no. 1, January 1961, pp. 322–335.
16. *National Radio Astronomy Observatory Annual Report*, July 1, 1959.
17. W. Findlay, "Operating Experience at the National Radio Astronomy Observatory," Annals of the New York Academy of Sciences vol. 116, pp. 25–37, June 26, 1964.
18. *Ibid.*, pp. 25–37.
19. National Radio Astronomy Observatory Annual Report, July 1, 1959.
20. John D. Kraus, *Radio Astronomy*, New York: McGraw-Hill Book Company, 1966.
21. D. S. Heeschen, "Observations of Radio Sources at Four Frequencies," *Astrophysical Journal*, vol. 133, no. 1, January 1961, pp. 322–335.
22. James F. Crews, "The Tatel 85-foot Telescope," *The Observer*, National Radio Astronomy Observatory, Green Bank, West Virginia, vol. 1 no. 5, March 31, 1964, p. 1.
23. F. D. Drake, "The Position Determination Program of the National Radio Astronomy Observatory," *Publications of the Astronomical Society of the Pacific*, vol. 72, no. 429, December 1960, pp. 494–497.
24. *Ibid.*, pp. 494–497.
25. D. S. Heeschen, "Observations of Radio Sources at Four Frequencies," *The Astrophysical Journal*, vol. 133, no. 1, January 1961, pp. 323–331.
26. *Ibid.*, pp. 323–331.
27. Kenneth L. Franklin, "The Discovery of Jupiter Bursts," included in *Serendipitous Discoveries in Radio Astronomy*, Proceedings of a Workshop held at the National Radio Astronomy Observatory at Green Bank, West Virginia on May 4,5,6, 1983, K. Kellerman and B. Sheets, (eds.) NRAO Publication: Green Bank, West Virginia, pp. 252–257.
28. J. S. Hey, *The Evolution of Radio Astronomy*, London: Elk Science, 1973, pp. 116–121.
29. F. D. Drake, "Radio Emission from the Planets," *Physics Today*, April 1961, pp. 30–35.
30. *Ibid.*, pp. 30–35.
31. "NRAO Observations of Venus" *The Observer*, National Radio Astronomy Observatory, Green Bank, West Virginia, vol. 2, no. 4, May 29, 1963, p. 4.
32. *National Radio Astronomy Observatory Annual Report*, July 1, 1959.
33. F. D. Drake, "Radio Resolution of the Galactic Nucleus," *Sky and Telescope*, 1959, pp. 2–3.
34. *National Radio Astronomy Observatory Annual Report*, July 1, 1959.
35. *Ibid.*
36. Hogbom and Shakeshaft, "Secular Variation of the Flux Density of the Radio Source Cassiopeia A," *Nature*, vol. 190, no. 4777, May 20, 1961, pp. 705–706.
37. *National Radio Astronomy Observatory Annual Report*, July 1, 1959.
38. C. M. Wade, "A Possible New Radio Galaxy in the Virgo Cluster," *The Observatory*, vol. 80, no. 919, 1959, pp. 235–236.
39. G. Swarup, "Radio Astronomy," Commission 40 Report covering surveys of radio sources, National Radio Astronomy Observatory Reprint Series B 534, 1981.
40. *National Radio Astronomy Observatory Annual Report*, July 1, 1959.
41. Frank D. Drake, "Project OZMA," included in *The Search for Extraterrestrial Intelligence*, Proceedings of a workshop held at the National Radio Astronomy Observatory, Green Bank, West Virginia, May 20, 21, 22, 1985, K. I. Kellerman and G. A. Seielstad, (eds.), pp. 17–26.
42. Frank D. Drake, "Project OZMA," pp. 17–26, Frank D. Drake, "How Can We Detect Radio Transmissions from Distant Planetary Systems?," *Sky and Telescope*, vol. 19, no. 3, January, 1960, pp. 3–7.
43. Frank D. Drake, "Project OZMA," p. 18.
44. Thomas R. McDonough, *The Search for Extraterrestrial Intelligence: Listening for Life in the Cosmos*, New York: Wiley Interscience, 1987, pp. 110–117.
45. Frank D. Drake, "Project OZMA," p. 19.
46. *Ibid.*, pp. 19–24.
47. *Ibid.*, pp. 22–24.
48. Giuseppe Cocconi and Philip Morrison, "Searching for Interstellar Communications," *Nature*, vol. 184, no. 4690, September 19, 1959, pp. 844–846.
49. Frank D. Drake, "Project OZMA," p. 20.
50. *Ibid.*, pp. 24–26.
51. *Ibid.*, pp. 24–26.
52. Thomas R. McDonough, *The Search for Extraterrestrial Intelligence: Listening for Life in the Cosmos*, New York: Wiley Interscience, 1987, pp. 110–117.
53. *Ibid.*, pp. 110–117.
54. *Ibid.*, pp. 110–117.
55. The NASA Microwave Observing Project for the Search for Extraterrestrial Intelligence (SETI), Publication of the SETI Institute, February 16, 1989.
56. Frank D. Drake, "Project OZMA," p. 19.
57. *National Radio Astronomy Observatory Annual Report*, July 1, 1959, p. 19.
58. *Ibid.*, p. 14.
59. "The Atmospheric Projects," *The Observer*, National Radio Astronomy Observatory, Green Bank, West Virginia, vol. 1, no. 10, September 30, 1962, p. 1.
60. J. W. Findlay, H. Hvatum, and W. B. Waltman, "An Absolute Flux-Density Measurement of Cassiopeia A at 1440 MHz," *Astrophysical Journal*, vol. 141, no. 3, April 1, 1965, pp. 873–884.
61. *National Radio Astronomy Observatory Annual Report*, July 1, 1959, pp. 12–15. J. W. Findlay, H. Hvatum, and W. B. Waltman, "An Absolute Flux-Density Measurement of Cassiopeia A at 1440 MHz," *Astrophysical Journal*, vol. 141, no. 3, April 1, 1965, pp. 873–884.
62. J. W. Findlay, H. Hvatum, and W. B. Waltman, "An Absolute Flux-Density Measurement of Cassiopeia A at 1440 MHz," pp. 873–884.
63. "The 40-foot Telescope," *The Observer*, vol. 1 no. 3, February 28, 1962, p. 1.
64. *Ibid.*, p. 1.
65. Benjamin K. Malphrus and Richard F. Bradley, "The NRAO 40-foot Radio Telescope Operators Manual," NRAO Internal Publication, edited by the NRAO Electronics Staff, July 1987, pp. 2–8.
66. *Ibid.*, pp. 2–8.
67. "Dr. Dave S. Heeschen Appointed Director of NRAO," *The Observer*, vol. 1 no. 11, October 31, 1962, p. 1.

Chapter 4

The NRAO 140-Foot Radio Telescope

A 140-foot all-purpose, high precision radio telescope was planned in the feasibility study for the proposed major instruments at the NRAO. Enormous aperture was placed at a premium by the planners in this envisioned age of gargantuan telescopes. There was a need for high resolution observations at short (centimeter) wavelengths. The short wavelength region of the radio spectrum, down to 1 centimeter, was essentially a frontier territory at this point in the young science. A telescope that could make high resolution observations at these high frequencies demanded radical design specifications and represented a tremendous engineering challenge. The high resolution desired required a large aperture, while the high frequency observations required a precision surface.

The NRAO planners decided to build the smaller, high precision telescope first, to gain experience before an attempt was made to construct the 300-foot and 600-foot telescopes. The exact size of the high precision telescope became an early subject of debate among the planners. The diameter of this precision telescope was not determined by physical observation requirements. The size was in fact determined as a compromise. The earlier consensus reached by the AUI committee, based on advising by numerous astronomers and engineers, was that the telescope should be larger than 100 feet (31.5 meters). The size was later changed to 40 meters to facilitate the expression of the aperture in wavelengths. Subsequently a "member of a governmental review group" suggested that the aperture of a national telescope be expressed in feet to facilitate comparison with other telescopes in the United States and Britain.[1] The second conversion of units also resulted in inflationary rounding off and the telescope aperture was set at 140 feet (42.67 meters or 200 wavelengths at 21 centimeters and roughly 1,400 wavelengths at 3 centimeters).

The size chosen by the steering committee was intensely debated among the astronomical community after the controversial decision was made to undertake a radical equatorial design. A cost study was made for a mounting design, equatorial or alt-azimuth for the 140-foot aperture. The study was based on information and estimates provided by J. Feld, H. C. Husband, and D. S. Kennedy. Alt-azimuth designs were generally less expensive to construct, but the cost of coordinate conversion was then a function of the precision desired and reliability was a problem as the reflector must be moved at variable speeds to track an object as it moves across the sky. The computer power required for the conversions was enormous compared to the state-of-the-art computers. The world's largest telescope at the time, the new Jodrell Bank 250-foot alt-azimuth telescope, had been plagued by pointing and tracking errors because the existing computer technology was insufficient to handle the conversions from alt-azimuth to equatorial coordinates. An equatorial design would alleviate the pointing and tracking errors as the polar axis is aligned with the North Star. Equatorial designs are relatively inexpensive for small reflectors, but as the size increases, the equatorial mount becomes more difficult and expensive to build. The engineering and design problems increase with size because of the large counterweights necessitated by the design. The associated bearings and other structures must also be larger and stronger to support the weight. The initial cost study failed to demonstrate that for a 140-foot reflector one type of mounting design would be less expensive to construct than the other.[2]

There was dissension among the astronomical community, however, as to the accuracy of the findings of the cost study. The 140-foot aperture size was in the middle of a structural "gray area" and thus a mounting design was not easily dictated. Grote Reber was a major opponent to the selection of the 140-foot aperture size. Reber argued that the 140-foot figure seems to have been arrived at as some kind of compromise. "Apertures up to about 100 feet should be made equatorial. At 200 feet this is impossible. Only an alt-azimuth mount is feasible. Within this range, 100 to 200 feet the problem is indeterminate and much time and effort may be expended trying to find a solution."[3] Reber and others felt that the selection of aperture size within this range was therefore a mistake. Nonetheless, the 140-foot size was retained and the mounting design was initially left open to the potential designers.

Figure 4.1 The Jacob Feld alt-azimuth design for the NRAO 140-foot (42.7-meter) telescope. The Feld design called for a parabolic reflector mounted on two central towers supported on a cross-bridge that rotated on a central azimuth bearing.

Planning and Design of the 140-Foot

The feasibility study committee for the NRAO set the size at 140 feet for the high precision telescope. The performance specifications were outlined in detail in the NRAO Planning Document and evolved from many ideas originated at committee meetings and from discussions with individual scientists. An initial draft of the design and performance specifications was adopted by the steering committee in May 1955. The basic design called for a focal length/diameter (f/d) ratio of 0.35 to 0.5; a surface tolerance of $\pm 3/8$ inch over the entire aperture and $\pm 1/4$ inch over the inner half aperture; rigidity and drive and control systems to be capable of an accuracy of 30 arcseconds; and complete hemispheric sky coverage and the ability to point below the horizon in some position.[4] Design proposals were subsequently requested from several engineering firms selected "because of their known competence and interest in devices comparable in size and complexity to the desired instrument."[5]

Three designs were developed by engineering firms that were solicited for consideration and inclusion in the planning document. The design proposals were requested from Jacob Feld, H. C. Husband Inc., and D. S. Kennedy and Company. Feld indicated that he could develop a 140-foot design which would allow an opportunity to test design solutions for the proposed 600-foot instrument, for which he performed the feasibility study. He was accordingly asked to submit a 140-foot design. H. C. Husband Inc. was asked

to develop a 140-foot design because of its experience in the design and construction of the 250-foot Jodrell Bank telescope. Kennedy and Company was asked to undertake a 140-foot design because of its experience in the construction of parabolic reflectors for the U.S. military and radio telescopes for several institutions in the United States. Development of the three designs began in November 1955. By that time the Steering Committee had reduced the surface tolerance specification to 1/4 inch over the entire surface and ruled out an equatorial mount as an alternative.[6]

The Feld design, seen in Figure 4.1, consisted of a parabolic reflector mounted on two towers supported on a cross-bridge that rotates on a central azimuth bearing. The towers were extended at the ends of the bridge and fitted with rollers that rest on a circular railroad track. The reflector was a thin shell supported only at its perimeter by a massive ring girder. The Feld design utilized 410 tons of structural steel and had a reflector weight of 219 tons.[7]

The Husband design (Figure 4.2) was similar to a design independently developed by Grote Reber in 1948. The design consisted of a massive reflector built into one side of a cylinder of structural steel. The length and diameter of the cylinder were nearly equal to the diameter of the reflector, providing much structural support. The cylinder was mounted between two altitude bearings at the top of two towers, as in the Feld design. In the Husband design, most of the weight of the reflector and cylinder was carried by four circumferential bands resting on rollers mounted on a cross-bridge. The entire unit rotated on concentric railroad tracks similar to the Feld design. The Husband design required 1,125 tons of structural steel and the reflector weighed 130 tons.[8]

The Kennedy design (Figure 4.3) consisted of a reflector built into a system of rear supporting trusses which is connected to a D-shaped turret. The entire assembly was carried by altitude bearings at the end of the turret. The reflector was balanced about the altitude axis by counterweights on large-radius gear sectors. The turret rested on an oil pad bearing ring mounted at the top of a concrete silo. Smooth motion in azimuth would be provided by the azimuth bearing mounted in the rigid silo. The Kennedy design used 778 tons of structural steel and the reflector weighed 146 tons.[9]

Although the three designs were dissimilar, all met the design and performance specifications and all were alt-azimuth telescopes, in accordance with the Steering Committee's requests. The selection of an alt-azimuth design was made from purely structural considerations. The alt-azimuth mount was found to be the simplest and least expensive design by each of the three designers. A question soon was raised whether the alt-azimuth would still be the simplest and least expensive design when the drive and control systems were considered.

The question of the mounting design became a subject of

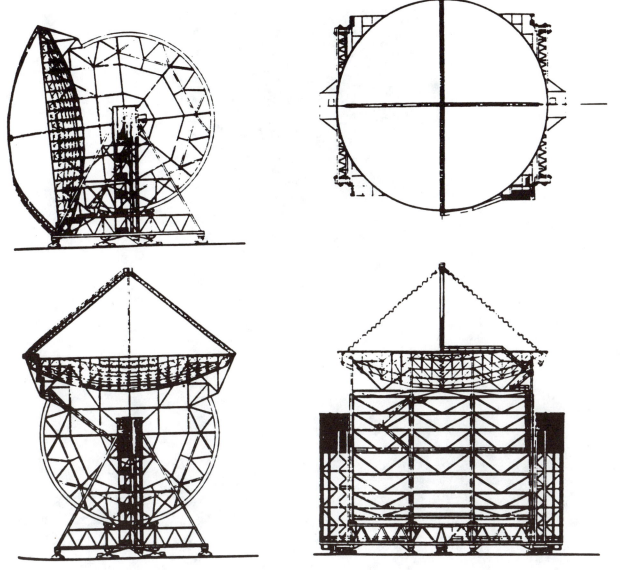

Figure 4.2 H. C. Husband Inc. of Sheffield, England alt-azimuth design for the NRAO 140-foot telescope. Two end elevation views (top and bottom left) show the reflector pointed toward the horizon and the zenith. A view from above (top right) shows the quadrapole feed support structure. The frontal view (bottom right) shows the massive structural cylinder which provided motion in declination.

much debate. Two groups developed, one in favor of an alt-azimuth design based on structural considerations and one in favor of an equatorial telescope based on observational considerations. Members of the NSF Advisory Committee and other astronomers interested in the program had reservations concerning the reliability in pointing and tracking of an alt-azimuth telescope. The Advisory Committee reviewed the matter at its March 27–28, 1956, meeting. It decided to recommend that an equatorial design based on the needs of the astronomers (particularly those in the committee and acting as advisers to the committee) be developed to meet the set design specifications.[10]

The question of aperture size was the first controversial decision. The equatorial mount was a second. These decisions along with the stringent performance specifications

set for the telescope dictated a radically different design concept.

The AUI Steering Committee pursued the development of a new design at the request of the NSF Advisory Panel. A decision was made to entrust the development of this design to Ned Ashton, designer of the NRL 50-foot high precision telescope, with an ad hoc group of scientists and engineers continuously reviewing the work. The ad hoc panel consisted of T. C. Kavanagh, who chaired the committee, P. Bijlaard, N. A. Christensen, A. M. Freudenthal, J. O. Silvey, D. Lindorff, E. J. Poitras, B. H. Rule, F. T. Haddock, E. F. McClain, J. G. Bolton, and H. E. Tatel.[11] The design development went through several subsequent stages. The Franklin Institute of Philadelphia designed the hydrostatic main bearings to accompany the basic telescope design. T. W.

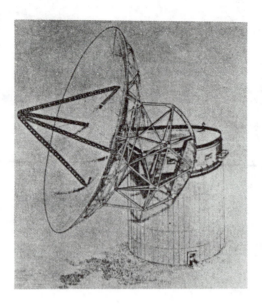

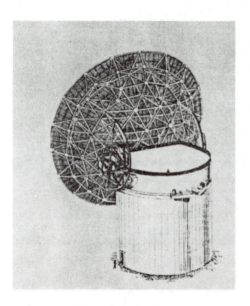

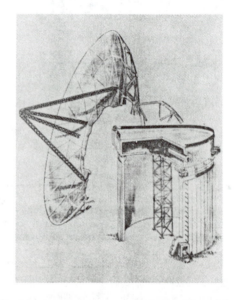

Figure 4.3   The Kennedy alt-azimuth design for the NRAO 140-foot telescope. The top view shows a general side elevation of the telescope design which consisted of a reflector built into a system of rear supporting trusses which was connected to a D-shaped turret mounted on a concrete silo containing the control room. A rear elevation view is shown at bottom left accentuating the superstructure. A cut-away view shown bottom right shows the torque tube connecting the altitude gear sectors.

Brown developed a design for the drive and control system. A final design provided by Ashton, although controversial, was approved by the committee in 1957.[12]

The NRAO 140-Foot Telescope Design

The design produced by Ashton called for a solid surface paraboloid supported by an aluminum superstructure equatorially mounted on a concrete base (Figure 4.4). The superstructure is attached to a massive yoke arm by means of a declination shaft and bearings. Motion in declination is provided by a gear that extends from the back of the superstructure and a declination axis that runs through the superstructure. Motion in right ascension is provided by a solid steel polar shaft that is parallel to the earth's axis of rotation. The polar axis shaft is supported on its spherical north end by four hydrostatic oil pad bearings. A steel yoke assembly is attached to the polar axis shaft by means of a

Figure 4.4 N. Ashton 140-foot equatorially mounted telescope design. The superstructure is attached to a massive yoke arm by means of a declination shaft and bearing. Motion in right ascension is provided by a solid steel polar axis shaft supported on its spherical north end by hydrostatic oil pad bearings. Huge declination and polar gears are utilized in the equatorial design. The structure is mounted on a concrete foundation which contains the control room and mechanical rooms. The design was considered radical from an engineering standpoint. *The original design blueprints were microfilmed by Micrographics II, of Charlottesville, Virginia, and photographically reproduced to produce this image.*

spherical bearing, the heart of the structural design. The yoke assembly, which provides counterbalance for the reflector, contains an enormous polar gear attached to the truss structure to rotate the telescope east and west, as well as the declination drives to rotate to the telescope north and south.

The scale described by the design is enormous. The 140-foot diameter paraboloid has a surface area of 3/8 of an acre. The reflector is comprised of 350 tons of aluminum, 35 tons of concrete ballast, and 5 tons of balancing blocks. The focus is 60 feet above the surface and carries 1/2 ton of receiver components. The polar shaft, made of machined steel, is 67 feet long and is surrounded by welded castings. The polar shaft assembly weighs 555 tons. An additional 170 tons of high density concrete are incorporated into the polar shaft assembly for ballast. The total moving weight of the telescope totals 2,500 tons and is designed to rotate on the spherical bearing. The spherical bearing is the largest in the world, 17-1/2 feet in diameter and weighing 400 tons. The spherical bearing supports the entire 2,500 ton moving weight while floating on a film of oil only 0.005 inch thick. Hydraulic pressure of up to 3,000 pounds per square inch (psi) maintains the oil film, supplied by 16 constant-volume, constant-speed hydraulic pumps. Sixteen other hydraulic pumps supply the tail bearing which floats on four adjustable hydrostatic cylindrical pads.[13] The foundation is 60 feet high and contains the control room, hydraulic and electrical equipment, and transformer vault.[14] Ashton's design for the instrument was structurally very complex and was expected to be a tremendous engineering challenge. The de-

sign had evolved into the largest equatorial telescope in the world.

In August 1957, bids were solicited for the fabrication and construction, and the detailed design, of the NRAO 140-foot telescope. The E. W. Bliss Company of Canton, Ohio, was awarded the initial prime fabrication and construction contract in June 1958. The Bliss Company subcontracted the field work to Darin and Armstrong, who began fabrication on site in August of that year.[15] The Electric Boat Division of General Dynamics Corporation was subcontracted to design and construct the hydraulic drive systems. The detailed designs and construction of the drive and control systems were based on the basic designs of T. W. Brown, but incorporated modifications imposed by the NRAO staff to improve the instrument's versatility. The subcontract for the drive and control system was awarded in April 1959.[16] The Electric Boat Division was the primary contractor then for atomic submarines. In fact, the drive skids for the 140-foot were the same hydraulic drive skids that were used on atomic submarines at the time.[17]

Construction Phase

The construction of the 140-foot radio telescope involved tremendous engineering challenges from the beginning. The design and construction were initially handled almost exclusively outside Green Bank. The design specifications were produced by the AUI planners and the construction contracted to the various companies. Several companies failed along the way. There were companies that could not

Figure 4.5 The partially complete 140-foot telescope foundation and the matched pair of 250 ton derricks used to erect the major telescope subassemblies. The foundation is made of 5,700 tons of 5,000 psi concrete and 140 tons of reinforcing steel anchored on rock 30 feet (9.15 meters) below the ground. The walls of the foundation are 3 feet thick. *Courtesy NRAO/AUI.*

can Hoist and Derrick Company (Figure 4.5) which would lift major subassemblies including the polar axis shaft, spherical bearing and tail bearing assembly, which weighed 390 tons, the 405 ton yoke and hub assembly, and the super-structure, which weighed 210 tons.[19] The foundation was formed and poured by Darin and Armstrong. It required 5,700 tons of 5,000 psi (pounds per square inch) concrete and 140 tons of reinforcing steel, anchored on rock 30 feet below the ground. The walls of the foundation are 3 feet thick. The interior foundation was finished by the contractors to house electrical and hydraulic primary components in the basement, the control room on the first floor, high pressure hydraulic pumps on the second floor, and additional drive and brake system hydraulic components on cantilevered arms extending from the third floor. An observation deck is located on the fourth level, which also supports the tail bearing for the polar shaft at the south end and the main drive and brakes for rotation in right ascension at the north end.[20]

Contracts for the fabrication of the major components were made early in 1958 and construction of these components began shortly thereafter. The first components arrived on site in 1958. The first major problem, the fabrication of the polar axis shaft, immediately surfaced. The 67-foot long, 12-foot diameter all-welded steel polar shaft proved too difficult to make, and its construction was abandoned by the E. W. Bliss Company. This was one in a series of ongoing engineering problems that began to mount by 1961. It became clear that the prime contractor would be unable to construct the telescope. The prime contract with E. W. Bliss and their subcontractor Darin and Armstrong was terminated in 1961 by mutual consent.[21] The engineering firm who completed the assembly of the telescope was Stone and Webster, with the NRAO engineering staff overseeing the process.[22]

The engineering and fabrication problems that surfaced were understandable in light of the fact that the components were enormous in scale and yet required precise tolerances. As a result of this problem, the polar axis shaft along with many other components, were delayed in shop fabrication. After fabrication of the original polar axis shaft was abandoned by the Bliss Corp., a second attempt was undertaken by the Westinghouse Corporation of Pittsburgh, Pennsylvania. A design that apparently met the specifications was completed and shipped to the site, but rejected because of brittle stress fracture that the chosen alloy was subject to during cold weather. Taking the brittle fracture problem into account, the Westinghouse Corporation fabricated a second declination axis shaft that was acceptable.[23] The new polar axis shaft was composed of machined 3-1/2 percent nickel steel casting that weighed 40 tons, welded to a tubular rolled plate straight section that weighed 90 tons which in turn was welded to 3-1/2 percent nickel conical transition casting machined to 85 tons.[24] The massive shaft assembly was machined to impressive toler-

make the contracts.[18] Several components were received that did not meet the fabrication specifications. Numerous delays were created by these problems. There were even problems with erection of the telescope after all components were fabricated and approved for assembly. By the time it was completed, the telescope had been in the construction stage for so long that many components were no longer state-of-the-art. The construction of the 140-foot telescope evolved into a virtual engineering nightmare, one that was reflected in reality by the failure of the Sugar Grove telescope, only 40 miles away. The construction phase was interrupted on several occasions by numerous construction, fabrication, and contractual problems, one of which ultimately ended in the termination of the prime contractor of the project. These problems were overcome as the NRAO engineering staff eventually assumed control of the project.

Construction of the foundation was begun in 1958 and completed in late 1959. The first erection at the site was a matched pair of giant 250 ton derricks, built by the Ameri-

ances of 0.0001 of an inch (0.0002 centimeter).[25] The second Westinghouse polar axis shaft was shipped to Green Bank on November 18, 1963, nearly a year after the expected date. The polar axis shaft finally met the performance specifications and was approved for erection in the telescope. The original Westinghouse shaft that was rejected remains unaccounted for in the literature and records. Entire assemblies that were found to be subject to brittle stress fracture were supposedly buried somewhere near the 140-foot site as they were too expensive to be shipped away.[26]

The declination shaft is one of the most complex subassemblies of the telescope. Its aluminum shell is adjustable about a steel central shaft. The quill assembly allows the antenna to be precisely adjusted to the true declination axis. The steel shaft is adjustable to bring the declination axis precisely perpendicular to the polar axis. After its arrival on site the second declination shaft was snaked under the erected gear girder and fitted into place.[27]

The spherical bearing on which the telescope rests and pivots is a critical component that affects the performance of the instrument. The 17-1/2 foot (5.34-meter) spherical bearing was cast of 3-1/2 percent nickel steel alloy by General Steel Industries at Eddystone, Pennsylvania, in January 1963. Because of its size, a special railroad car was cast to transport the spherical bearing to the Westinghouse plant in Pittsburgh where it was machined. The original casting from which the sphere was machined weighed 350 tons, the largest nickel steel casting ever poured. Following rough machining the casting was examined by radiograph, magnetograph, and reflectoscope. Final machining was completed at Westinghouse in April 1964. The completed sphere weighed 150 tons and was machined to within 1/8 inch of design diameter and within 0.003 inch of sphericity (Figure 4.6). The surface was highly polished to between 10 and 20 microinch finish.[28]

Shipping the spherical bearing to Green Bank was a tremendous task in itself. The components were shipped by rail to Bartow, West Virginia, and there transferred to tractor trailer trucks to continue to Green Bank. The seven bridges between Bartow and Green Bank were shored to permit the passage of the major components. Concrete mats were hoisted onto each bridge to allow the component to pass and then transported to the next bridge to be hoisted into place. Transportation of the spherical bearing across these bridges proved difficult, actually sticking once. To unstick the lowboy, it was jacked up and runners placed underneath. An additional truck attempted to push from the rear, as a mobile crane lifted from above. The convoy made it up the hills en route with two trucks pulling, one pushing, and the mobile crane winching from the top.[29] The arrival of the spherical bearing represented the last and most difficult transportation of the major components.

The first and longest duration field assembly was the aluminum truss superstructure which was welded piece by

Figure 4.6 Machining of the spherical journal to be positioned at the end of the polar shaft. The spherical bearing is 17-1/2 feet (5.4 meters) in diameter and cast of 400 tons of 3-1/2 percent nickel steel. The spherical bearing is designed to support the full 2,700 ton moving weight of the telescope while floating on a film of oil only 0.005 inch (0.0127 cm) thick. *Courtesy NRAO/AUI.*

piece upside down at the site (Figure 4.7). The structure was welded from 5456 aluminum, the heaviest known aluminum rolled sections. Special techniques of field welding were developed for heavy aluminum sections and a special group of welders were trained and qualified on site. A stiff, highly compliant structure that would not deflect more than 3/8 inch in winds up to 135 mph was created. The first large subassembly fitted to the assembled superstructure was the declination gear girder with counterweight box section. This 20-ton aluminum fabrication was lifted into place and attached to the superstructure truss structure. The girder holds the bronze declination gear and 36 tons of concrete and steel counterweighting.[30]

It was desirable from a structural point of view that the yoke hub weldment be fabricated in as large a section as possible and under controlled conditions in order that the 137-ton nickel steel casting welded to 6-inch plate could be heat treated and radiographed, then precision machined to fit the spherical bearing. The size of the hub weldment had been set by the narrowest railroad passage en route. The yoke hub weldment, which bolts to the north end of the polar shaft, required a special railroad car to bring it from the fabricator (Westinghouse) in Pittsburgh to the station at Bartow, West Virginia, where it was offloaded and trucked to Green Bank. The car permitted the load to be suspended between two identical end sections that could be pulled apart. The load could therefore be lowered to allow clear-

Figure 4.7 The all-welded 5456 aluminum superstructure in assembly on site. A field tolerance of ± 3/16 inch (0.48 centimeter) across any dimension was maintained. An argon/helium atmosphere was utilized in the welds. The completed weldment weighed 180 tons. Erection of the superstructure involved a 20 percent overload on both derricks but was successful. *Courtesy NRAO/AUI.*

Figure 4.8 Section of the polar gear. Separate gear segments designed and manufactured by Philadelphia Gear were hung to the girder plates as the structure was rotated via a small temporary gear system. The assembled main polar drive gear is 84 feet (25.62 meters) in diameter. *Courtesy NRAO/AUI.*

Figure 4.9 Erection of the superstructure. Erection of the 180-ton structure was tricky because of its size and the position of the derricks. A first attempt was unsuccessful. Part of one of the derricks had to be cut away to provide more clearance. The second attempt was successful. The surface panels are visible in the left foreground. The focal feed support being constructed on the ground is visible in the background at the left. *Courtesy NRAO/AUI.*

ance at tight spots. The yoke hub weldment was an enormous subassembly, weighing slightly under 200 tons.[31]

After the arrival of the major subassemblies of the yoke, partial assembly of the yoke began. The Pacific Crane and Rigging Company completed a final fit-up of the yoke components, which were fabricated by Sun Shipbuilding, with the exception of the yoke hub section that was built and machined by Westinghouse. Final welding of the subassemblies was completed on the ground. When erected and completed, the entire yoke is a single weldment of 1,125,000 pounds of steel and 1,300,000 pounds of high density concrete counterweighting made of high density ulmanite (iron) ore pumped with ferrophosphorous mortar to yield a high density concrete, 250 pounds per cubic foot. [32]

The next major mechanical challenge after the erection of the yoke assembly was to attach the 28 segments of the 84-foot diameter permanent main driving gear. The 28 segments of the polar gear were designed to be attached to the periphery of the counterweight section of the yoke. The

gear girder plates were welded in place and torch cut to the proper radius as the structure was rotated about its polar axis by the temporary small gear system. The separate gear segments were hung to the girder plates as the structure was again rotated (Figure 4.8). Alignment had to be precise for the segments to form a single gear. Despite thermal expansion and contraction due to daily temperature variations and periodic vertical oscillations, tolerances of 2-1/2 thousandths of an inch (0.006 centimeter) were maintained. All gear sections were designed and manufactured by Philadelphia Gear.[33]

The polar drive and main polar brake units were installed after the polar gear was complete. The three main polar brakes were designed and fabricated by the Goodyear Tire and Rubber Company. These horseshoe type hydraulic spring brakes are the largest ever known to have been constructed.[34] Once set on the polar gear, the three brakes apply a total braking force of 630,000 pounds (286,000 kilograms). The brakes are designed to automatically lock on the gear's sides in the event of a power failure. Anti-back-

Figure 4.10 Final erection of the superstructure. The superstructure was secured by bolts which were entered into the declination bearing housings. The surface panels could then be attached to the superstructure. *Courtesy NRAO/AUI.*

lash motors were also installed in the reduction gear housing. For normal braking during operation of the telescope, each of the reduction gear units has internal disc brake units. The main polar drive installed was a 50 horsepower hydraulic power package designed and constructed by the Electric Boat Division of General Dynamics Corporation. The two reduction gear trains connected the power unit to two drive pinions thereby driving the instrument in right ascension.[35]

By the fall of 1964, all components of the telescope had arrived on site. Fabrication of the aluminum superstructure was complete. The most difficult erection, that of the antenna superstructure, was at hand. Many had hypothesized that the lift, as planned, was impossible.[36] Steel pickpoints, calculated to divide the load between the two derricks, were built into the superstructure. The superstructure was raised and simultaneously rotated 45° from the horizontal position in which it had been welded, by means of the two derricks and a holdback crane (Figure 4.9). The lift continued to 50° where a snap at the south pickpoint occurred. At 55° the north pickpoint shifted 2 inches. The structure shuddered, but held. At 75° the holdback crane went into overload and

a 1-3/8-inch holdback cable separated. The superstructure dropped 4 feet and the bottom dug into the ground. The derricks, the main load lines, and the weldments held, and no structural damage occurred. The next day, the superstructure was lowered to the ground. The pickpoints were relocated for better balance and the lift was again attempted. The south pickpoint had allowed too much imbalance for control of the lift. The second attempt went smoothly at first. The lift continued to 90° where it was pinwheeled to align the declination axis shaft directly with its yoke arm bearings. The lift went successfully almost to the final position. Clearance between the derricks had, however, run out with just a few feet to go. More clearance was provided by cutting away a portion of the south derrick. The booms were at the limit of their hoisting capacity, but the bearings were in position to engage the keys into the yoke arm keyways. The fits were made and bolts were entered into the declination bearing housing. The superstructure was thereby secured on its bearings (Figure 4.10).[37]

The last major construction job was erection of the surface panels. The 60 aluminum surface panel sections were designed and manufactured by the D. S. Kennedy Company Division of Electronic Specialties, the same company that had provided the original unused Kennedy design for the 140-foot telescope. Each panel contained four support points at which jack screws were secured on spherical bearings for rotation at the panel end. The screws were placed in clearance holes for lateral adjustment when the panel was attached to the superstructure.[38] The design of the surface panels was insightful. The original design called for adjustment of the panels to be motor controlled. High precision jack screws were therefore incorporated into the design. Each quarter turn resulted in a thousandth of an inch of lateral adjustment. Although the idea of active surface panel measurement and adjustment was impossible to achieve with the existing technology, the design allowed high precision adjustments to be made mechanically.[39] Readjustment was therefore possible as surface measurement techniques improved.

The installation of the surface panels was begun at the bottom of the superstructure to retain control from above (Figure 4.11). The two 250-ton derricks were dismantled simultaneously as the surface panels were installed. The surface panels were rigged on the ground to the proper predetermined angle and plane. The surface panels were lifted and placed, one by one, at their proper position on the face of the superstructure. The antenna panel subassemblies were fabricated to maintain a surface tolerance within 0.030 inch (0.08 centimeter) rms. Each panel was adjusted after attachment to the superstructure. Precise placement of each panel was critical. Edge clearances between panel sides and ends were measured and maintained with each placement to ensure this accuracy. Placement and fitting of the final panels would critically depend on the precise placement of the initial panels. Fine adjustment of the panels was based on the theodolite method. The surface measured to within 0.030 inch rms of the desired paraboloid.[40] After the final adjustments were made, the panels were painted with a white heat-diffusing paint.

As soon as enough surface panels had been installed to reveal the focal feed support mounting holes, erection of the four-legged feed support was undertaken. The 65-foot (20-meter) focal feed support structure had been welded into one unit on the ground by the E. W. Bliss Company. The feed support was swung up and installed by two truck cranes assisted by tag lines.[41] The focal feed support was designed to be adjustable using tapered shims under each of the four legs to compensate for deviation of the focal point from the calculated point. The doughnut-shaped receiver support structure was designed to support a thousand pounds of receiving equipment. The focal axis of the antenna is defined by the vertical axis of the feed support aligned with the instrument support stand above the apex of the paraboloid. Both the focal feed support and the instrument support stand are adjustable to focus the telescope.

After the focal feed and instrument support structures were erected and the last surface panel was placed, the telescope was driven to zenith and the declination drive and brakes installed (Figure 4.12). With these final installations in February 1965, the erection of the telescope was complete.[42] The entire instrument, with the exception of the polar and declination gears, was painted with a highly reflective white paint to minimize thermal distortion due to the absorption of solar heat (Figure 4.13). With these tasks complete, installation of the receiving systems could begin.

The assembly of the telescope was finally finished in the spring of 1965. Although it took more than 6 years to complete, the 140-foot telescope performed beautifully. The telescope was eventually brought up to date with state-of-the-art components and receivers. The final product was the largest equatorial telescope with the lowest noise temperature receiver of any significant bandwidth in the world.[43] This status has been maintained ever since. These characteristics, as well as the high precision surface, operable down to 1 centimeter, have made the 140-foot one of the world's most demanded and productive radio telescopes of any kind.

140-Foot Telescope Receiving Systems

The telescope was initially used as a prime focus instrument, but later converted to Cassegrainian focus with optional prime focus capability. The original receivers were installed in a standard front-end box mounted to the prime focus front-end support. A remote control focusing system was designed to permit continuous axial focusing by mechanically moving the front-end box up to 90 centimeters along the focal axis. The focal feed support design for the

Figure 4.11 Erection of the surface panels. Installation of the panels was performed while simultaneously dismantling the derricks. Installation of the surface panels was begun at the bottom to retain control from above. *Courtesy NRAO/AUI.*

140-foot telescope was insightful. Along with its continuous focusing ability, the feed support was designed to accommodate a standard front-end box in which various receivers could be mounted. The design of the mount and standard front-end box was developed by the NRAO electronics staff. The basic design idea was that the feed and receiver systems required for specific astronomical research programs could be easily installed and then removed for the next research program. Receivers could also be interchanged between the 140-foot and 300-foot telescopes. Another tremendous advantage of the design is that multiple feed and receiver systems can be operated simultaneously, providing simultaneous observation at a number of frequencies. The arrangement also allowed the front-end box to be rotated 450° around its axis of symmetry to permit polarization studies.[44] These design capabilities along with the high precision surface, accurate tracking afforded by the equatorial mount, which allowed long integration time observations, and the various low noise receivers, made the 140-foot a very versatile instrument.

The standard front-end box was designed to contain the feed and preamplifiers and other front-end components of each radiometer. The back-end components for the various receiving systems were housed in the control room in the foundation along with the position control and various data collection hardware. The first receiver installed was an NRAO designed and constructed radiometer operating at 6 cm which utilized tunnel diode amplifiers. A 2-cm switched crystal-mixer radiometer and an 11-cm radiometer utilizing

Figure 4.12   Final assemblies of the 140-foot telescope. The telescope was driven to the zenith position after the surface panels were installed. Installation of the declination drive and brakes were permitted in this position. *Courtesy NRAO/AUI.*

parametric preamplifiers were added in 1966. By 1967 receivers operating at 0.9, 18, and 21 cm developed by the NRAO electronics staff were also in active use.[45] Receivers developed by 1971 for use with the 140-foot included a tunable 5–10 GHz receiver, a 24-GHz receiver, a 2-cm cooled paramp receiver, a new cooled parametric 6-cm receiver and a 3-cm receiver.[46] These radiometers were modified and adapted for specific research programs.

The 21-centimeter receiver used for Zeeman effect research reflects the basic design and components common to most of the radiometers. A simplified schematic block diagram of the 21-cm receiver is shown in Figure 4.14. A shallow waveguide directed the radiation to a pair of crossed dipoles directly fed to a hybrid to produce two outputs of opposite circular polarization. A switch injected a radio signal from a noise generator for calibration purposes. The switch was controlled by a 416-channel digital spectrometer used as two separate autocorrelators of 192 channels each,

known as receivers A and B. The two receivers could be set to different overall bandwidths. The receivers were very sensitive over the bandwidth range and produced a very low system temperature. The state-of-the-art NRAO receivers such as the 21-centimeter radiometer used in combination with the digital spectrometer resulted in a powerful tool for radio spectroscopic research. Significant research was performed with this receiver, most notably Gerrit Verschuur's 1968 discovery of the Zeeman effect.[47] The original 21-cm receiver was replaced with a dual-channel cooled 21-cm radiometer in 1972.[48]

In 1968 the 416-channel digital spectrometer developed by the NRAO electronics staff was brought on-line for use with the various front-end receivers. The digital spectrometer, also known as the 416-channel autocorrelation receiver, was developed by A. M. Shalloway, R. Mauzy, J. Greenhalgh, and S. Weinreb of the electronics division in 1968.[49] The 416-channel autocorrelator receiver operated at a wide

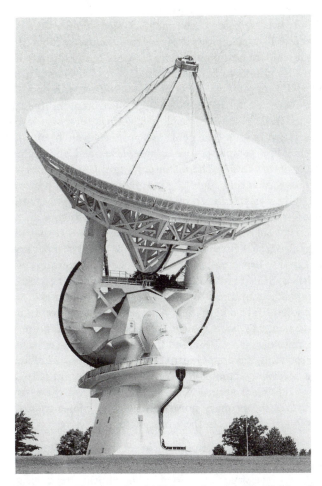

Figure 4.13 The completed 140-foot telescope. The highly reflective white paint minimizes thermal distortion due to the absorption of solar heat. The telescope was converted to a Cassegrain focus instrument in 1974 as seen here. *Courtesy NRAO/AUI.*

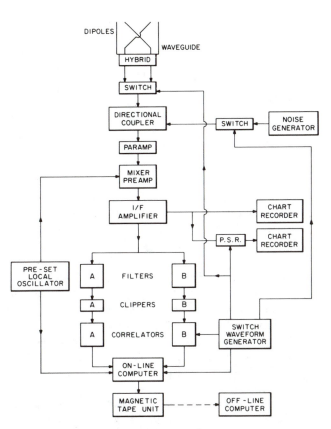

Figure 4.14 Simplified schematic diagram of the NRAO 21-cm receiver. This receiver, which utilized parametric amplifiers, was used in the discovery of the Zeeman effect.

bandwidth and could simultaneously examine fine-scale frequency structure, making it ideally suited for research in spectroscopy. Research that required scanning over various frequencies previously involved using narrow band filter receivers that had to be slowly scanned across the desired profiles. Little baseline could be obtained with this method and the resulting sensitivity was poor. The two receivers of the autocorrelator of 192 channels each could be set at different overall bandwidths and reveal fine structure at each channel. An overall bandwidth of sensitivity ranging from 120 Hz to 62 kHz was achieved. Because the two receivers of the autocorrelator were designed to act independently, the data from the two could be added together to increase the effective integration time. Long total integration times, often 20 to 40 hours, could be undertaken to greatly increase the effective sensitivity. Signals detected in one receiver could also be verified in the other at a different frequency, thereby eliminating spurious signals generated by the autocorrelators themselves.[50]

Data acquisition, recording, and display, as well as telescope monitoring, were provided by means of an on-line

DDP-116 computer. The DDP-116 displayed either a reference spectrum determined by the signal transform over the desired channels or a quotient spectrum calculated from the signal transform compared to a reference transform. The quotient was generally then converted to antenna temperature in degrees by an off-line computer.[51] Typical representation of the data plotted antenna temperature thus derived versus channel or velocity (a representation of frequency). The computer calculated the spectral frequency of the receiver corrected for the specified local standard of rest (LSR). The LSR velocity was defined by the solar standard of motion 20 kms^{-1} toward 18^h, $+30°$.[52] The sophisticated data collection and display system enhanced the versatility of the 140-foot by enabling efficient analysis and interpretation of large amounts of data over a wide range of frequencies.

Testing the Performance Characteristics of the 140-Foot Telescope

The performance characteristics of the 140-foot telescope were extensively tested upon its completion and installation of the first receiving systems in 1965. Characteristics tested included the surface accuracy over the full range of pointing positions, aperture efficiency, beam efficiency, and pointing and tracking accuracy. Radio measurements of these characteristics were typically made at wavelengths of 11, 6,

Table 4.1 Measured 140-Foot Performance Characteristics

Wavelength (cm)	11.1	5.99	1.95
HPBW (arcminutes)	11.8	6.4	1.9
Aperture Efficiency	59%	53%	43%
Beam Efficiency	78%	81%	43%
First Sidelobe Intensity (decibels below main beam)	< −28	< −24	< −17

and 2 cm. The first and most basic measurement was that of the axial focal length. The focal length was found to deviate only 10 millimeters over the entire range of observing declinations and hour angles. The lateral focus position was measured and adjusted by rotation of the front-end box. This is a critical adjustment because the maximum antenna temperature observed from an object will remain constant only if the center of rotation of the front-end box coincides with the electrical axis of the paraboloid. Measurements at the three frequencies of the surface deviations at positions within 20° of zenith showed that the effective rms value of the deviations from an ideal paraboloid was only 0.92 ± 0.17 millimeter. At other declinations and during tracking, the rms deviations increased to 1.44 millimeters. The final adjustment of the surface panels was evidently very precise. The pointing errors of the telescope were also measured at the three specified frequencies. A study by Pauliny-Toth found that the instrumental pointing error was less than 2 seconds of time in right ascension at any hour angle and less than 2.5 minutes of arc for any setting of the telescope in declination from −30° to +90°.[53]

Measurements of the beam efficiency and aperture efficiency were made at the three specified wavelengths. Cassiopeia A, Taurus A, and Cygnus A were used as calibration sources for these measurements. The beam efficiency was determined from the integrated antenna temperature taken during observations of these sources. The aperture efficiency was then calculated from these results. The resulting figures for the beamwidth (HPBW), beam efficiency, aperture efficiency, and first sidelobe intensity are given in Table 4.1.[54]

Evolving Technology at the 140-Foot Telescope

Ever since its completion in 1965, the NRAO 140-foot telescope has essentially defined state of the art in centimeter radio astronomy instrumentation. New developments and improvements have continually been incorporated into the 140-foot telescope system. The evolution of receiver technology as evidenced in the 140-foot instrumentation is the single largest factor responsible for the vastly improved sensitivity over the years. Great strides were made in the first stage amplification of the receivers. The system noise

was lowered with each new development. The effective sensitivity was increased each time the system temperature was lowered. The technology evolved from the use of room temperature parametric amplifiers to room temperature transistor amplifiers to field effect transistor (FET) amplifiers. The FET amplifiers were later cooled to further reduce the noise level. Better FETs called high electron mobility transistors (HEMTs) were developed with lower noise levels. The HEMTs were then cooled to produce even lower noise levels. Maser amplifiers were later developed and defined the state of the art. The technology involved in the maser receivers is cutting edge. The masing medium is a precisely cut ruby crystal injected by high power microwave exciting devices. Powerful superconductors are used to contain the electric field and therefore concentrate it within the ruby crystal. Masers provide extremely low noise temperature and a very broad bandwidth. The system temperature was reduced from hundreds of degrees kelvin, in the case of parametric amplifiers to tens of degrees kelvin, in the case of cooled FETs and masers. At least a factor of ten in improvement has been realized. The system temperature of the state-of-the-art FET and maser amplifiers is currently approaching the temperature of liquid helium, 4.2 kelvin.[55] The receiver technology has evolved to the point that the noise levels are approaching absolute zero. At these temperatures the primary noise contribution comes from the motion of electrons in the wires and circuitry.

Another major development in receiver technology pioneered by the NRAO is the multiple feed receiver. Multiple beams on the sky are provided by the multiple feed design. During the 1970s the NRAO electronics staff developed numerous multiple receivers. Notable examples are the three-feed 11-cm and the four-feed 21-cm receivers. The three-feed 11-cm allowed the observer to collect three strips of data on the sky at one time. Three feed horns were arranged side by side. The two outer horns were mounted at 45° to the center horn for polarization measurements. The central horn had two feeds to measure right and left circular polarization. The four-feed 21-cm produced four beams on the sky to obtain similar polarization information. State-of-the-art NRAO receivers developed in the 1980s have even more beams on the sky and lower noise temperature. A very productive seven-feed cooled receiver, with a central horn surrounded by six horns in a circular pattern, for example, was developed in 1984.[56] The state-of-the-art receiver technology developed at the NRAO over the years has advanced to more beams on the sky and lower noise temperatures.

The improvements in back-end receiver technology have also been significant. The primary early example is the development of the 416-channel autocorrelator. A 384-channel autocorrelator was developed in 1972.[57] Other examples include developments in the areas of data collection and reduction. Numerous modifications and improvements of the on-line analysis programs were implemented over the years.

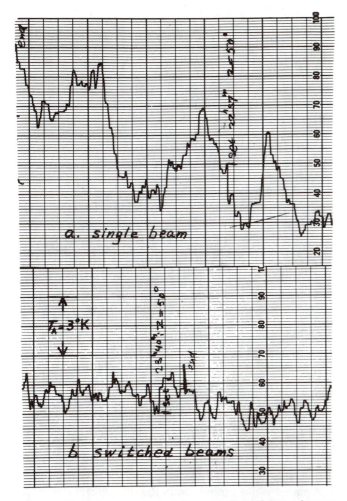

Figure 4.15 Data of sky noise taken with the 140-foot telescope at 1.95 cm. The top figure shows receiver input switched between the feed and calibration load. The bottom figure shows the effect of beam switching between the source and background sky. The effective reduction in noise level afforded by beam switching is evident.

Upgrades in computer technology and data manipulation programming have been continual. Central processing unit (CPU) upgrades in the central computing facility have been implemented as well as the on-line computer upgrades. Image processing and analysis systems have been developed to facilitate representation and interpretation of the observations. The POPS (People-Oriented Parsing Service) command-line interpreter was developed over a 20 year period by the NRAO staff as the basis for a number of the major data reduction systems used by NRAO observers. Many programmers have contributed to upgrading POPS and adapting it to different computers and operating systems since its original development. POPS has become the brain behind single-dish analysis packages used at Green Bank and Tucson as well as for the 140-foot telescope control system and the heavily used AIPS package. The Astronomical Image Processing System (AIPS) developed in 1982 is a widely used system, primarily implemented in VLA and

VLBI data processing, reduction, and analysis.[58] Improvements in the back-end receiving systems have been continually implemented on the 140-foot since its construction to complement the rapidly advancing front-end receiver technology.

Other early innovations in technology developed for the 140-foot involved beam switching and frequency switching of the radiometers. Many of the early radiometers employed on the 140-foot incorporated switched feeds. In observing a spectral line the radiometer could be rapidly switched between the line frequency and a slightly different one by switching between feeds. Because atmospheric radiation is essentially the same for nearby frequencies and its changes are slow with respect to the switching rate, sky noise fluctuations can be virtually eliminated. Several radiometers could also be switched between the feed and an artificial load or between two feed horns, one of which was laterally displaced by several HPBWs. Fluctuations in the system noise temperature could be significantly reduced by switching to provide continual calibration. The effect of feed switching and beam switching is shown in Figure 4.15. Beam switching between two horns pointed at slightly different regions of the sky significantly reduced the system noise by virtually eliminating sky-noise fluctuations. Beam overlap occurs in the first few kilometers of the atmosphere where the noise fluctuations originate. Beam switching at centimeter wavelengths therefore provided reduced noise levels, more stable baselines, and improved sensitivity.[59]

From the 140-foot telescope's completion in 1965, a series of improvements in the performance was implemented. The receiver systems, automation of the telescope, and the control systems and back ends were improved. A fundamental change in the telescope design was implemented in 1974. The 140-foot was converted from a prime focus telescope to Cassegrain configuration. Receiving systems by the early 1970s had improved to the point that noise radiated from the ground around the telescope made a significant contribution to the system temperature. In the Cassegrain configuration the feed horns point upward to a subreflector (Cassegrain secondary) instead of downward to the dish and ground. The sky is much colder at radio frequencies and adds little to the system noise.[60] The Cassegrain configuration focuses the radio waves by means of a hyperbolic surface. The focus of the Cassegrain instrument is near the vertex of the 140-foot main dish where an equipment room containing the front-end equipment is mounted (Figures 4.16 and 4.17). The Cassegrain configuration has a much longer effective focal length allowing the telescope beam to be precisely directed by tilting the subreflector. The new vertex box could also support more equipment and weight with less deviation of the focal point due to gravitational sagging of the feed legs.

The subreflector has been a focus of development and improvement of the performance of the 140-foot telescope.

Figure 4.16 Conversion of the 140-foot to a Cassegrain focus instrument, showing the vertex box which housed the receivers mounted in a standard NRAO front-end box.

Figure 4.17 Conversion of the 140-foot to a Cassegrain focus instrument showing the installation of the vertex box. The conversion allows the 140-foot to be used either as a Cassegrain or prime focus instrument.

The first innovation developed in the subreflector design was a nutating subreflector that pointed the beam alternately to the object being observed and to the sky beside it. The nutating subreflector was installed in 1974. The subreflector could be wobbled back and forth alternately pointing the beam to the object and the sky three times per second.[61] Atmospheric effects were significantly minimized and the effective sensitivity increased.

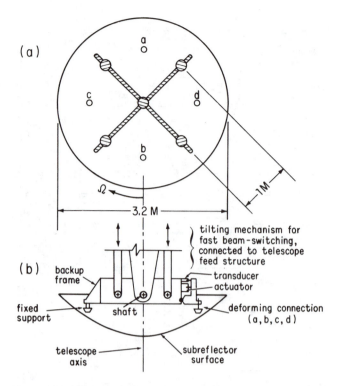

Figure 4.18 Deformable subreflector. The top view shows a back view which indicates the position of the actuators used to mechanically deform the shape of the subreflector to compensate for gravitational deformations of the main dish. A cross-sectional view (bottom) shows the basic mechanical design.

Another major innovation in the subreflector design developed for the 140-foot was a deformable subreflector installed in 1979. The deformable subreflector was a nutating design whose shape could be mechanically altered to compensate for gravitational deformations of the main dish. The deformable subreflector was made of a shell of fiberglass covering a honeycomb aluminum framework. Four actuators shaped the surface using an electromechanical mechanism servo-controlled by the on-line computer (Figure 4.18). The deformable subreflector accounted for considerable improvements of the beam shape and efficiency, particularly at pointing positions far east or far south where the efficiency was increased by nearly a factor of three for the maximum available deformation. The beamshape became more narrow and symmetrical. At certain declinations the aperture efficiency increased by factors between two and three.[62]

Innovations developed by the NRAO staff have continued to maintain the 140-foot telescope at the cutting edge of instrumentation technology for the 20-plus years since its construction. Important innovations to the 140-foot instrumentation not previously mentioned include holographic measurement and subsequent adjustment of the surface panels and the design and development of low noise feeds. The holographic measurement of surface panels was

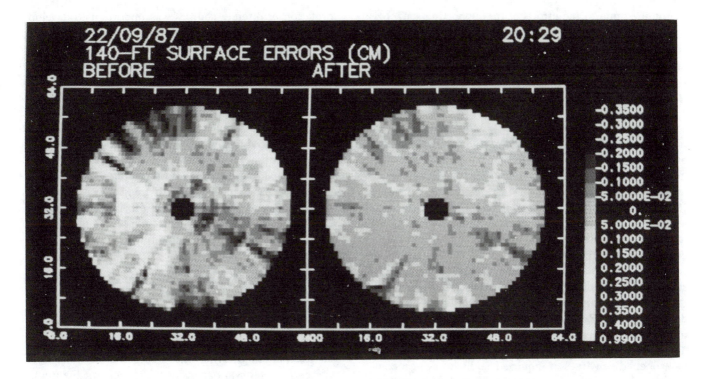

Figure 4.19 Results of the holographic survey of the surface of the 140-foot telescope. Subsequent adjustment of the surface panels allowed the telescope to be used down to 36 GHz.

a scheme devised by Martin Ryle at Cambridge. The process has been adapted by Ron Maddelena and the NRAO electronics group to adjust the surface panels of the 140-foot on several occasions. The process uses a satellite as a reference source to make direct radio frequency measurements of the phase distribution across the aperture. Two receivers are utilized to observe the reference source. A receiver with a 10° beam is mounted on the front-end box at the prime focus facing the sky. A second receiver with a 1.5 arcminute beam is mounted at the prime focus facing the dish. The incoming signal from the satellite travels different distances to reach the respective receivers. The signal is directly received by the receiver pointing skyward. The signal received by the second receiver travels from the source to the dish and is reflected to the prime focus. The signals are then combined to produce an interference pattern. The pattern for an ideal paraboloid is calculated and compared to the measured pattern. An inverse Fourier transform of the measured signals reveals the surface deviations. Surface deviations from the ideal paraboloid produce phase deviations from a plane wave in the aperture field distribution. The phase deviations appear in the far-field pattern using the complex Fourier transform relation. The amplitude and phase of the far-field pattern are measured revealing the phase perturbations in the aperture distribution and thereby the surface deviations.[63] Precise measurements can be made in this manner that far exceed the accuracy of physical surveyor's measurements. Results of the 1987 holographic surface survey are shown in Figure 4.19. Adjustments of the panels are based on this precise information. As a result of these adjustments the telescope which was good to 24 GHz is now good to 36 GHz.[64]

The development of low noise feeds to improve aperture efficiency of the telescope is another area where the NRAO has excelled. The feeds are designed and shaped to provide the maximum and most symmetrical illumination of the dish possible. The size and length of the feed control how the illumination spreads out—the illumination pattern. The ideal pattern is essentially a doughnut shape created by illuminating the edges more intensely than the middle to make more efficient use of the available aperture. A dual hyper-mode feed developed by Rick Fisher in the early 1980s to accompany a modern low frequency 21-cm receiver provided a very significant increase in aperture efficiency. Typical feeds built prior to 1980 measured 50 percent efficiency. The dual hyper-mode feed increased the efficiency to 57 percent. In combination with the new cooled 21-cm receiver built by Roger Norrod and the NRAO electronics group, the effective sensitivity of the receiver system was doubled.[65]

Nearly every aspect of the 140-foot telescope has been improved over the years. From the feeds and front-end receivers to back-end instrumentation, the technology has steadily evolved. The 140-foot telescope represented state of the art in centimeter radio astronomy instrumentation in 1965 when it was completed. That status has since been maintained due to the efforts of the supporting staff. It is widely believed that the NRAO 140-foot telescope currently represents the best equipment in the world for filled aperture research in centimeter radio astronomy.

140-Foot Telescope Programs

Dedication ceremonies for the 140-foot telescope were held on October 13, 1965. Dedication ceremonies took place after the major telescope performance characteristics had been tested and found to meet or exceed the design specifications. Galactic and extragalactic research programs were instigated by the staff and visiting scientists. Extragalactic astronomy programs included surveys and studies of the neutral hydrogen distribution in other galaxies. The major contributions of the 140-foot telescope have, however, been in galactic astronomy particularly in molecular spectroscopy. The 140-foot telescope has been used to make major discoveries including discovery of recombination lines, the discovery of the Zeeman effect, and the discovery of many molecules in space, including the first organic molecule. These discoveries are fundamental in the sense that they have opened new lines of inquiry. The discovery of recombination lines provided a new tool to observe the structure of galactic and extragalactic neutral hydrogen, lending insight into the structure of a wide spectrum of objects. The discovery of the Zeeman effect proved the existence of interstellar magnetic fields and opened the door to the observations of the magnetic fields associated with cosmic phenomena. The discovery of many molecules in space has been credited to observers using the 140-foot telescope. The 140-foot greatly contributed to the revolution in molecular spectroscopy in the 1960s and 1970s. A new line of inquiry, astrochemistry, has subsequently evolved. The discovery of the first organic molecule in space, formaldehyde, with the 140-foot also represents a significant event in the development of another new line of scientific inquiry, that of exobiology.

Research in Extragalactic Astronomy

Early surveys of discrete sources conducted in 1965–1970 represented an important aspect of research with the newly constructed 140-foot telescope. The 140-foot was used to make source surveys, measuring characteristics such as polarization, variability, and flux density of the sources surveyed. A discrete source survey at 20 cm by M. M. Davis as early as 1967 confirmed 69 new sources that were nonexistent in source catalogs at longer wavelengths.[66] K .I. Kellerman, I. Pauliny-Toth, and W. C. Tyler had measured the 11-cm flux density of over 400 sources by 1967.[67] Kellerman and Pauliny-Toth had also measured the flux density of 480 sources at 6 cm by 1968.[68] Kellerman and Pauliny-Toth made these observations as part of a much more ambitious research program aimed at determining the spectra of extragalactic sources at centimeter wavelengths. Spectral in-

formation of discrete sources was important to develop a theoretical understanding of the underlying physics involved in the production of the intense radiation observed. Many unidentified sources turned up in these early surveys. In the 1968 survey, for example, 89 unidentified sources were observed.[69] The sources were correctly assumed to be extremely distant galaxies beyond the limit of the Palomar sky survey, the most comprehensive optical survey then existent. As the instrumentation technology evolved, the number of sources in the surveys greatly increased. The 300-foot telescope would prove to be more suited to weak source surveys because of its large aperture and transit nature. The 140-foot, however, was very useful for high frequency surveys. The early surveys represent important basic science that provided measurements upon which theories and further research could be based.

The 140-foot telescope has been extensively used in observations of the neutral hydrogen in other galaxies. Structural details are often revealed as in the case of observations of the Andromeda galaxy (M31) in 1967 at 11 cm by R. C. Cooley and M. S. Roberts.[70] Maps of the neutral hydrogen distribution of Andromeda revealed far more detail than earlier observations at lower frequencies. Figure 4.20 shows a 1967 (incomplete) contour map of M31 that reveals complex source structure as well as other radio sources in the area not associated with the structure of the galaxy. The radio-emitting areas were found to be far more extensive than the optical region. A nucleus and hot spots are also evident in the Andromeda galaxy which is very similar and also very close to the Milky Way. The Andromeda galaxy and the Milky Way are now considered twin galaxies, each with smaller satellite galaxies, the two anchoring the smaller galaxies of the local cluster. Techniques have been developed for estimating the parameters and kinematics of the neutral hydrogen within galaxies. Extragalactic studies with the 140-foot have provided cosmological information as well as clues to the structure of our own galaxy and galactic neighborhood.

Other extragalactic researchers have observed the full range of galaxies from normal to quasars, and searched for small scale anisotropy of the cosmic microwave background. Variations in the flux density of quasi-stellar sources have been measured. This data provides indications of the size and structure of these mysterious objects as well as clues as to the powerhouse of these phenomena. Attempts were also made to measure small scale fluctuations that have been predicted to exist in the temperature of the cosmic microwave background. The anisotropy is an indication of how galaxies formed from the extremely smooth distribution of matter in the early stages of the universe. Early upper limits to the small scale anisotropy in the cosmic microwave background were set by this research performed with the 140-foot telescope and later with the NRAO Very Large Array (VLA).[71] Evidence for the small scale anisotropy provided by the NASA Cosmic Microwave Background Explorer

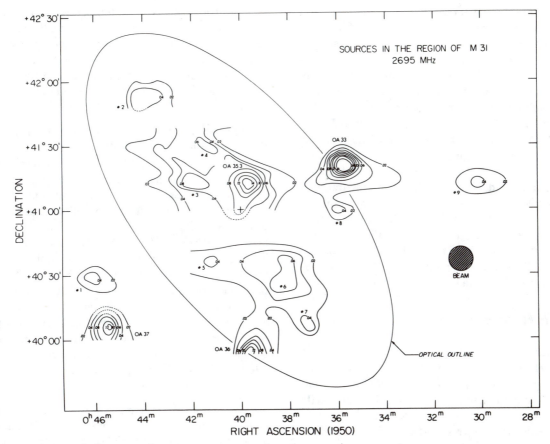

Figure 4.20 Contour map of the radio intensity of the Andromeda galaxy (M31). This incomplete data taken in 1967 with the 140-foot telescope at 2,680 MHz shows the radio-emitting regions in comparison with the optical galaxy (oval outline). *Courtesy M. Roberts and Science,* © *1967 American Association for the Advancement of Science.*

(COBE) was announced in April 1992. The COBE findings experimentally verify the hot big bang model of the creation of the universe and provide important insight into the formation of galaxies. The COBE data essentially represents the primordial blueprint that determined the structure and evolution of the modern universe.[72]

Research in Galactic Astronomy and the Discovery of the Recombination Lines

Almost immediately after the telescope construction and testing phases were complete, contributions to galactic astronomy began to be made. The first experiment performed with the 140-foot telescope resulted in the discovery of a recombination line of ionized hydrogen (HII).[73] The possibility of observing recombination lines of ionized hydrogen in the radio spectrum had been suggested by Kardashev in 1959. Attempts to observe the hydrogen recombination lines had been made by Russian groups and also by the NRAO group using the Tatel 85-foot telescope. Positive results had not been achieved by these investigations. On July 9, 1965, the first unambiguous detection was provided by Höglund and Mezger while using the NRAO 140-foot telescope. They detected the H109 line in Orion A and M17

using a cooled parametric amplifier radiometer.[74] The observation was the first time that a closed cycle cryogenic system was used to cool a parametric amplifier for astronomical observations. An unprecedented system noise temperature of only 80 K was obtained with this radiometer.[75] The great advance in sensitivity afforded by the cooled paramp receiver was instrumental in the discovery. The discovery of the H109 recombination line provided a new means to observe galactic HII and opened the door to the discovery of other recombination lines of hydrogen as well as the recombination lines of other elements such as helium and carbon.

The recombination occurs as electrons in the HII clouds lose momentum in their motion and are recaptured by the hydrogen ions. Although the density of HII clouds is extremely low, free electrons are eventually captured by the hydrogen ions. Spectral lines are produced that correspond to the quantum level at which the recombination occurs. The recombining electron can occupy any of the energy levels of the atom or ion, but if it recombines at any quantum level other than the ground state, one or more down transitions occur resulting in the emission of radiation. The initial and final energies of the electron are well defined resulting in a bound-bound transition that emits a spectral line. Recombination lines fall in the optical band for transitions originating at quantum levels near the nucleus. Radio frequency recombination lines occur when highly excited electrons are captured at high quantum levels far from the nucleus. Thus the H109 recombination line occurs as an electron recombines with a hydrogen ion at a specified quantum level ($n = 109$) producing a spectral line at 5,008.9 MHz.[76] Both free electrons and electrons bound in atoms produce radiation in HII regions. A free electron in the HII region will either be recaptured by an ion or scattered by the ion's Coulomb field and radiate by the bremsstrahlung process. This free-free interaction produces continuum radiation. Three distinct types of spectral lines in the cosmic radio radiation were therefore observed by 1967: the 21 cm hyperfine structure radiation emitted by HI regions; molecular spectral lines from OH produced by HI and HII regions; and the recombination lines produced in HII regions.

HI and HII Regions

The discovery of recombination lines represented a new tool with which to observe the distribution of ionized hydrogen in the galaxy. The spiral arms of galaxies are composed of stars and interstellar matter which is primarily made up of hydrogen. Regions of interstellar matter such as star forming regions and the regions in the vicinity of hot O and B stars are ionized by the ultraviolet radiation associated with these objects. These regions are designated HII regions because roughly 90 percent of the ions are of hydrogen. Only continuum radiation and line radiation from hydrogen at 21 cm and from the hydroxyl (OH) molecule at 18 cm from these regions had previously been observed. The discovery of recombination lines revealed the structure of HII regions and provided a wealth of astrophysical information when considered in combination with the continuum and line radiation previously observed from these regions.

HI regions represent areas of cold or neutral hydrogen. HII regions represent areas of ionized hydrogen and are often referred to as emission nebulas. Extended emission nebulas often contain compact HII regions within them which represent regions of star formation. HI clouds are tenuous cold clouds pervading the interstellar space within galaxies. A typical HI cloud ranges from two to a few hundred light-years in extent. Typical densities range from 3 to 100 atoms per cubic centimeter of space. Their relative density is extremely low compared, for example, to the density of air at the earth's surface which is about 10^{19} molecules per cubic inch.[77] Observations of HI clouds provide tremendous insight into the large-scale structure of the galaxy whereas observations of HII regions provide insight into the early stages of stellar evolution as well as the structure of galactic spiral arms. Spectral line and recombination line observations have helped define parameters of these regions

such as their distance, size, temperature, density, and constituent elements and have therefore greatly enhanced models of the underlying physics of these phenomena. The 140-foot telescope, because of its high precision surface, sensitive radiometers, and ability to observe fine-scale detail over a wide range of frequencies, proved to be ideal for these observations.

Observations of HII regions revealed many recombination lines for hydrogen as well as other elements. These line observations of HII regions provided much insight into the physical structure and mechanics of these phenomena. The radio emission associated with an emission nebula is typically far more extensive than the optical region. Figure 4.21 shows 1967 data taken with the 140-foot telescope of the HII region NGC 6357.[78] The contours of constant radio brightness are superimposed onto an optical photograph (negative). A small region of optical nebulosity just below the intense radio region at the top right is the only portion of the nebula even barely visible at optical frequencies. This example demonstrates the dramatically different picture of HII regions afforded by radio frequency observations of recombination lines. Much research performed with the NRAO 140-foot telescope was subsequently devoted to galactic HI and HII regions.

Observations of HII regions revealed much about the underlying physics of these phenomena. HII regions are heated by ionization of the plasma by ultraviolet radiation released by hot stars forming in the interior of the emission nebula. The cloud is cooled by radiation occurring during the recombination of ionized hydrogen. The electron temperature of the nebula depends on the surface temperature of the star or stars in the interior. Typical cloud tempera-

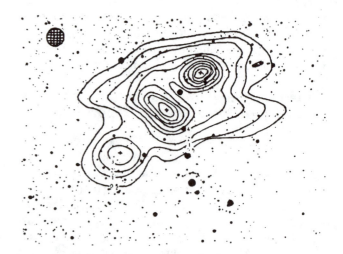

Figure 4.21 Radio contours of the HII regions NGC 6357. This map was compiled from data taken in 1967 at 6 cm with the 140-foot telescope by Mezger and Palmer. The radio contour map is superimposed over a negative photograph of the region. The extension of HII regions relative to the visible counterpart (a small nebulosity near the upper right radio hot spot) is illustrated. *Courtesy P. G. Mezger.*

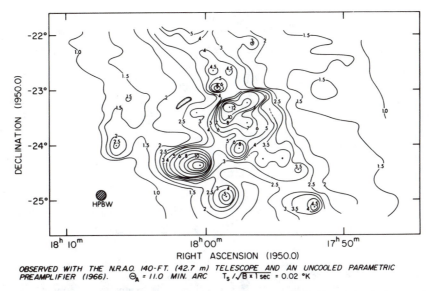

Figure 4.22   Radio contour map of the Orion region. This data was taken at 11 cm with the 140-foot telescope in 1966. The extension and complexity of the giant emission nebula in Orion are evident in this data. *Courtesy P. G. Mezger.*

tures range from 10,000 K in compact clouds to 1,000 K in diffuse emission nebula.[79] The observed radio brightness is directly related to the electron temperature. Electron temperatures are measured by comparing two observable quantities, two spectral lines, or a spectral line and the adjacent continuum. The ratio of a hydrogen recombination line to the adjacent free-free continuum radiation, for example, provides a measure of electron temperature. Radio contour data of galactic emission nebula therefore reveals important information about electron temperature as well as information about the physical parameters of the nebula. The large-scale distribution of galactic HII regions has been studied by examining the Doppler shift of recombination lines providing kinematic information regarding the motion and distance of these clouds. This method was applied by Mezger and Höglund to determine large-scale galactic structure by undertaking an extended H109 line survey of all detectable galactic sources. The survey provided vast improvements in the continuum observations of many sources because of the resolution and sensitive radiometers. Figure 4.22 shows a contour map of the Orion nebula revealing great detail in the emission centers. Giant emission nebula such as this one often contain 10 or more young main sequence O and B stars in clusters in the interior.[80] The giant HII regions surveyed seem to correspond well with the position of galactic HI regions. The giant HII regions were found to help define the spiral arm structure of the galaxy. By 1967 several hydrogen recombination lines had been observed at wavelengths ranging from 4 to 59 cm. Recombination lines of helium and carbon had also been discovered in these HII regions with the 140-foot by this time. HI and HII regions have been extensively studied by Höglund and Mezger, T. K. Menon, G. Westerhout, W. Sullivan, C. Gordon, G. Verschuur, B. Burke, T. Wilson, and many others with the 140-foot.[81] These observations of HII re-

gions have subsequently enhanced our knowledge of star formation and stellar evolution as well as the large-scale structure of our galaxy.

Discovery of the Zeeman Effect

Another important aspect of HI regions was revealed by the NRAO 140-foot telescope. The first measurement of a magnetic field associated with hydrogen clouds was realized with the discovery of the Zeeman effect by Gerrit Verschuur in 1968 using the NRAO 140-foot telescope. This discovery proved the existence of interstellar magnetic fields and opened an important new line of inquiry. Interstellar clouds are permeated by magnetic fields. The motion of free electrons in the cloud is slightly altered by these fields. Spectral lines of elements in these clouds are split in two because of the presence of these magnetic fields. It was first suggested in 1959 that this phenomena, called the Zeeman effect, could be observed in interstellar clouds and that measurements of the Zeeman effect would indicate the strength of the magnetic fields. Not until 1968, however, did Gerrit Verschuur detect interstellar magnetic fields in several HI clouds in the Perseus spiral arm in the direction of Cassiopeia A by observing the splitting of spectral lines.[82]

Verschuur used the 140-foot telescope with the 21-cm paramp radiometer and the 416-channel digital spectrometer to discover the Zeeman effect. The digital correlator was switched between left- and right-hand polarizations. These signals were recorded and later combined by computer to present the difference between the two polarizations for each scan and one of the polarization signals to provide a reference spectrum. The scans were taken with long integration times to increase the effective sensitivity. The scheme was successful. The absorption spectra of Cass A showed

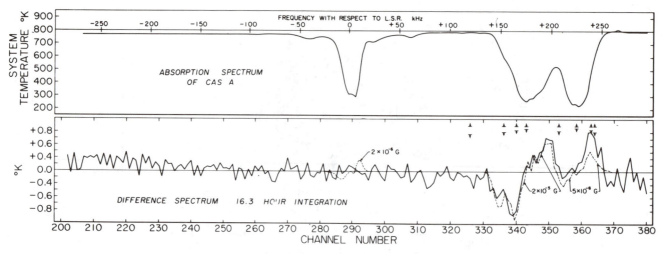

Figure 4.23 The absorption spectrum (top) and difference spectrum (bottom) of right-hand minus left-hand polarization of Cass A. This data taken by Verschuur with the 140-foot telescope in 1968 represents 16.3 hours of integration time. The Zeeman effect is evident in the splitting of the absorption line at the far left. Predicted curves for the values of the magnetic field strengths are plotted (dotted line) in the difference spectrum for comparison. *Courtesy G. Verschuur.*

two widely separated components due to the presence of a magnetic field. The Cass A data (Figure 4.23) was taken over a 16.3-hour integration time and plotted as the difference between right- and left-hand incident polarization. Zeeman splitting effects are manifested as the derivative of the observed absorption lines. Predicted curves for various field strengths were generated for comparison (dotted lines in Figure 4.23).[83] The Cass A data exhibited classic Zeeman splitting that corresponded very well with the theoretical values.

The Zeeman effect has been observed extensively in HI regions and other interstellar clouds throughout the galaxy. Field strengths of between 2 and 20 microgauss have been observed. The earth's magnetic field, by comparison, is 1/10 of a gauss. The research in the Zeeman effect at radio frequencies has resulted in the measurement of magnetic fields one hundred-thousandth of the earth's magnetic field in interstellar clouds 10,000 light-years away.[84] The Zeeman effect has since been observed in OH lines of molecular clouds. Heilis and Troland first observed the Zeeman effect in emission lines of OH. Subsequent OH Zeeman data combined with the HI Zeeman data in the late 1970s indicated that magnetic field strengths increase with density in interstellar clouds.[85] The observed field strengths are strong enough to be an important variable in the evolution of the cloud and therefore in the formation of stars. Research in magnetic field strengths has provided important clues into the collapse of interstellar clouds and the evolution into protostars.

Research in Molecular Spectroscopy and the Evolution of Astrochemistry

The 1951 discovery of the 21-cm line emission from atomic hydrogen by Ewen and Purcell represented an im-

portant step in the evolution of radio astronomy—the introduction of spectral line studies. Although three diatomic molecules CH, CH+, and CN had been discovered by optical astronomers during the 1930s, the ensuing search for other interstellar molecules was not successful until 1963. The discovery of the hydroxyl (OH) absorption line by S. Weinreb, A. Barrett, M. Littleton Meeks, and J. Henry in 1963, however, opened the door to the discovery of a diverse and complex interstellar molecule population. More than 90 interstellar molecules are now known to exist, many of which have been discovered with the NRAO 140-foot telescope.

Spectral line radiation is produced by atoms or molecules that radiate or absorb energy at discrete wavelengths. Atoms and molecules have a specific quantum structure that dictates the transitions between energy states. Spectral lines are produced by hyperfine transitions, down-transitions, recombination, and rotational transitions. Each atom or molecule radiates a specific spectral line pattern due to its unique quantum structure. The spectral line patterns may therefore be used to identify elements and molecules that constitute the interstellar medium. The spectral lines are broadened by motions in the gas cloud and Doppler shifted if the cloud has a motion relative to earth. The composition, motions, intensity, and distribution of interstellar clouds may therefore be deduced from spectral line measurements. The spectroscopic observations provide much insight into the conditions present in the interstellar medium as implied by the formation and preservation of complex molecules.

In the years following the 1951 Ewen and Purcell discovery, many attempts were made to observe molecular lines. The 1963 discovery of hydroxyl by Weinreb et al. using the Lincoln Laboratory's 84-foot Millstone Hill telescope was primarily attributed to recent improvements in receiver and autocorrelator technology. Other attempts were limited in

primarily attributed to recent improvements in receiver and autocorrelator technology. Other attempts were limited in sensitivity and also by the limited theoretical understanding of the interstellar medium. Charles H. Townes had suggested searching for spectral line emission of several interstellar molecules in 1955. The search was, however, limited to diatomic molecules because it was considered unlikely that polyatomic molecules could survive the conditions in the interstellar medium. It was not until 1968 that radio astronomy groups at Berkeley and the National Radio Astronomy Observatory at Green Bank began to search for complex interstellar molecules. The discoveries came quickly. The group at Berkeley had detected ammonia (NH_3) and water (H_2O) with the Hat Creek 20-foot telescope by late 1968. In 1969, the first organic molecule in the interstellar medium was discovered with the NRAO 140-foot telescope. By 1970, it was clear that a "Pandora's box of possibilities concerning the chemistry of interstellar space," had been opened and a revolution of our understanding of the interstellar medium was inevitable.[86]

A wide variety of complex molecules has been observed in the interstellar medium since 1968. Discovery of the known interstellar molecules has primarily been achieved at radio frequencies. The NRAO 140-foot and 36-foot instruments lead the field in the discovery of the known interstellar molecules. Table 4.2 lists all the interstellar molecules known to exist. The contributions of the NRAO 140-foot and later the NRAO 36-foot millimeter-wave telescope constructed in 1968 at Kitt Peak, Arizona, are evident in the number of discoveries of molecules accredited to each instrument. The 140-foot alone was used in the discovery of formaldehyde, cyanoacetylene, methyl alcohol, formamide, acetaldehyde, and formic acid as well as numerous new isotopes of other previously discovered molecules.[87] Only the NRAO 36-foot millimeter-wave telescope has surpassed the number of molecules discovered by the 140-foot telescope. The number and complexity of the interstellar molecules lend tremendous insight into the evolution of matter into stars, planetary systems, and ultimately, at least in one case, life forms. The abundance of these chemical factories in interstellar clouds was not previously expected or even imagined as intense ultraviolet radiation and the presence of cosmic rays were thought to be detrimental to the preservation of complex molecules. Theories of the chemical evolution of the interstellar medium were disrupted and insight provided into the evolution of life on earth by the discovery of complex interstellar molecules.

The first organic molecule discovered in space, formaldehyde (H_2CO) was discovered by B. Zuckerman, P. Palmer, L. Snyder, and D. Buhl using the 140-foot telescope in 1969. Zuckerman, Palmer, Snyder, and Buhl discovered the absorption spectrum of formaldehyde in dark dust clouds. The formaldehyde molecules in these dark clouds apparently absorb radiation from the 3 K microwave background. The lower energy state of the formaldehyde molecule is over-

populated caused by decay of the formaldehyde to low energy levels after excitation by collisions with hydrogen atoms. The cloud therefore absorbs energy and produces an absorption line in the formaldehyde spectrum. The 140-foot telescope was used to detect the $l_{11}-l_{10}$ ground state rotational transition of formaldehyde at 4,830 MHz.[88] Formaldehyde was found in numerous galactic and extragalactic sources indicating that organic molecules are not uncommon in the interstellar medium.

The discovery of formaldehyde initiated a series of detections of organic molecules. Organic, carbon-based, molecules represent the ingredients necessary for the origin, evolution, and perpetuation of life. It was a striking discovery to find an abundance of these molecules in the harsh conditions of the interstellar medium. The interstellar medium is bathed in high energy radiation that tends to disassociate complex organic molecules. The abundant ultraviolet and other forms of high energy radiation combined with the incredibly sparse density of matter (typically 1 atom per cubic centimeter, although individual regions may have densities that greatly depart from this average) allowed little speculation that complex molecules could survive in the interstellar medium. The discovery of organic molecules and of the many other complex molecules has resulted in a new understanding of the interstellar medium. These discoveries opened the door to a new scientific discipline—the science of astrochemistry. The organic constituents of molecular clouds in the interstellar medium reveal a prebiological astrochemistry of interstellar clouds. The implications of the findings to the evolution of life on planet earth are evident. The implications of these discoveries to the prebiological evolution of life became even more profound with the discovery of glycine, the simplest amino acid, in the interstellar medium in 1994. This discovery proved that not only organic molecules, but even amino acids—molecules that are essential building blocks of life—have evolved in the harsh environment of the interstellar space. The discovery indicates a propensity for the formation of the ingredients necessary for the evolution of life in the universe. Observations of molecular clouds provide insight into evolution at three levels: the evolution of molecules in interstellar clouds, the evolution of interstellar clouds into stars and planetary systems, and the subsequent evolution of life on the planets.[89]

The known interstellar molecules are typically found either in dense interstellar clouds or in shells around stars. Typical interstellar molecular clouds are found around HII regions and are thought to play a role in the evolution of the tenuous interstellar hydrogen gas into stars. Although the mechanism is not well understood, dust grains within these clouds are thought to effect the preservation of complex molecules in the presence of ultraviolet radiation. Many of the interstellar molecules have also been found in extragalactic sources. Observations of extragalactic molecules indicate that the chemistry involved in our neighborhood of

Table 4.2   Atomic and Molecular Species in the Interstellar Medium

Symbol	Molecule	Year of Discovery	Wavelength	Telescope
Two atoms				
CH	methylidyne	1937	visible	Mount Wilson
CN	cyanogen radical	1940	visible	Mount Wilson
CH^+	methyladyne ion	1941	visible	Mount Wilson
OH	hydroxyl radical	1963	18.0 cm	Lincoln Labs 84-foot
CO	carbon monoxide	1970	2.6 mm	NRAO 36-foot
H_2	molecular hydrogen	1970	ultraviolet	NRL Rocket Camera
CS	carbon monosulfide	1971	2.0 mm	NRAO 36-foot
SiO	silicon monoxide	1971	2.7 mm	NRAO 36-foot
SO	sulfur monxide	1973	3.0 mm	NRAO 36-foot
NS	nitrogen sulfide radical	1975	2.6 mm	NRAO 36-foot
SiS	silicon sulfide	1975	2.8 mm	NRAO 36-foot
C_2	diatomic carbon	1977	infrared	Mt. Wilson
NO	nitric oxide	1978	2.0 mm	NRAO 36-foot
HCl	hydrogen chloride	1985	1.3 cm	NASA Kupier AO
PN	phosphorus nitride	1987	1.3 cm	FCRAO 14-m
NaCl	sodium chloride	1987	1.8 mm	IRAM 30-m
AlCl	aluminum chloride	1987	1.8 mm	IRAM 30-m
KCl	potassium chloride	1987	3 mm	IRAM 30-m
AlF	aluminum fluoride	1987	3 mm	IRAM 30-m
SiC	silicon carbide	1989	1.3, 2, 3 mm	IRAM 30-m
CP	phosphorus carbide	1989	1.3 mm	IRAM 30-m
SiN	silicon nitride	1990	1.4 mm	NRAO 12-m
NH	nitrogen hydride	1991	ultraviolet	KPNO 4-m

Symbol	Molecule	Year of Discovery	Wavelength	Telescope
Three atoms				
H_2O	water	1968	1.3 cm	Hat Creek 20-foot
HCO^+	formyl ion	1970	3.4 mm	NRAO 36-foot
HCN	hydrogen cyanide	1970	3.4 mm	NRAO 36-foot
HNC	hydrogen isocyanide	1971	3.3 mm	NRAO 36-foot
OCS	carbonyl sulfide	1971	2.7 mm	NRAO 36-foot
H_2S	hydrogen sulfide	1972	1.8 mm	NRAO 36-foot
C_2H	ethynyl radical	1974	3.4 mm	NRAO 36-foot / Mitaka 6-m
N_2H^+	diazenylium	1974	3.2 mm	NRAO 36-foot
SO_2	sulfur dioxide	1975	3.6 mm	NRAO 36-foot
HCO	formyl radical	1976	3.5 mm	NRAO 36-foot
HNO	nitroxyl radical	1977	3.7 mm	NRAO 36-foot
HCS^+	thioformylium	1980	1.1 - 3.5 mm	Bell Labs 7-meter
SiC_2	silicon dicarbide	1984	1-3 mm	NRAO 36-foot
H_2D^+	(unnamed)	1985	infrared	NASA Kupier Obs.
C_2S	(unnamed)	1986	3.7 mm	IRAM 30-m
C_2O	dicarbon monxide	1989	6.5 mm	Nobeyama

Table 4.2 Atomic and Molecular Species in the Interstellar Medium (*continued*)

Four atoms

NH_3	ammonia	1968	1.3 cm	Hat Creek 20-foot
H_2CO	formaldehyde	1969	6.2 cm	NRAO 140-foot
$HNCO$	isocyanic acid	1971	3.4 mm	NRAO 36-foot
H_2CS	thioformaldehyde	1971	9.5 cm	Parkes 210-foot
C_2H_2	acetylene	1976	infrared	KPNO 4-m
C_3N	cyanoethynyl radical	1976	3.4 mm	NRAO 36-foot
$HNCS$	isothiocyanic acid	1979	3.0 mm	Bell Labs 7-meter
$HOCO^+$	protonated carbon dioxide	1980	3 mm	Onsala 20m/Bell 7m
C_3H	propynylidyne	1984	3.9 mm	NRAO 36-foot, Osala 20-m, Bell 7-m
C_3O	tricarbon monoxide	1984	1.7 cm	NRAO 140-foot
$HCNH^+$	protonated HCN	1984	4.1 mm	NRAO 12-m
H_3O^+	protonated water	1986	1 mm	NRAO 12-m
C_3S	tricarbon monosulfide	1986	4 mm	IRAM 30-m
$HCCN$	(unnamed)	1991	3 mm	IRAM 30-m

Five atoms

$HCOOH$	formic acid	1970	18.0 cm	NRAO 140-foot
HC_3N	cyanoacetylene	1970	3.3 cm	NRAO 140-foot
CH_2NH	methanimine	1972	5.7 cm	Parkes 210-foot
NH_2CN	cyanamide	1975	3.0 mm	NRAO 30-foot
H_2CCO	ketene	1976	2.9 mm	NRAO 36-foot
C_4H	butadiynyl radical	1978	2.6 cm	NRAO 36-foot
CH_4	methane	1978	3.9 mm	NRAO 36-foot
SiH_4	silane	1984	infrared	IRFT (NASA)
C_3H_2	cyclopropenylidene	1985	1.6 cm	NRAO 140-foot
CH_2CN	cyanomethyl radical	1987	1.5 cm	FRACO 14-m, Nobeyama, NRAO 43-m Onsala
C_4Si	(unnamed)	1989	19.6 cm	Nobeyama 45-m
H_2C_3	propadienylidene	1990	1.4 cm	IRAM 30-m

Six atoms

CH_3OH	methyl alcohol	1970	36.0 cm	NRAO 140-foot
CH_3CN	methyl cyanide	1971	2.7 mm	NRAO 36-foot
NH_2CHO	formamide	1971	6.5 cm	NRAO 140-foot
CH_2SH	methyl mercaptan	1979	3.0 mm	Bell Labs 7-meter
C_2H_4	ethylene	1980	infrared	Kitt Peak
C_5H	pentynylidyne radical	1986	12.6 cm	IRAM 30-m
CH_3NC	methyl isocyanide	1987	2-3 mm	IRAM 30-m
$HCCCHO$	propynal	1989	1.7 cm	Nobeyama 45-m, NRAO 140-foot
H_2C_4	butatrienylidene	1990	2.3-3.7 mm	IRAM 30-m

Seven atoms

CH_3C_2H	methyl acetylene	1971	3.5 mm	NRAO 36-foot
CH_3CHO	acetaldehyde	1971	20.0 cm	NRAO 140-foot
CH_3NH_2	methylamine	1974	3.5 mm	NRAO 36-foot
CH_2CHCN	vinyl cyanide	1975	21.8 cm	Parkes 210-foot
HC_5N	cyanodiacetylene	1976	2.8 cm	Algonquin 150-foot
C_6H	hexatriynyl radical	1986	21.6 cm	IRAM 30-m

Table 4.2   Atomic and Molecular Species in the Interstellar Medium (*continued*)

Eight atoms				
CH_3OHCO	methyl formate	1975	18.6 cm	Parkes 210-foot
CH_3C_3N	methyl cyanoacetylene	1983	1.5 cm	NRAO 140-foot

Nine atoms				
CH_3CH_2OH	ethyl alcohol (ethanol)	1974	3.2 mm	NRAO 36-foot
$(CH_3)_2O$	dimethyl ether	1974	3.3 mm	NARO 36-foot
CH_3CH_2CN	ethyl cyanide	1977	3.0 mm	NARO 36-foot
HC_5N	cyanotriacetylene	1976	3.0 cm	Algonquin 150-foot
CH_3C_4H	methyl diacetylene	1984	1.5 cm	Haystack R.O. 36-m

Ten atoms				
$(CH_3)_2CO$	acetone (unconfirmed)	1987	14.7 cm	IRAM 30-m
NH_2CH_2COOH	glycine (unconfirmed)	1994	2.79 mm	BIMA Array

Eleven atoms				
HC_9N	cyano-octatetrayne	1977	2.9 cm	Algonquin 150-foot

Thirteen atoms				
$HC_{11}N$	cyanotetracetylene	1981	1.3 cm	Haystack R.O. 36-m

space is common to extragalactic sources.[90] The principle of isotropy, the theory that the universe is essentially the same everywhere, has been maintained by these investigations. Important cosmological information has therefore been provided by radio spectroscopic observations. The 140-foot has played an important role in the evolution of molecular spectroscopy at radio frequencies which, in turn, has caused a revolution in our understanding of the interstellar medium and resulted in the genesis of astrochemistry.

Giant Molecular Clouds and Cosmic Masers

Two bizarre cosmic phenomena have been discovered through spectral line studies at radio frequencies: giant molecular clouds (GMCs) and cosmic masers. GMCs and cosmic masers have subsequently been intensely investigated with the 140-foot. The existence of GMCs was eventually realized from the interpretation of numerous data in the early 1970s. The evolution of HII regions to fully developed molecular clouds and star formation regions is apparent in the data. The diffuse HII regions, HII regions with compact sources, GMCs, and clouds with O and B stars are thought to represent different stages of star formation.[91] GMCs are the most massive objects known to exist in the galaxy, up to 10 million solar masses. GMCs are almost entirely made up of molecular hydrogen and carbon monoxide with small amounts of many other molecules interspersed throughout the cloud. GMCs are typically 150–250 light-years in diame-

ter and are surrounded by a tenuous enveloping cloud of atomic hydrogen. From one to a few dozen dense cores are contained within the interior of a GMC. These cores represent the birthplace of stars.[92] These protostars are invisible at optical frequencies because of the obscuring gas and dust. The classic example of a GMC is associated with the Orion Nebula. Data taken with the 140-foot telescope by Maddalena, Bally, and Marziti (Figure 4.24) and compiled from numerous observing programs shows the complexity and extent of the GMC in Orion and Monoceros.[93] Images of same region observed by Maddalena et al. at a much higher frequency (115 GHz) with the Columbia University 1.2-m telescope show a different view of the same features of the GMC. In Figure 4.25, the visible stars of the famous Orion constellation have been superimposed onto the radio view. This striking image shows that the extent of the GMC far exceeds that of the visual component.[94] Numerous molecules were initially discovered in this GMC. Stars are born in the interior of GMCs and eventually consume or ionize the cloud. Much of the matter is ejected from the vicinity of protostar. Remaining materials may form accretion disks and planets. Many GMCs have been observed to have compact central objects and bipolar flows of ejecta emanating from an accretion disk which often appears as a dark band across the object. Investigations of GMCs have provided important insight into the early stages of stellar evolution which, prior to their discovery, were not well understood. In 1965 an anomalous emission of the OH line was discovered simultaneously by a group at Berkeley (Weaver, Dieter, and

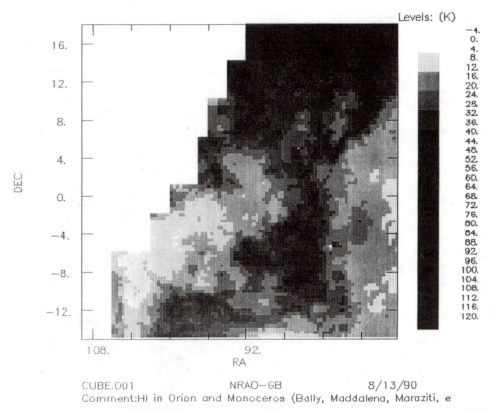

Figure 4.24 Radio contour map of the GMC in Orion and Monoceros. This image was produced from data compiled in 1990 by Maddalena, Bally, and Maraziti, using the 140-foot telescope. The GMC in Orion has been extensively observed. Many molecules were observed for the first time in this GMC. *Courtesy R. Maddalena.*

Williams) and a group at Harvard (Gunderman, Goldstein, and Lilley). The line emission was originally called "mysterium" because of its intensity, narrow line width, and unexpected ratios to other lines. The existence of cosmic masers was eventually realized to be responsible for this peculiar spectral line data.[95] The observed intensities are many hundreds of times the expected value. Cosmic masers produce these intense spectral lines by interstellar amplification by the maser (microwave amplification by stimulated emission of radiation) principle, the same principle as the laser. The maser amplifies radiation from molecular clouds using energy supplied by a star or stars in the interior. The energy source must be 10,000 times more luminous than the sun to pump the maser. The molecules absorb energy from the stars; this energy pumps the molecules into excited states. Other pumping sources include collisions between amplifying molecules and interstellar hydrogen atoms. Although the pumping source may be spread over a considerable energy range, the absorbed energy is radiated at specific spectral line frequencies. A population inversion is created in the masing medium in which a majority of the molecules are pumped to an excited state. The excited molecules then radiate at a specific frequency. The observed effect is the tremendous amplification of a single microwave emission line. Figure 4.26 shows the spectral line emission

from the water maser in W49N (June 1978 VLBI data).[96] A spectrum of lines is observed as the frequency is shifted by motion of the cloud relative to the observer. Maser emission has been detected from several molecules including water, silicon monoxide, formaldehyde, and methyl alcohol. Powerful masers have been observed in other galaxies that appear to be 10 times stronger than galactic masers. These megamasers, as they have been termed, are low gain masers but occupy a substantial fraction of the entire galaxy. Masers were initially considered to be unambiguous indicators of star formation. Subsequent observations using the NRAO 140-foot and VLBI instruments have also found cosmic masers in planetary nebula and in the expanding shells of highly evolved red giant stars, which are observed as long-period variables.[97] Researchers using the 140-foot have been actively involved in investigations of GMCs and cosmic masers and have revolutionized our understanding of the molecular evolution of the interstellar medium as well as stellar evolution in general.

Very Long Baseline Interferometry

The history of radio astronomy has been one of a quest for ever increasing sensitivity and angular resolution. The early instruments could barely distinguish between cosmic

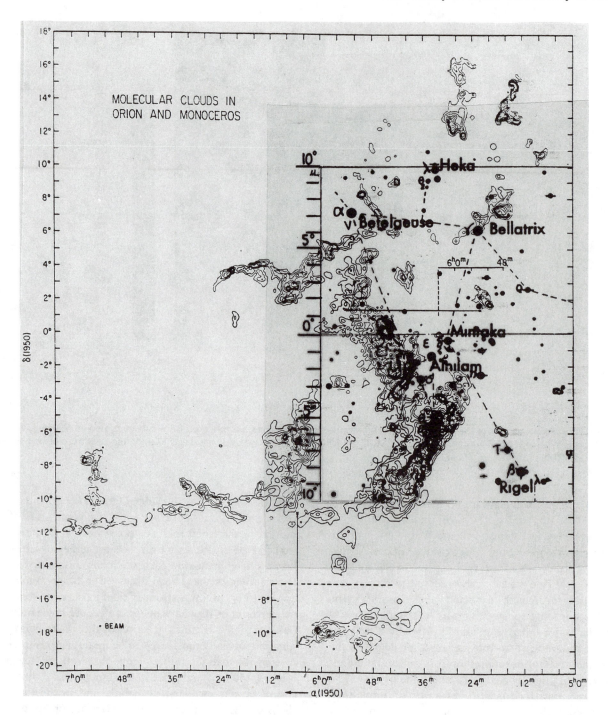

Figure 4.25   The same region shown in Figure 4.24 as observed by Maddalena et al. at a much higher frequency (115 GHZ) with the Columbia University 1.2-m telescope. This image shows a different view of the same features of the GMC. The visible stars of the Orion constellation have been superimposed onto the radio view. Maddalena references Kuhr et al., Baars et al. and Maddalena et al., for the source list from which the contour map is generated. *Courtesy R. Maddalena.*

sources widely separated in the sky. The basic problem was that the angular resolution of a radio telescope, the ability to distinguish detail, is proportional to the observed wavelength divided by the effective diameter of the aperture. For example, a 100-meter telescope observing at a wavelength of 21 cm has a resolution of only 8 arcminutes, worse than

that of the unaided human eye. To achieve a resolution at this wavelength comparable to the largest optical telescopes the telescope reflecting surface would have to be hundreds of kilometers in diameter. The construction of structures of this magnitude was, of course, impractical. An answer to the problem of angular resolution was simultaneously provided

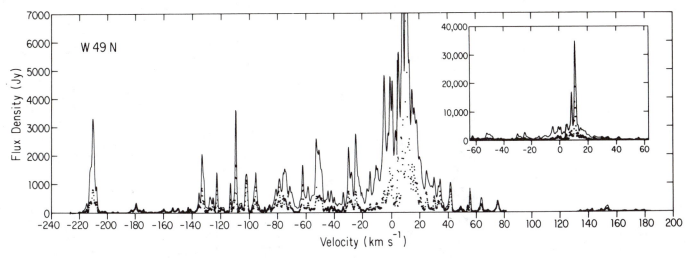

Figure 4.26 Spectral line emission from the water maser in W49N. This June 1978 VLBI data shows the tremendous intensity of line emission produced by masers. Observers were R. C. Walker, D. N. Matsakis, and J. A. Garcia-Barreto.

of radio frequency interferometry, a method of combining two or more radio telescopes together electronically to effectively produce the diameter of one large telescope was developed by Martin Ryle, D. D. Vonberg, Anthony Hewish, and a major British group that gathered at the Cavendish Laboratory of Cambridge University after World War II. A similar technique was pioneered by McCready, Pawsey, and Payne-Scott in Australia. Pawsey et al. used an ingenious instrument, the sea interferometer, which consisted of only one antenna placed on a bluff overlooking the ocean. The antenna's reflection on the water's surface provided the second element of the interferometer system. The first interferometer fringes in radio astronomy were produced in January 1946 with this instrument.[98] Although the two groups were aware of each others' research, the techniques were developed essentially independently.

The technique of radio frequency interferometry is based on the optical work of Michelson who used the technique to measure the angular diameters of nearby and bright stars. Michelson combined wavefronts from two mirrors directed at a single aperture to produce a brightness distribution of stellar images of such high precision that angular diameters could be measured.[99] The British and Australian groups applied the technique to radio frequency observations. When the radio signals from two antennas separated by some baseline distance are combined, the wavefronts interfere, either constructively or destructively. The combined information produces interference fringe patterns. The data collected by the pair inherently contains a time or phase delay due to the separational distance between the two. A wavefront incident upon the antennas arrives at one before the other and it is this time delay ΔT that represents the basic measurement. The longer the baseline length, the greater the precision of the determination of the orientation of the baseline relative to the source. The resulting angular reso-

lution for the most widely separated telescopes is on the order of 0.001 arcsecond. This procedure provided a powerful tool for the observation of discrete radio sources. Figure 4.27 shows a basic schematic diagram of a radio frequency interferometer. The essential independence of the error in determining ΔT from the length of the baseline separating the two antennas is the key to the potential usefulness of this technique for many scientific applications. A determination of the distance between two antennas a few kilometers apart with an error on the order of a few centimeters is impressive, but would have little scientific advantage. The fact that this error can be kept to a few centimeters, although the antennas are separated by intercontinental distances, is the advantage of the technique.

The first interferometer systems developed during the late 1940s consisted of telescopes physically connected by cable. Observations made with this first generation of radio interferometers proved that many sources were unresolved and that much greater detail could be observed in the resolved sources. Baselines increased and angular sizes resolved decreased dramatically over the next 10 years as new techniques were employed. Microwave links were established between telescopes via repeater stations utilizing high frequency phase locking systems at each site to increase baselines far beyond the limits of a physical connection by cable. This technique was first used by Australian and British radio astronomers to achieve baselines of more than 100 kilometers and a resolution of greater than 1 arcsecond.[100] Microwave links pushed baselines to a few hundred kilometers and greatly improved angular resolution. Even at this point in the evolution of the technology astronomers had long discussed the possibility of completely eliminating the direct electrical connection between interferometer elements by separately recording the signals at each end on magnetic tape and later comparing the two.

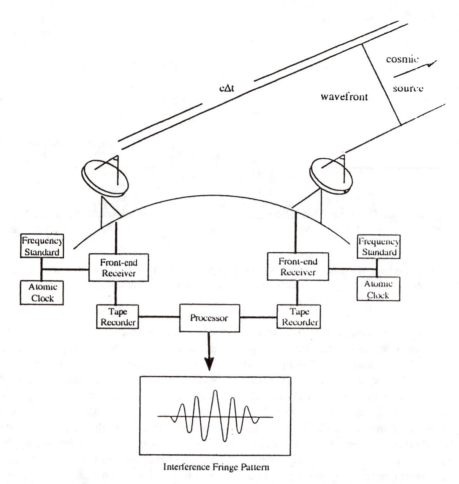

Figure 4.27 Schematic diagram of radio frequency interferometer using independent time and frequency standards. In long baseline interferometry the signals are recorded independently at the two antennas and later combined to produce fringes.

interferometer elements by separately recording the signals at each end on magnetic tape and later comparing the two. The possibility of independent-local oscillator tape recording interferometry was considered in the U.S.S.R. as early as 1961.[101] At the time, however, sufficiently stable frequency standards and wide-band tape recorders needed for high sensitivity were not available.

The development of very precise clocks, stable frequency standards, and high precision wide-band tape recorders in the late 1960s eliminated the need for a physical link between the two telescopes. A tape-recording, frequency-independent interferometer was first used by Brown, Carr, and Block at the University of Florida in January 1967 to observe the Jupiter bursts, intense radio frequency radiation associated with the planet's magnetosphere and its interaction with the Jovian moon Io.[102] The University of Florida team used time signals broadcast from the National Bureau of Standards station WWV for frequency stability and time synchronization. Although a very early important experiment, the detection of the incredibly weak emission from other cosmic sources did not occur until stable atomic fre-

quency standards and precise, high speed magnetic tape recorders were developed. The first results with independent-clock and frequency standards that provided sufficient frequency stability and time synchronization necessary to observe extragalactic sources were independently obtained by Canadian and U.S. teams in 1967.[103] Many discrete sources were still unresolved by these instruments. Longer baselines were needed to improve the resolution of the observations. The major problem in achieving long baselines was that the elements needed to be connected in some manner. This limitation was overcome by the Canadian and American groups who developed a means of recording the signals simultaneously at each antenna on high speed video recorders together with time signals generated by atomic clocks. The tapes could be later loaded into a correlator and be combined to observe the interference fringes. The baseline distance between the two elements causes phase differences in the signals reaching the two aerials. When the signals are combined interference fringe patterns are produced that may be used to synthesize an effective resolution equivalent to the baseline distance. Figure 4.28 shows the basic VLBI instrumentation design as well as the difference

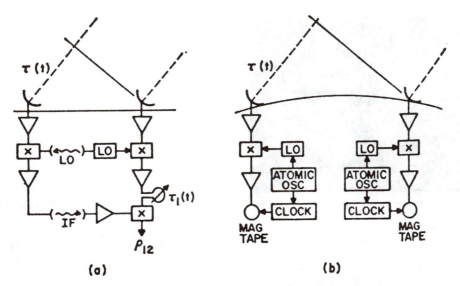

Figure 4.28   Simplified schematic diagram of a conventional two element interferometer with a radio link (a) and VLBI system (B) with independent atomic standard oscillators. VLBI techniques have provided tremendous gains in resolution.

between VLBI techniques and conventional interferometry.[104] The Canadian group was first to successfully implement these VLBI investigations. They attained a baseline of more than 3,000 km or 5 million wavelengths which gave a resolution of 0.02 arcsecond. Many discrete sources still remained unresolved. The American group at the NRAO used essentially the same technique but digitized the signals at each telescope before recording. The digital technique allowed more precise preservation of timing the observations.

Tremendous increases in resolution came quickly after these groups demonstrated that a physical connection was no longer necessary. Using the digital technique, the NRAO was able to make breakthroughs in the obtainable resolution. A major breakthrough in resolution resulted when a continental baseline was established between Hat Creek, California, and NRAO at Green Bank, West Virginia, that was nearly 20 million wavelengths long at 18 cm producing an angular resolution of 0.01 arcsecond.[105] Intercontinental baselines followed shortly thereafter. The first major successes at intercontinental VLBI were achieved in 1968 using baselines stretching from the United States to Sweden, and Australia. One of the earliest experiments produced a baseline of 6,319 kilometers (3,925 miles) from Green Bank to the Onsala Space Observatory near Gothenburg, Sweden, exceeded 100 million wavelengths at 6 cm. The spatial resolution achieved in this experiment was comparable to the angular size of a postage stamp as viewed from 6,319 kilometers—the distance between Gothenburg and Green Bank. A landmark VLBI experiment was performed in 1969 using the NRAO 140-foot telescope and the Soviet 72-foot (22-meter) telescope in Crimea. The joint Soviet-American experiment produced a resolution of 0.001

arcsecond breaking all records up until that time.[107] Since that time, the 140-foot has been used in conjunction with telescopes in the United States, Sweden, and Australia in extremely productive VLBI experiments. These and other subsequent observations producing resolutions of 0.001 arcsecond proved that numerous extragalactic sources were even smaller than these fantastically small dimensions. Such intercontinental interferometers actually mimic a telescope whose diameter approaches that of the earth itself.

VLBI has proven to be a powerful technique for providing resolution previously inconceivable in radio astronomy. Observations of the hydrogen in other galaxies require the resolution afforded by VLBI. Many discrete radio sources require VLBI techniques to resolve the objects. VLBI observations have revealed the internal structure of many compact objects. Source structure as small as 0.0004 arcsecond has been revealed through VLBI observations.[108] VLBI techniques have been used to observe galaxies, masers, quasars, and spectral lines in extragalactic sources at a routine resolution of 0.001 arcsecond. VLBI observations of quasars have been particularly enlightening as no other observation techniques reveal source structure within these phenomena. The 140-foot telescope has been extensively used as a VLBI element to observe quasars. Quasars have often been found to have jets of ejecta that are only revealed through extremely high resolution. Material in the jets of several quasars appears to be traveling many times the speed of light. This superluminal velocity is thought to be a projection effect in sources with components moving along the line of sight at velocities very near the speed of light. VLBI data of the quasar 3C 273 using the 140-foot as an element indicates superluminal motion exceeding ten times the speed of light (Figure 4.29).[109] Although observations at

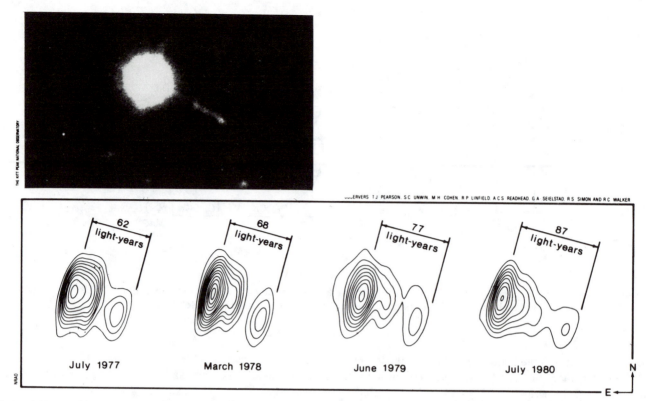

Figure 4.29   VLBI observations of 3C 273, a relatively nearby quasar (T. J. Pearson, S. C. Unwin, M. H. Cohen, R. P. Linfield, A. C. S. Readhead, G. A. Seielstad, R. S. Simon, and R. C. Walker). The NRAO 140-foot telescope served as one of the elements in the VLBI system for these observations. The jet extending from the quasar appears to be moving 10 times the speed of light. *Courtesy G. A. Seielstad.*

reached—the diameter of the earth. In the late 1970s astronomers at the NRAO began planning an array of ten 25-meter telescopes placed at locations across the United States, including Hawaii and Puerto Rico, to be used exclusively for VLBI research. Construction of the antennas of the very long baseline array (VLBA), as it has been termed, was begun in 1984 and completed in 1993. VLBI and the VLBA are the only known techniques in which details of the oldest, most distant objects in the universe may be revealed.

Radio Astronomy and Geophysics

The contributions of research performed at the NRAO and other observatories to the present understanding of celestial phenomena are considerable. The use of radio telescopes in research has greatly enhanced fundamental astrophysics by allowing astronomers to observe a wide variety of objects and astrophysical processes. The achievements in astrophysics of radio astronomy techniques have been widely appreciated. The usefulness of research in radio astronomy for observing these astrophysical phenomena was anticipated and deliberate, and intricately involved with the evolution of instrumentation in the field. The evolution of the instrumentation techniques, however, led to an unexpected use of radio astronomy instrumentation—the application of radio astronomy techniques in geodesy and geophysics.

The surprising marriage between the diverse fields of radio astronomy and geophysics has led to major contributions in the areas of geophysics and geodynamics. Terrestrial applications of radio astronomy techniques, particularly VLBI, has produced observations of distant quasars that have been used to determine tectonic plate motions of a few centimeters per year, to provide fundamental measurements of polar motion and earth rotation, and to study the complex interaction between global atmospheric circulation and the rotation of the earth. Applications of VLBI to geophysics and geodynamics have produced high precision measurements in three major areas: crustal dynamics, earth rotation, and polar motion. Coordinated telescopes on separate continents can measure the orientation of distant radio sources such as quasars so accurately relative to the earth's surface that VLBI measurements have become the standard measures of the motion of the earth's poles and variations in the length of the day (LOD). The Chandler wobble, precession, and nutation have all been measured to a precision of an order of magnitude or more greater than measurements made with conventional methods. Measurements using the same VLBI techniques have provided extremely detailed observations of crustal dynam-

ods. Measurements using the same VLBI techniques have provided extremely detailed observations of crustal dynamics. This research has provided the level of accuracy required to directly measure tectonic motion of crustal plates and the gravitational deformation of the solid earth. The implications of this research to geoscience are profound. The precise measurements afforded by VLBI techniques have allowed geoscientists to refine basic models of geophysics and geodynamics. Several ambitious programs using VLBI observation techniques that link radio telescopes on separate continents together were carried out by various groups including the NRAO, NASA, and other U.S. and international observatories during the 1970s and 1980s.

The geodetic and geophysical potentials of VLBI were recognized in the late 1960s. The idea of using VLBI techniques to measure crustal dynamics, earth rotation and polar motion first occurred to Irwin Shapiro in 1966 as the technology to produce phase stable frequency standards had come into existence.[110] In astrophysical research utilizing interferometers, the position and structure of the radio sources are measured very precisely. The technique may also be used in reverse; the position of the sources may be assumed known and the baseline orientation is then measured. The technique was employed for this purpose in 1968, shortly after Shapiro and others (Gold in 1967; MacDonald in 1967) recognized the potential.[111] Research over the subsequent 25 years has met and even exceeded these early optimistic expectations.

This new generation of VLBI telescopes and the quest for higher angular resolution led to innovative uses of VLBI techniques, particularly very precise measurements of crustal dynamics, the earth's rotation and polar motion. Several innovative research initiatives been undertaken by various groups in these areas have produced a new level of accuracy beyond conventional methods.

Developing an understanding of the crustal dynamic processes of the earth has been a long and arduous process in the history of geophysics. Plate tectonic motion is an incredibly slow process; the average rate for plate movement is on the order of 1 to 5 cm per year. The measurement of this minute crustal motion was well beyond the ability of conventional geodetic technology prior to the 1970s. By 1979, however, the techniques of SLR (satellite laser ranging) and VLBI had proven to exhibit the high accuracy capabilities necessary for geodetic measurements and NASA established the Crustal Dynamics Project (CDP) for the application of these space techniques to study the dynamic motions of the earth. The CDP utilized fixed telescopes and mobile systems developed specifically for the project. The method of applying VBLI techniques to crustal movement is conceptually quite simple. The orientation of the baselines is changed by tectonic deformation of the crust and by displacements of the telescopes caused by differential

gravitational tidal deformation of the solid earth. The tectonic movement of crustal plates has been directly measured by this program since its initiation. Long-term measurements have allowed the direct observation of continental drift. Periodic distensions due to the differential attraction of both the sun and the moon are also measured by the CDP. These vertical motions as well as the motions of the major tectonic plates were measured periodically by the CDP through 1990. Principal applications of these measurements have been to determine global plate motion, regional deformation, and plate stability.

A second major objective of the CDP as well as of other major geodetic initiatives including POLARIS, MERIT, and IRIS has been the measurement of the earth's rotation and polar motion. VLBI measurements are used to determine the rotational motion of the baseline in space, and as a result have been used to measure basic parameters of rotational motion of the earth. Polar motion, UT1 (the earth's rotational orientation at any given epoch), irregularities in the speed of rotation, length of day (LOD), precession, nutation, and the Chandler wobble have all been measured to a new level of accuracy using VLBI techniques. The earth's speed of rotation and therefore the LOD had been proven to vary in a complex manner by the 1950s as atomic frequency standards became available. Knowledge of the earth's rotational dynamics was at that point limited by astronomical observations used to determine UT1. The explosion of technology stimulated by space exploration in the 1960s and 1970s produced two different, independent techniques for monitoring UT1: lunar laser ranging (LLR) and VLBI. VLBI measurements of LOD and UT1 represented remarkably improved measurements of these phenomena. Resulting measurements determined UT1 with a formal standard of error of ± 0.05 millisecond of arc by the mid-1980s.[112] This precise measurement of UT1 allowed comparable measurements of LOD. This data is, in turn, applied to models of precession, nutation, and the Chandler wobble and greatly increases the accuracy of these models of the earth's rotational characteristics. Complex periodic and nonperiodic variations in UT1 and LOD have been observed. The application of VLBI techniques have allowed the observation and measurement of a new range of geophysical effects in the earth's rotation.

Innovations in VLBI techniques and the application of these techniques to geodesy and geophysics have succeeded in rapid progression since the inception of the technique in the 1950s. The design and construction of new instruments attest to this evolution of the technology. The Very Large Array (VLA) completed in 1980, the Very Long Baseline Array (VLBA) completed in 1993, and the proposed European VLBI system represent the world's major instruments permanently devoted to VLBI and aperture synthesis.

The contributions made to the fields of geophysics and

geodynamics by NRAO and other researchers using VLBI techniques are considerable. The application of VLBI techniques to crustal dynamics and the earth's rotation and polar motion over long time periods and with great precision has provided measurements essential to fundamental geoscience. Such measurements are key to understanding the history of the earth's crust and interior, of deformation events, and of polar motion and rotation. The application of radio interferometry to geophysics has not changed the accepted vision of the earth but rather enhanced models of crustal motion and characteristics of rotation through greatly improved precision in measurement of these systems.

Endnotes

1. Otto Struve, R. M. Emberson, and J. W. Findlay, "The 140-foot Radio Telescope of the National Radio Astronomy Observatory," *Publications of the Astronomical Society of the Pacific*, vol. 72, no. 429, December 1960, pp. 439–457.
2. R. M. Emberson, "The Telescope Program for the National Radio Astronomy Observatory at Green Bank, West Virginia," *Proceedings of the Institute of Radio Engineers*, vol. 46, no. 1, January 1958, p. 30, T. C. Kavanagh, private communication, January 4, 1957.
3. R. M. Emberson, "The Telescope Program for the National Radio Astronomy Observatory at Green Bank, West Virginia," p. 30.
4. Planning Document for the Establishment and Operation of a National Radio Astronomy Observatory, prepared for the National Science Foundation by Associated Universities Inc., New York, New York, 1956, p. 33.
5. *Ibid.*, p. 33.
6. *Ibid.*, p. 33.
7. *Ibid.*, pp. 33–34.
8. *Ibid.*, pp. 33–36.
9. *Ibid.*, pp. 33–37.
10. *Ibid.*, pp. 33–37.
11. R. M. Emberson, "The Telescope Program for the National Radio Astronomy Observatory at Green Bank, West Virginia," p. 30.
12. R. M. Emberson, "National Radio Astronomy Observatory," *Science*, vol. 130, no. 3385, November 13, 1957, pp. 1307–1318.
13. Maxwell M. Small, "The New 140-foot Radio Telescope," *Sky and Telescope*, vol. 30, no. 5, November 1965, pp. 267–274.
14. "Technical Description: 140-foot Radio Telescope," National Radio Astronomy Observatory, Internal Publication, 1987.
15. Maxwell M. Small, "The 140-foot Radio Telescope Dedication: October 13, 1965," National Radio Astronomy Observatory, Internal Publication, 1965, pp. 2–22.
16. *National Radio Astronomy Observatory* July 1, 1959, National Radio Astronomy Observatory, Green Bank, West Virginia, operated by Associated Universities, Inc. under contract with the National Science Foundation.
17. James F. Crews, Interview with B. Malphrus, at the National Radio Astronomy Observatory, Green Bank, West Virginia, March 3, 1989.
18. *Ibid.*
19. "'Guy Derricks Tested and Approved at the 140-foot Telescope," *The Observer*, National Radio Astronomy Observatory, vol. 1, no. 10, September 30, 1962, p. 6.
20. Maxwell M. Small, "The 140-foot Radio Telescope Dedication: October 13, 1965."
21. *Ibid.*, p. 4.
22. James F. Crews, Interview with B. Malphrus, March 3, 1989.
23. "The 140-foot Radio Telescope: Construction Story," NRAO film, produced by Peter B. Good, December 1964.
24. Maxwell M. Small, "The 140-foot Radio Telescope Dedication: October 13, 1965," "The 140-foot Telescope," *The Observer*, vol. 3, no. 2, November 27, 1963, p. 15.
25. "The 140-foot Radio Telescope: Construction Story," NRAO film, 1964.
26. Jay Lockman, "The History of the 300-foot Telescope or Astropoli-

tics," Lecture given at the National Radio Astronomy Observatory at Green Bank, West Virginia, on July 20, 1988.
27. "The 140-foot Radio Telescope: Construction Story," NRAO film, 1964.
28. "Spherical Bearing," *The Observer*, vol. 3, no. 6, April 30, 1964, p. 1.
29. "The 140-foot Radio Telescope: Construction Story," NRAO film, 1964.
30. *Ibid.*
31. *Ibid.*
32. *Ibid.*
33. Maxwell M. Small, "The 140-foot Radio Telescope Dedication: October 13, 1965."
34. "The 140-foot Radio Telescope: Construction Story," NRAO film, 1964.
35. *Ibid.*
36. *Ibid.*
37. Maxwell M. Small, "The 140-foot Radio Telescope Dedication: October 13, 1965."
38. "The 140-foot Radio Telescope: Construction Story," NRAO film, 1964.
39. James F. Crews, Interview with B. Malphrus, March 3, 1989.
40. Maxwell M. Small, "The 140-foot Radio Telescope Dedication: October 13, 1965."
41. "The 140-foot Radio Telescope: Construction Story," NRAO film, 1964.
42. Maxwell M. Small, "The New 140-foot Radio Telescope," *Sky and Telescope*, vol. 29, no. 11, November 1965, pp. 272–273.
43. J. Richard Fisher, Interview with B. Malphrus, at the National Radio Astronomy Observatory, Green Bank, West Virginia, March 3, 1989.
44. W. M. Baars and P. G. Mezger, "First Observations at Short Wavelengths with the 140-foot Radio Telescope," *Sky and Telescope*, vol. 31, no. 1, January 1966.
45. William E. Howard III, "From the Director's Office," *The Observer*, vol. 10, no. 2, March 1970 p. 3.
46. *National Radio Astronomy Observatory Annual Report*, July 1966–June 1967, *Astronomical Journal*, vol. 72, no. 9, November 1967, pp. 1118–1122.
47. G. L. Verschuur, "Measurements of Magnetic Fields in Interstellar Clouds of Neutral Hydrogen," *Astrophysical Journal*, vol. 156, June 1969, pp. 863–873.
48. D. L. Thacker, "Dual-Channel Cooled 21-cm Radiometer," *NRAO Electronics Division Internal Report* 124, 1972.
49. A. M. Shalloway, R. Mauzy, J. Greenhalgh, and S. Weinreb, "The NRAO 416-Channel Autocorrelation Spectrometer," NRAO *Electronics Division Internal Report* 75, 1968.
50. G. L. Verschuur, "Measurements of Magnetic Fields in Interstellar Clouds of Neutral Hydrogen," pp. 863–873.
51. *Ibid.*, pp. 863–873.
52. R. P. Sinha, "Survey of Neutral Hydrogen in the Galactic Center Region," *Astrophysical Journal*, Supplement, vol. 37, November 17, 1978, pp. 403–463.
53. W. M. Baars and P. G. Mezger, "First Observations at Short Wavelengths with the 140-foot Radio Telescope."
54. *Ibid.*
55. Richard Fleming, Interview with B. Malphrus.
56. Richard Fleming, Interview with B. Malphrus.
57. Richard Fleming, Interview with B. Malphrus.
58. *National Radio Astronomy Observatory Annual Report* 1982, *Bulletin of the American Astronomical Society*, vol. 4, no. 1, 1982, pp. 139–149.
59. *Ibid.*, pp. 397–399.
60. W. M. Baars and P. G. Mezger, "First Observations at Short Wavelengths with the 140-foot Radio Telescope."
61. John W. Findlay, "The National Radio Astronomy Observatory," *Sky and Telescope*, vol. 48, no. 6, December, 1974, pp. 352–363.
62. Sebastian von Hoerner and Woon-Yin Wong, "Improved Efficiency with a Mechanically Deformable Subreflector," *IEE Transactions on Antennas and Propagation*, vol. AP-27, no. 5, September 1979, pp. 719–724.
63. Ronald J. Maddalena, "Results of the 1987 Holographic Survey of the 140-foot Radio Telescope," NRAO Internal Document, 1987.
64. J. D. Kraus, *Radio Astronomy*, 2nd edition, Cygnus-Quasar: Powell, Ohio, 1986.
65. James F. Crews, Interview with B. Malphrus, March 3, 1989.
66. J. Richard Fisher, Interview with B. Malphrus, March 3, 1989.

67. *National Radio Astronomy Observatory Annual Report*, 1967, pp. 1118–1122.
68. I. K. Pauliny-Toth and K. I. Kellerman, "Measurements of the Flux Density and Spectra of Discrete Radio Sources at Centimeter Wavelengths. II. The Observations at 5 GHz (6cm)," *The Astronomical Journal*, vol. 73, no. 10, Part 1, December 1968, pp. 953–968.
69. *Ibid.*, pp. 1118–1122.
70. R. C. Cooley and M. S. Roberts, "Observations of the Andromeda Galaxy at 11 Centimeter Wavelength, *Science*, vol. 156, no. 3778, May 26, 1967, pp. 1087–1088.
71. Juan M. Uson and David T. Wilkinson, "Search for Small-Scale Anisotropy in the Cosmic Microwave Background," *Physical Review Letters*, vol. 49, no. 19, November 8, 1982, pp. 1463–1465.
72. Corey S. Powell, "The Golden Age of Cosmology," *Scientific American*, vol. 267, no. 1, July 1992, pp. 17–22.
73. *National Radio Astronomy Observatory Annual Report*, July 1965–June 1966, *Astronomical Journal*, vol. 71, no. 9, November 1966, pp. 798–802.
74. B. Höglund and P. G. Mezger, "Detection of the H109 Recombination Line," *Science*, vol. 150, 1965, p. 339.
75. P. G. Mezger and P. Palmer, " Radio Recombination Lines: A New Observational Tool in Astrophysics," *NRAO Reprint Series A*, 1960.
76. *Ibid.*, p. 60.
77. Gerrit L. Verschuur, *The Invisible Universe Revealed: The Story of Radio Astronomy*, New York: Springer-Verlag, 1986, pp. 107–113.
78. P. G. Mezger and P. Palmer, " Radio Recombination Lines: A New Observational Tool in Astrophysics," 1960.
79. *Ibid.*
80. *Ibid.*, p. 60.
81. *National Radio Astronomy Observatory Annual Report*, 1967, pp. 1118–1122.
82. *Ibid.*, pp. 1118–1122.
83. Gerrit L. Verschuur, "Positive Determination of an Interstellar Magnetic Field by Measurement of the Zeeman Splitting of the 21-cm Hydrogen Line," Physical Review Letters, vol. 21, no. 11, September 9, 1968, pp. 775–778.
84. Gerrit L. Verschuur, *The Invisible Universe Revealed: The Story of Radio Astronomy*, 1986, pp. 107–113.
85. R. M. Crutcher, I. Kazes, and T. H. Troland, "Magnetic Field Strengths in Molecular Clouds," *Astronomy and Astrophysics*, vol. 181, 1987, pp. 119–126.
86. D. Buhl and L. Snyder, "From Radio Astronomy Towards Astrochemistry," *Technology Review*, April 1971, pp. 126.
87. B. E. Turner, "Interstellar Molecules- A Review of Recent Developments," *Journal of the Royal Astronomical Society of Canada*, vol. 68, no. 2, January 1974, pp. 55–88.
88. D. Buhl, L. Snyder, B. Zuckerman, and P. Palmer, "Microwave Detection of Interstellar Formaldehyde," *Physical Review Letters*, vol. 22, no. 13, March 31, 1969, pp. 679–681.
89. B. E. Turner, "Interstellar Molecules- A Review of Recent Developments," January 1974, pp. 55–88.
90. D. Buhl, "Molecules and Evolution in the Galaxy," *NRAO Reprint Series A*, 1972.
91. L. Blitz and M. Kutner, ed., *Extragalactic Molecules*, Proceedings of a workshop held at the National Radio Astronomy Observatory at Green Bank, West Virginia, on November 2–4, 1981.
92. B. E. Turner, "Interstellar Molecules—A Review of Recent Developments," January 1974, pp. 55–88.
93. Ronald J. Maddalena, Unpublished observation of the GMC complex in Orion, taken with the 140-foot telescope, 1993, shows the extent of GMC in Orion.
94. Ronald J. Maddalena, Mark Morris, J. Moscowits, and P. Thaddeus, "The Large System of Molecular Clouds in Orion and Monoceros," *Astrophysical Journal*, vol. 303, April 1986, pp. 375–391.
95. R. Norris, "Cosmic Masers," *Sky and Telescope*, March 1986, pp. 248–252.
96. Data taken with the 140-foot telescope (Figure 4.33) shows the spectral line emission from the water maser in W49N (June 1978 VLBI data). Observers were R. C. Walker, D. N. Matsakis, and J. A. Garcia-Barreto.
97. K. I. Kellerman, and G. W. Swenson, "An Intercontinental Array- A Next-Generation Radio Telescope." *Science.*, vol. 188, 1975, pp. 1263–1268.
98. Richard Thompson, James, M. Moran, and George W. Swenson, *Interferometry and Synthesis in Radio Astronomy*. Malabar, FL: Krieger Publishing Company, 1991, p. 15.
99. A. A. Michelson, "On the Application of Interference Methods to Astronomical Measurements." *Astrophysical Journal*. 51 (1920):257–262.
100. K. I. Kellerman, and G. W. Swenson, "An Intercontinental Array- A Next-Generation Radio Telescope." pp. 1263–1268.
101. *Ibid.*
102. Richard Thompson, James, M. Moran, and George W. Swenson, *Interferometry and Synthesis in Radio Astronomy*. p. 20.
103. K. I. Kellerman, and G. W. Swenson, "An Intercontinental Array- A Next-Generation Radio Telescope." pp. 1263–1268.
104. Mark Gordon, "VLBA- A Continent-Size Radio Telescope," *Sky and Telescope*, June 1985, pp. 487–490.
105. *Ibid.*
106. K.I. Kellerman, and G. W. Swenson, "An Intercontinental Array- A Next-Generation Radio Telescope." pp. 1263–1268.
107. Marshall H. Cohen, "Introduction to Very Long Baseline Interferometry," *Proceedings of the IDEE*. vol. 61, no. 9, September 1973, pp. 1192–1197.
108. K. I. Kellerman, "Intercontinental Radio Astronomy," *Scientific American*, vol. 226, February 1972, pp. 72–83.
109. VLBI data of the quasar 3C 273 using the 140-foot as an element is seen to exhibit superluminal motion exceeding ten times the speed of light as observed periodically from July 1977 to July 1980. Observers were T. J. Pearson, S. C. Unwin, M. H. Cohen, R.P. Linfield, A. C. S. Readhead, G. A. Seielstad, R. S. Simon, and R. C. Walker.
110. Benjamin K. Malphrus, Interview with Irwin Shapiro, Center for Astrophysics, Harvard University, Cambridge, Massachusetts, July 30, 1993.
111. T. Gold, "Radio Method for the Precise Measurement of the Rotation Period of the Earth." *Science*, vol. 157, 1967, pp. 302–305.
112. William E. Carter and Douglas S. Robertson, "High-Frequency Variations in the Rotation of the Earth." *IEE Transactions on Geoscience and Remote Sensing*. GE-23, vol. 4, 1985, pp. 369–372.

Chapter 5

The NRAO 300-Foot Radio Telescope

The story of the NRAO 300-foot telescope begins in 1958 with the history of the 140-foot telescope. The in-progress 140-foot was a modest telescope even by 1958 standards. The Jodrell Bank 250-foot telescope was making important discoveries and the Australian Parkes 210-foot telescope was about to come on-line. The 140-foot design was selected essentially as a compromise between Tuve and Berkner. The idea was to quickly construct this telescope to gain experience before attempting a more ambitious project. By 1958, it was realized that the 140-foot was in trouble. Construction was far over budget and far behind schedule. Only the pedestal had been built. Major components and subassemblies had been fabricated and received on the site and then rejected due to susceptibility to brittle stress fracture. Design problems were beginning to be realized, causing many delays in the construction. The specifications imposed by the design committee had been altered throughout the design stage. The designers continually upgraded the specifications, laboring under the assumption that the first and possibly the only national telescope should be a versatile instrument that would meet the needs of many of the nation's radio astronomers. The upgraded requirements specified that the yoke arms had to be longer, which meant that the spherical bearing had to be larger, and the foundation deeper. The design committee imposed these requirements which multiplied many of the design specifications.[1] The design evolved from a simple essentially off-the-shelf telescope to a radical design that involved numerous engineering risks. Many wondered if it could be built. Such was the state of affairs in 1958. The national observatory, a user institution, was essentially without users because there was no major instrument.

The 140-foot was initially handled from AUI headquarters in New York which was also a source of contention. The observatory staff at Green Bank had little input into the design and construction of the 140-foot. The NRAO staff, in particular John Findlay and Dave Heeschen, saw a need to quickly build a telescope whose concept, design and construction they could oversee. The staff began to perceive a need for an instrument to attract visitors to the national facility prior to the completion of the 140-foot and the proposed 600-foot telescopes. The science of radio astronomy needed an interim telescope. A research instrument was needed to fill a gap of 5 to 10 years until the 600-foot telescope came on line. The concept therefore began to evolve for the 300-foot telescope.

From the beginning, the concept design was for a simple instrument, one that could be built quickly and would not evolve into an instrument of enormous complexity like the 140-foot. The performance specifications were subsequently decided upon by the NRAO scientific staff. The scientific staff foresaw an instrument with a very large collecting area that would produce a high level of sensitivity. It was desired that the telescope provide good sky coverage, work at 21 cm, and be relatively inexpensive to construct. A transit design was decided upon early in the design stage. A large paraboloid could be transit mounted at a fraction of the cost of a fully steerable telescope. Good sky coverage can be provided by a transit instrument as the earth rotates. Objects are, however, only visible once a day and for a short time with a transit design. John Findlay of the NRAO scientific staff claimed that the new instrument could be built fairly quickly and inexpensively and that its full scientific value could be realized before progress in the science would defeat its purpose.[2] The transit design seemed to meet the performance and cost criteria despite an inability to track objects across the sky.

Design of the 300-Foot Telescope

In 1959 the NRAO scientific staff began a preliminary study of the design of the 300-foot radio telescope. After considering various design possibilities, the basic design of the 300-foot was decided in late October 1960. Among the various designs considered was a spherical telescope for which a study was made in 1959.[3] John Findlay of the NRAO scientific staff oversaw the initial basic design, which by mid-1960 had evolved into a transit-mounted paraboloidal instrument. In early November 1960, consulting contracts were signed with Robert D. Hall and seven associate consulting engineers for the specific design. Bob Hall, formerly head of the antenna division of the Blaw-Knox Company, was selected as the primary design engineer because of his experience in the design of radio telescopes. The Blaw-Knox Company, which had built the 85-foot telescope, was taken over by a company that dissolved the antenna division. The entire antenna division group went to work for Rohr Aircraft, a West Coast corporation. Bob Hall and the antenna division were in the process of transferring to Rohr when they undertook the 300-foot design. The engineering team met for a few days in Hall's basement and produced the 300-foot concept design prior to their relocation.[4] The completion of the detailed design and the production of the design drawings were turned over to E. R. Faelten, a consulting engineer in Buffalo, New York, in December 1960.[5] He subsequently served as the engineer for the construction phase.

The Rohr and Faelten designs called for a 300-foot paraboloid mounted on two 87-foot towers shown in Figures 5.1 through 5.5. The design changed very little between the Rohr design drawings and the detailed Faelten engineering drawings. The 300-foot (paraboloid diameter) surface was designed to be constructed of aluminum mesh (Figure 5.3) supported by a structural steel superstructure. The f/d ratio of the paraboloid designed was 4.3. The design of the surface panels was selected in the winter of 1961. The strategy was to wait until the superstructure was complete so that its exact dimensions and characteristics could be measured. The superstructure (Figure 5.4) could not be fabricated to very precise tolerances since it was decided that standard structural steel would be used and the members bolted together. It was then decided what type surface panels would be used and how they would be attached. In many respects, the 300-foot was designed and fabricated at the same time.[6]

The enormous surface area was 7,850 square feet or 1.8 acres (2,394 square meters). A mesh as opposed to a solid surface was allowable as the telescope was originally designed to operate at wavelengths of 21 cm and longer. The major advantages of a mesh surface are that it is much less expensive to manufacture and the superstructure need not be as strong as one supporting a solid surface. The support towers, bearings, and bearing housings in turn need to support less loading weight and torque (seen in Figure 5.5). The telescope design achieved acceptable surface accuracy at a low cost by mounting 5/8-inch squarex aluminum mesh surface panels to a superstructure made of standard structural steel. The panels were designed to be adjustable by means of a precision screw that attached the panels to supporting members of the superstructure. The surface tolerance was set at 1 inch (2.54 centimeters) rms over the entire surface.

The telescope was designed to withstand 65-mph (105-kilometer) winds at stow but precise operation was limited to winds of 15 mph (24 kilometers) and under. Reduced precision operation was possible with this design at 25 mph (40 kilometers). The superstructure was also designed to support the surface and a uniform loading of 10 pounds (4.5 kilograms) per square inch snow over the whole reflector surface at stow. To facilitate quick fabrication, all the connections between major components were designed to be bolted together as opposed to welded. Motion in declination was achieved through a declination drive circle mounted to the superstructure which utilized a quadruple chain of 3-inch pitch attached to the periphery of the semicircle. The declination drive circle was driven by a two speed induction motor with a gear reducer producing drive speeds of 2-1/2 to 10 degrees per minute. A 50-ton counterweight was designed to be attached to the declination drive circle at the midway point of its periphery. The surface, superstructure, declination wheel, and counterweight represented the moving structure of the telescope. The entire 500-ton moving weight was supported by four 17-inch spherical roller bearings, two positioned at each end of the horizontal axis, at the top of the control towers. Two feed support legs were designed to extend from the surface. The feed support legs were attached to the superstructure and supported by guy wires attached to the surface. A focus mount supported by the feed legs which held up to 1/4 ton of receiving equipment was placed at the focal point 128-1/2 feet (39 meters) above the surface. The maximum height above ground of this structure was 225 feet (68.6 meters).[7] The design of the 300-foot telescope was innovative. A tremendous collecting area and full sky coverage was achieved at a low cost with the transit-mounted mesh surface paraboloid. The 300-foot design represented the undertaking of the construction of the largest telescope in the world at an important transformational stage in the science. The 300-foot design would allow for quick construction and therefore meet the immediate needs of the formative stages of the rapidly advancing science of radio astronomy. The scale of instrumentation reached a new level with the construction of a new generation of enormous telescopes such as the 300-foot, the Jodrell Bank telescope, and the Parkes instrument.

A cost estimate was submitted to the NSF as a supplement to the observatory FY budget of 1961. NSF had approved funding of the design in 1960. In February 1961, proposals for the fabrication and erection of the telescope were invited from 21 companies, 12 of which submitted bids. The main contract for the construction was approved on April 7, 1961, by the NSF. A prime contract for the construction, fabrication, and erection was awarded to Bristol Steel and Iron Works of Bristol, Virginia, on April 21, 1961. Other contracts were awarded for fabrication of the various components. A contract for construction of the foundation was awarded to B. F. Parrott and Company of Roanoke, Virginia. Design and fabrication of the main bearings were contracted to Lake Erie Machinery Corporation of Buffalo, New York. The design and construction of the drive system were con-

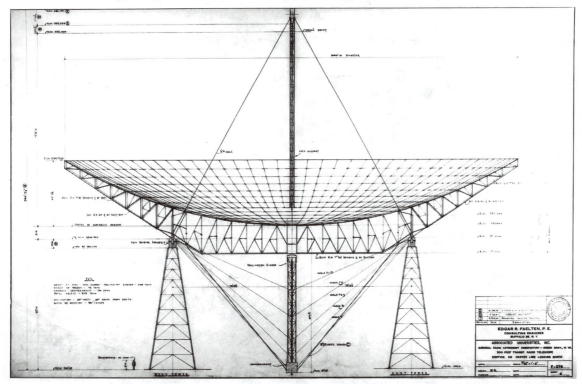

Figure 5.1 Rohr 300-foot (91.5-meter) telescope design drawing blueprints, showing side view. *NRAO image produced by Micrographics II, Charlottesville, Virginia, 1961. Courtesy NRAO/AUI.*

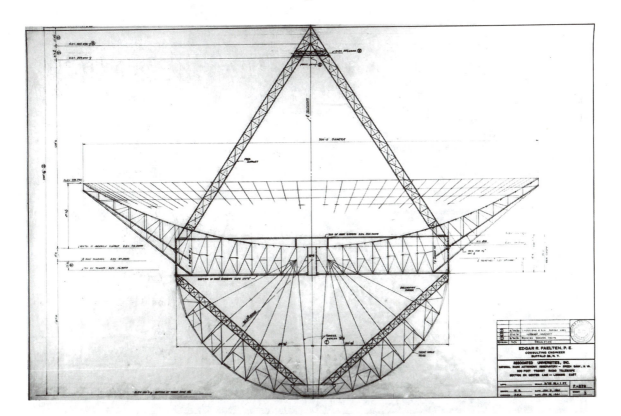

Figure 5.2 Rohr 300-foot telescope design drawing blueprints, showing frontal view. *NRAO image produced by Micrographics II, 1961. Courtesy NRAO/AUI.*

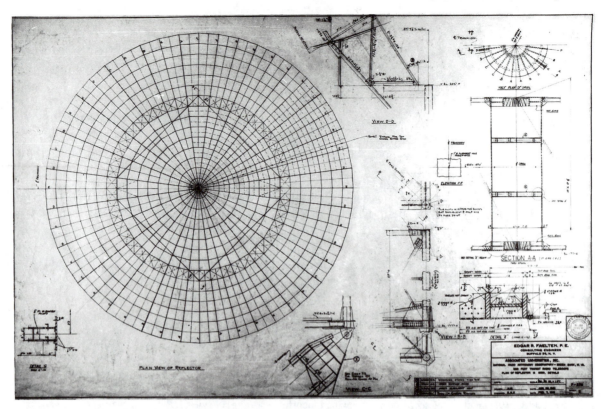

Figure 5.3 Original surface designed for the NRAO 300-foot telescope The panels were constructed of 5/8-inch squarex aluminum mesh. *NRAO image produced by Micrographics II, 1961. Courtesy NRAO/AUI.*

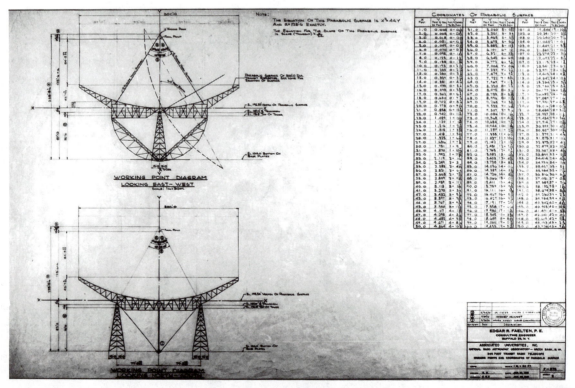

Figure 5.4 Rohr 300-foot telescope design drawing blueprint for the instrument superstructure. The superstructure was designed to be constructed to typically good standards for large structures. Higher precision adjustment of the paraboloid was designed to be achieved by installing adjustable panels. *NRAO image produced by Micrographics II, 1961. Courtesy NRAO/AUI.*

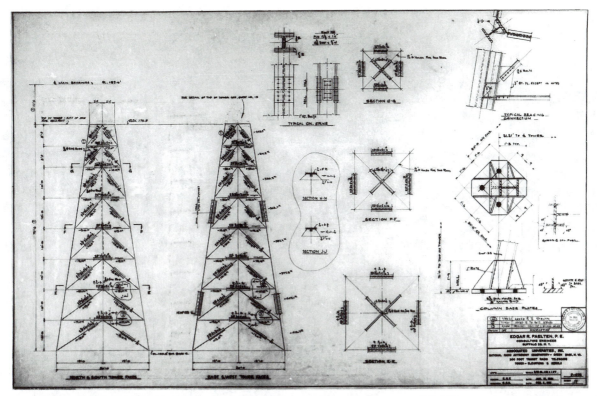

Figure 5.5 Rohr 300-foot telescope design drawing blueprint for the telescope support towers. *NRAO image produced by Micrographics II, 1961. Courtesy NRAO/AUI.*

tracted to Link-Belt Company of Philadelphia, Pennsylvania. Lockheed Electronics Company of Metuchen, New Jersey was contracted to design and fabricate the position indicator.[8] Other companies were subcontracted by the major contractors for details such as field painting of the instrument.

Construction Phase

On April 27, 1961, a groundbreaking ceremony took place for the NRAO 300-foot radio telescope (Figure 5.6). NRAO Director Otto Struve broke the ground to initiate the construction. Struve served as director throughout the construction phase of the 300-foot telescope. David S. Heeschen succeeded Struve as director in October 1962 shortly after the construction was complete. The construction phase of the 300-foot telescope went very much according to plans. The entire instrument was designed and built from the ground up in 23 months and for a cost of only $850,000. The 140-foot telescope by comparison took over 6 years to build at a cost of over $5 million.[9]

Fabrication and erection of the steel began shortly after a contract was signed with Bristol Steel in April 1961. The Pittsburgh Testing Laboratory was contracted to perform shop inspections of the fabrication. The main foundation contract was signed with B. F. Parrott on May 20. The twin tower foundations, one of which is seen in Figure 5.7, were

completed by August 14. The steel components had been fabricated and received on site by this time. Steel erection started on August 14 immediately after the support tower foundations were completed. The towers were erected early in September. Figure 5.8 shows the nearly completed towers on September 8. The drive foundation was also completed by September 29. The steel superstructure went up quickly. The basic box structure, which would support the paraboloid, was visible in the structure by mid-October (Figure 5.9). The feed support legs were erected and attached to the superstructure at this stage of the construction. Much of the superstructure had been completed by early November (Figure 5.10). By December 12, 1961, the steel erection had been completed (Figure 5.11).[10]

In the months of January through March, the dimensions of the superstructure were measured. The dish-shaped frame was surveyed by the staff with a theodolite mounted above the apex of the dish and with surveyors steel tape. The elevation angles of 288 points were measured. In this manner positions on the structure could be measured to 1/5 inch (0.5 centimeter). The information collected in this survey was used to base the decision of the design of the surface panels. The superstructure could deviate several inches from the ideal parabolic curve within the design specifications, reflecting normally accepted tolerances of good steel structures of the period. The position of the surface panels would be adjusted to compensate for these deviations to

1. Hein Hvatum	12. M. Howell	23. Jim Elliot	34. Richard Emberson
2. John Findlay	13. Clifford McLaughlin	24. David Heeschen	35. Bill Kuhlken
3.	14.	25. Campbell Wade	36. Sidney Smith
4. Edgar Faelton	15. John Hawkins	26.	37. Carl Davis
5. John Ralston	16. Glen Grandon	27. Connie Mayer	38.
6. Jim Tilley	17. Robert Elliott	28. Gerard Kuiper	39. Marc Vinokur
7. Bedford Taylor	18. George Swenson	29. Bill Waltman	40.
8.	19. Frank Drake	30. Bernie Burke	41. Jamie Sheets
9. Frank Callendar	20. Naomi Daniels	31. Bernie Lindstrom	42. Roger Lynds
10. Robert Aldrich	21.	32.	43. Mike Waslo
11. Otto Struve	22. Beaty Sheets	33. Joe Carter	44. Torliev Orhaug

Figure 5.6 Groundbreaking ceremony for the NRAO 300-foot radio telescope held on April 27, 1961. NRAO Director Otto Struve broke the ground to initiate the construction. *Courtesy NRAO/AUI.*

Figure 5.7 Foundation for one of the twin support towers of the NRAO 300-foot telescope, July 25, 1961. *Courtesy NRAO/AUI.*

Figure 5.8 Nearly completed support towers of the NRAO 300-foot telescope, September 8, 1961. *Courtesy NRAO/AUI.*

Figure 5.9   Completed box girder of the NRAO 300-foot telescope superstructure on October 17, 1961. *Courtesy NRAO/AUI.*

Figure 5.10 Erection of the NRAO 300-foot telescope superstructure. By late October 1961, much of the superstructure had been erected. *Courtesy NRAO/AUI.*

Figure 5.11   Complete superstructure of the NRAO 300-foot radio telescope. By December 12, 1961, the 300-foot telescope superstructure was completed. The design strategy was to undertake the design of the surface panels only after measurements had been made of the accuracy of the superstructure. *Courtesy NRAO/AUI.*

Figure 5.12 Installation of the NRAO 300-foot telescope surface panels. The surface panels were designed by the NRAO scientific staff and constructed of 5/8-inch squarex aluminum mesh. Installation of the panels began on May 14, 1962, and was completed on August 30, 1962. Courtesy NRAO/AUI.

produce a surface structure with tolerances within the performance specifications. The surface was planned and designed by the NRAO scientific staff during the winter of 1961 and into the spring of 1962. The original 300-foot design called for an instrument operable at 21 cm and longer wavelengths so an inexpensive surface would suffice. The planning and design stages entailed radio frequency measurements of various wire meshes and digital computer work related to the problem of affixing the surface to the steel cantilever members of the superstructure.[11] The radio frequency measurements included measuring the radio reflectivity of the mesh at various frequencies. Squarex aluminum mesh of 5/8 inch by 0.91 inch was selected because of its excellent mechanical and electrical characteristics. The electrical characteristics are particularly important as it is primarily the electric field of electromagnetic radiation that interacts with the reflector surface. Panels were fabricated by affixing the rather flexible mesh to lightweight members. The panels were installed by workers walking along the cantilevers of the superstructure at stow position. The panels were attached to the steel cantilevers by 16,000 Nelson studs of various lengths based on the superstructure survey

results. The studs were shot welded to the cantilevers and nuts were set on the threads at the proper elevation. The lightweight members of the panels were placed over the studs and bolted into position with a second series of nuts.[12] Figure 5.12 shows the installation of the surface panels. Installation of the panels began on May 14, 1962, and was completed on August 30. The initial cost of the original surface was $125,000, or less than $2 per square foot.[13]

The drive system was delivered by the Link-Belt Company on April 3, 1962. Observatory personnel installed the drive, which was completed by April 10. Figure 5.13 shows the drive system wheel, gears, and chain drive. The control building was completed in June 1962. The control system, position indicators, cabling, focal point hoist, and electronic back-end equipment had been completed during the winter. The instrument was essentially complete by mid-September 1962 (Figure 5.14). The first receivers and feeds were installed by James F. Crews on September 20. At this time, the instrument was taken over by the NRAO scientific staff and the telescope passed from the construction to the operational stage. The first observations were made on September

Figure 5.13 Drive unit of the NRAO 300-foot telescope. The drive unit was built by the Link-Belt Company of Philadelphia, Pennsylvania. Observatory personnel installed the drive unit. The unit was delivered on April 3, 1962, and was installed by April 10, 1962. *Courtesy NRAO/AUI.*

21, 1962, less than 23 months after the design was undertaken.[14]

Performance Characteristics of the 300-Foot

The first setting of the 300-foot surface was based on measurements made by the staff with surveyors tape and based on transit observations of strong radio sources. In December 1962, a photogrammetric survey was performed by the Instrument Corporation of Florida. Photogrammetric surveys were made of the 300-foot and 85-foot telescopes as described in Chapter 4. The photogrammetric surface survey of the 300-foot telescope was made at three pointing positions: zenith and declinations of 30° and 51° 21'. A summary of the results are given in Table 5.1. The detailed results of the photogrammetric survey are seen in contour maps of equal areas of surface deviation in Figure 5.15. The average rms departure of the three positions was 1.09 inches.[15] This value was very close to the 1 inch rms departure from the best fit paraboloid described by the design specifications. The focal length of the best fit paraboloid decreases as the telescope points from the zenith toward the

horizon. The telescope was refocused by manually adjusting the feed support based on this information.

The aperture efficiency at the two operating frequencies was determined by Frank Drake, C. M. Wade, and P. G. Mezger from the measurements of the surface panels. The theoretical calculated aperture efficiencies were also determined as described in Chapter 4.[16] The measured values compared favorably with the theoretical values given in Table 5.2.[17]

Pointing errors of the 300-foot telescope were determined from observation of bright radio sources. Test observations to measure the pointing errors were made on October 1, 1962. The pointing of the telescope was calibrated by observing bright radio sources whose positions were known and comparing these values to the positions given by the telescope's pointing indicator. The errors were determined by plotting the difference of the peak signal minus the true right ascension (Figure 5.16) and the peak signal minus the true declination (Figure 5.17) as the telescope was moved from 20° south declination to 70° north.[18] The measurements indicated that the structure deflects considerably as the telescope is tilted from north to south. Movements of

Figure 5.14 The complete NRAO 300-foot telescope, 1962 version. The instrument was designed in November 1960. Construction was completed and the first observations took place on September 21, 1962, 23 months later. *Courtesy NRAO/AUI.*

Table 5.1 Results of the Photogrammetric Survey of the Surface of the 300-Foot Telescope

Position of Telescope (Declination)	Greatest Departure from Best Fit Paraboloid	RMS Departure from Best Fit Paraboloid	Focal Length of Best Fit Paraboloid
0°	3.3 cm	1.07 cm	128.145 ± 0.012 feet (39.084 ± 0.003 m)
30°	2.8 cm	1.27 cm	127.980 ± 0.011 feet (39.033 ± 0.003 m)
51° 24'	3.4 cm	0.95 cm	127.782 ± 0.009 feet (38.973 ± 0.002 m)

the feed supports were also caused by gravitational forces which partially accounted for the observed pointing errors.

Measurements of the antenna gain, beamwidth, and beam shape were undertaken by Frank Drake, C. M. Wade, and the NRAO staff shortly after the completion of the 300-foot. The radiation pattern of the beam was measured by moving the telescope in declination during the transit of a strong radio source. The beamwidth at 750 MHz was measured to be 18'.5 ± 0.01 and 10'.0 ± 0.1 at 1,400 MHz. The beam efficiency was found to be 0.74 at 750 MHz and 0.51 at 1,400 MHz.[19] Measurements of the performance characteristics of the completed 300-foot telescope indicated that the instrument met the design specifications very well. The effectiveness and versatility of the 300-foot telescope was implied even in the first measurements of its original performance characteristics.

Evolving Technology at the 300-Foot

The 300-foot telescope that was turned over to the NRAO scientific staff on September 20, 1962, was essentially a bare-bones instrument. Many receiver front-end and back-end systems were then developed by the NRAO staff

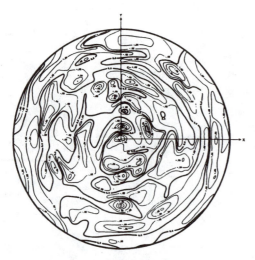

A. Pointing toward the zenith

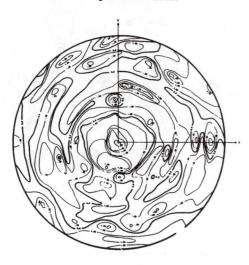

B. Pointing to 30° declination

C. Pointing to 51° 24' declination

Figure 5.15 Photogrammetric measurements of the surface of the NRAO 300-foot telescope. The contour interval is 0.2 foot (6.1 mm).

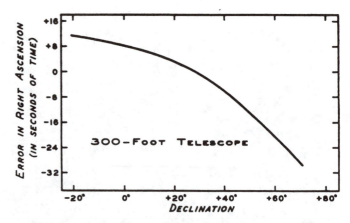

Figure 5.16 Pointing errors in right ascension of the NRAO 300-foot telescope. The pointing errors were measured in October 1962 by observing bright radio sources with precisely known positions as the telescope was moved from −20° to +70° declination. The known positions were compared to positions given by the telescope position indicator for calibration. Errors were determined by plotting the difference ot the peak signal minus the true right ascension of the sources.

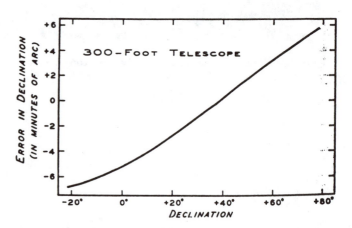

Figure 5.17 Pointing errors in declination of the NRAO 300-foot telescope. The pointing errors were measured inn October 1962. Errors were determined by plotting the difference of the peak signal minus the true declination of the sources.

Table 5.2 Aperture Efficiency of the 300-Foot Telescope

Weighted RMS Surface Errors	Frequency	Measured Aperture	Calculated Aperture Efficiency
1.20 cm	740 MHz	59%	58%
	1,400 MHz	40%	41%

Figure 5.18 The upgraded version of the NRAO 300-foot telescope as seen in 1972. An upgrade program to improve the strength and integrity of the 300-foot superstructure was undertaken in 1966. New, stronger feed support legs and a new focal feed support were added. (The first traveling feed, installed in 1969, is seen in this version.) *Courtesy NRAO/AUI.*

that greatly enhanced its performance. The telescope from the beginning performed research at the frontier of the science of radio astronomy. The enormous collecting area combined with state-of-the-art receiver technology produced a powerful research instrument. Improvements in virtually all subsystems were continually incorporated into the 300-foot telescope throughout its history. The telescope was continually upgraded and many components redesigned, including structural improvements, replacement of the surface, new feed support designs and feed legs as well as continual improvments in front-end and back-end receiver systems. These improvements in instrumentation and receiver technology at the NRAO over the years kept the instrument at the forefront of research in radio astronomy for the 26 years of its existence.

Structure

The first major improvements were undertaken in 1966. By this time, the telescope had endured several severe winters and the structure began to show some problems. Small cracks and failures appeared in some of the connections of the steel members of the reflector structure. The observatory engineering group and the Rohr Corporation group investigated the problem. They developed a plan to strengthen the superstructure and replace the feed support and feed legs with stronger ones that would allow more receiver equipment to be installed at the focus. Many light steel structural members were replaced with heavier ones, particularly where light members were attached to heavy members. Another structural improvement included replacing major components in the superstructure. The original design called for steel members running from the concrete counterweight to a point close to the elevation bearings in the superstructure. The members technically were designed to only carry tension loads. To reduce expenses, 3-inch bridge strand cables were used instead of steel members. The 1966 improvements replaced the cables with heavy steel members which greatly increased the integrity of the structure. The feed support legs were replaced with new, heavier feed support legs and a new focus mount that would support a standard front-end box.[20] The 1966 upgrade program was

Figure 5.19 The new 300-foot telescope surface installed in 1970. The 3/8-inch (0.95-centimeter) aluminum mesh panels allowed the telescope to operate at wavelengths down to 6 cm. *Courtesy NRAO/AUI.*

successful in improving the structural integrity of the 300-foot telescope. Figure 5.18 shows the upgraded 1966 instrument in which the new feed support legs and focal feed support are visible.

Reflecting Surface

The strengthening program along with severe winters, however, resulted in considerable damage to the surface. The surface panels were removed in 1966 and rolled with a road roller to reflatten them. This action, however, reduced the integrity of the panels and they sagged. The surface worked well for a while but the panels began to sag within a few days.[21] The original surface accuracy could no longer be achieved. The NRAO engineering division analyzed the telescope structure and predicted that if the panels were replaced with a good reflector surface that the telescope could perform "usefully at 11 cm and could even be expected to have a useful efficiency over a limited range of elevation angles at wavelengths as short as 6 cm."[22] Plans were therefore made to resurface the instrument and funds requested in

1969. The new panels were designed by the NRAO group in 1969. The panels were 3/8-inch aluminum mesh with lightweight supporting members mounted on adjustable studs, similar to the original panels but more rigid and providing a higher surface accuracy. The fabrication of the new surface was done by Radiation Systems, Incorporated, of McLean, Virginia. The surface panels were installed by the Micro-T Construction Company from July to early December 1970.[23] The new surface (Figure 5.19) was far more precise. The accuracy of the new panels was 4 mm rms. The accuracy of the total instrument, including the superstructure, measured 8 mm rms at any telescope position.[24] Measurements of the performance of the new surface indicated that the telescope would be operable at 6 cm. The new surface cost in the neighborhood of half a million dollars, nearly as much as the original telescope, but higher frequency observations were now feasible. The effeciency at 5 GHz was 35 percent with the new surface whereas the antenna was essentially useless above 1.4 GHz with the old surface.[25] The performance and versatility of the 300-foot telescope were greatly improved with the new surface.

Figure 5.20   The original focal feed support and mount as seen in 1964. The original feed support system was limited to a load of 500 pounds (227 kilograms) and did not provide a means of automatic focusing. *Courtesy NRAO/AUI.*

Focal Feed Support and Mount

The focal feed support and mount component of the telescope was improved on several occasions through the years. The original focal support (Figure 5.20) was limited to a load of 500 pounds and did not provide a means of automatic focusing. The 1966 upgrade included the installation of a large vertex cabin with a motor driven focus mount that allowed mounting a standard front-end box at the focus. A method of tracking sources for a limited time by moving the feed off the telescope axis by means of a traveling feed mechanism was developed by the NRAO staff. Limited tracking of sources was an innovative improvement in the performance of a transit instrument. The first traveling feed was installed in July 1969 (Figure 5.21). The 50-foot track was designed to automatically position the receivers and feeds accurately enough to provide tracking observations at frequencies up to 500 MHz. A clever strategy for tracking with a traveling feed was developed. The feed in theory would have to be linear for observations at the equator and very curved for observations near the pole to

follow the paths of objects across the sky. Instead of altering the shape of the feed track, the telescope was moved north and south to compensate for the curved motion of celestial objects. The motion was computer controlled for the proper declination. Tracking for several minutes was provided with the 1969 traveling feed. In April 1971, a receiver mount fabricated by the Sterling-Detroit Company was installed replacing the 1966 focus mount. The Sterling mount was compatible with the standard NRAO front-end box and had motor driven mechanisms that could move the box in an axial direction for focusing, 30 inches in an east-west direction for tracking, and could rotate the box for polarization measurements. The Sterling mount allowed tracking of sources at frequencies of up to 5 GHz for up to 4 minutes.[26]

A second, improved traveling feed was installed in 1980. The 1969 traveling feed could only support 50 pounds of receiving equipment over the 50-foot track length. A stronger, more accurate traveling feed system was installed that allowed the installation of cryogenic receivers. The 1980 traveling feed is seen in Figure 5.22. The improved traveling feed system allowed operation of cryogenic receivers up to 1,000 MHz.[27] The development of traveling feed systems added a new dimension of versatility to the 300-foot telescope performance.

As improvements were made in the structure, surface, and feed systems of the 300-foot telescope, the aperture efficiency was increased. Feed designs also contributed to significant improvements in aperture efficiency. A prime example is a feed built for a new low frequency 21-cm receiver during the early 1970s. A dual hyper-mode feed, developed by J. Richard Fisher, more efficiently illuminated more of the surface by spreading out the illumination pattern. The dual hyper-mode feed measured 57 percent efficient, a large improvement over the state-of-the-art feeds which typically performed at 50 percent. The improved aperture efficiency combined with the new highly sensitive cooled 21-cm receiver resulted in effectively improving the sensitivity of the receiver by a factor of two.[28] The evolution of aperture efficiency at the 300-foot is evident when this value is compared to the 40 percent measured efficiency of the 1962 instrument and the typical efficiency of the instrument in 1972, which averaged 50 percent at 21 cm (Figure 5.23).[29]

Receiver Systems

The technology perhaps most responsible for the performance capabilities of the 300-foot telescope is that of receiver systems. The evolution of receiver technology described in Chapter 4 has been considerable. Innovations in receiver technology developed by the NRAO electronics group were continually incorporated into the 300-foot receiver systems. The 1962 telescope was equipped with two crystal mixer amplifier receivers operating at 750 MHz and 1,400 MHz. A Jasik dual frequency feed was incorporated in the front-end receiver system to simultaneously utilize

Figure 5.21   The first traveling feed installed in July 1969. The NRAO staff designed and constructed the 1969 traveling feed to provide limited tracking ability. A 50-foot track automatically positioned the receivers and feeds accurately enough to provide tracking observations at frequencies up to 500 MHz. *Courtesy NRAO/AUI.*

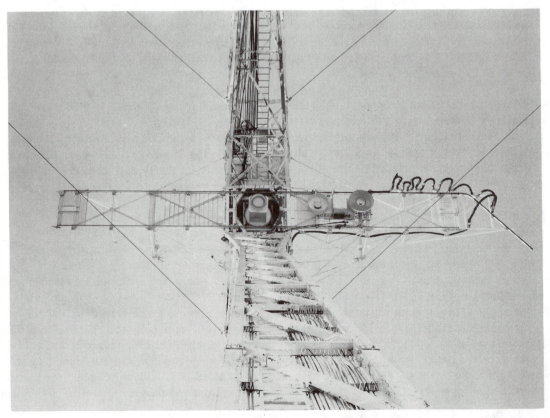

Figure 5.22   The 1980 traveling feed. This stronger and more accurate traveling feed allowed the first use of cryogenic receivers with a traveling feed. *Courtesy NRAO/AUI.*

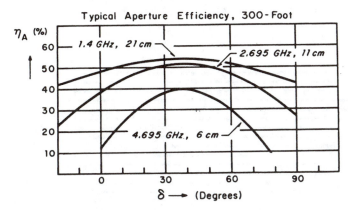

Figure 5.23 Typical aperture efficiency of the 300-foot telescope at frequencies of 1.4 GHz, 2.695 GHz, and 4.695 GHz in 1972. The aperture efficiency was increased to 57 percent at 1.4 GHz as a result of the developement of a new cooled 21-cm receiver and a dual hyper-mode feed installed in 1972.

the two receivers. The 750 MHz receiver produced a noise temperature of about 1,000 K and the 1,400 MHz operated at 800 K.[30]

By 1964 many improvements had been incorporated into the original dual receiver system. A tunnel diode preamplifier and improved mixer/IF amplifier were installed in the 750 MHz receiver that reduced the system temperature to 450 K. A parametric amplifier built by Microwave Physics Corporation was installed in the 1,400 MHz receiver that reduced its noise temperature to 190 K.[31] Other early receivers include a 1964 parametric amplifier receiver tunable from 1,390 to 1,420 MHz which was extensively used for HI research. A four-channel 21-cm receiver was brought on-line in October 1967 that produced a system temperature of 180 K and had four feeds giving four beams on the sky. This 21-cm receiver was upgraded several times, including replacement of the parametric amplifiers with transistor amplifiers in 1982.[32] In the period from the late 1960s to the early 1970s, several low frequency receivers were developed, covering the frequency range from 50 to 1,000 MHz. Transistor amplifiers were used for frequencies below 500 MHz and tunable paramps were used above 500 MHz.[33]

Higher frequency receivers were developed after 1970 to accompany the new surface. In May 1971, an 18-cm receiver was installed which was also the first cryogenic receiver used with the 300-foot telescope. This cooled 18-cm receiver utilized a closed cycle refrigeration system with NRAO-designed paramps and produced a noise temperature of only 65 K. A three-feed 11-cm receiver was also installed in 1971. The three-feed 11-cm receiver used room temperature degenerate paramps which gave a system temperature of 120 K.[34] A comparison of the noise factors of these two contemporary receivers shows the capabilities of cryogenically cooled receivers as early as 1971.

In 1972 the cooled 21-cm receiver was installed. The cooled 21-cm receiver was similar in design to the cooled

18-cm receiver. The cooled 21-cm receiver had a noise temperature of 50 K. The cooled 21-cm receiver was widely used until 1983. A six-feed 25-cm receiver was installed in October 1976 that was tunable over the frequency ranges from 1.0–1.4 GHz and 4.4–4.8 GHz. The first cryogenic receiver to be used with the traveling feed was the tunable 300–1,000 MHz receiver installed in May 1981. This receiver utilized cooled FET amplifiers. Balanced cooled FET amplifiers were incorporated into this receiver in 1987. In 1983, a 1.3–1.8 GHz receiver using cooled FET amplifiers and an innovative dewar waveguide input system and high-efficiency hybrid-mode feeds was installed that produced a system temperature of 22 K. A seven-feed 4.8 GHz was brought on-line in 1986 that utilized 14 cooled FET amplifiers.[35] The seven-feed receiver was ideal for survey programs because of its low operating temperature and its seven beams on the sky. The 300-foot telescope could make swaths of the sky with seven beams of the seven-feed receiver. Large areas of sky could quickly be covered with extreme sensitivity with this receiver system. Figure 5.24 shows total power observations of the sky with the 300-foot and seven-feed receiver.[36] The amount of quality information is evident in the data which includes the two polarizations of each feed at two back-end channels.

The evolution of receiver technology at the NRAO placed the group at the forefront of research from the late 1960s to the present day. The developments in receiver technology incorporated into the 300-foot receiving systems greatly increased the sensitivity of the instrument over the years by reducing the operating temperature. The evolution of receiver sensitivity of the 300-foot systems is seen in Figure 5.25 which gives sensitivity in terms of system noise temperature. Operating temperatures of the 300-foot receivers were reduced from over 800 K in 1962 to 15 K in 1987.[37] To a first order approximation the sensitivity of a radio telescope depends on its size and the sensitivity of its receivers. By these measurements the 300-foot had the same sensitivity of the entire VLA.[38] The sensitive receivers of the 300-foot telescope combined with its enormous collecting area and versatile receiver back ends produced a very sensitive instrument with multiple pencil beams on the sky.

The receiver systems were used in association with a variety of receiver back ends developed by the NRAO electronics group. Although many early observations were recorded with only a strip chart recorder, a digital recording system was available early in the 300-foot telescope history. The digital recording system consisted of a voltage-to-frequency converter, a frequency counter, and a paper tape punch, which was later replaced with magnetic tape. A solid-state standard receiver design developed by 1965 featured a broad IF range. These solid-state back-end receivers were used for continuum observations through the early 1980s. A spectrometer was available by 1963 for observations with the 300-foot telescope. The spectrometer consisted of a 20-channel 95 kHz/channel receiver which utilized a filter bank centered at 5 MHz. The output signals of

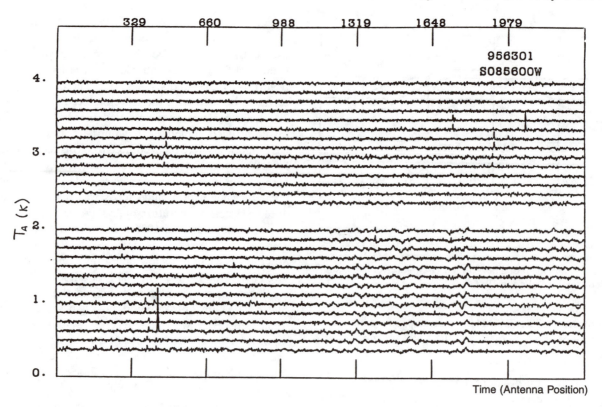

Figure 5.24 Total power observations using the 300-foot telescope and seven-feed receiver J. J. Condon, 1986). The seven-feed 4.8 GHz receiver was ideal for sky survey as it produced seven beams simultaneously on the sky and provided polarization information for each. *Courtesy J. J. Condon.*

the filter channels were sampled every 10 seconds and sent to the digital recording system. The first of a series of autocorrelation spectrometers was brought on-line in 1964. The Model I A/C had one IF channel at 5 MHz and produced a 100 point A/C function with a bandwidth of 2.5 MHz every 10 seconds. The data was directly recorded on magnetic tape and later Fourier transformed by computer. Other autocorrelation spectrometers, the Model II A/C (1968) and the Model III A/C (1971), were developed with more channels and wider bandwidths. The Model III also featured a synchronization system for pulsar observations. A digital continuum receiver was brought on-line in 1978 which used digital circuitry with a microcomputer to achieve improved flexibility in data acquisition and recording, and storage.[39] The development of innovative receiver systems has been a trademark of the NRAO electronics group. The NRAO has made many contributions to the evolution of receiver electronics and has historically defined the state-of-the-art in radio astronomy instrumentation since the early 1960s. Table 5.3 summarizes the significant events in the technical evolution of instrumentation encorporated into the 300-foot telescope.[40] The many sensitive receivers and back-end systems available made the NRAO 300-foot telescope a versatile instrument. The NRAO 300-foot telescope served as a powerful tool for many types of scientific programs and made major contributions to the science of radio astronomy.

300-Foot Telescope Programs

Researchers with the 300-foot telescope have made major contributions to the science of radio astronomy. A list of the discoveries is tremendous. Hundreds of sources, primarily distant extragalactic objects, were observed for the first time. The design of the telescope was ideal for mapping and conducting surveys of large areas of sky, so the nature of the accomplishments of the 300-foot telescope is primarily statistical. The transit design of the instrument and its pencil beams on the sky combined with its overall sensitivity provided for an ideal survey instrument. Breakthrough discoveries in radio astronomy are also credited to the 300-foot telescope. The discovery of the pulsar in the Crab Nebula in 1968, for example, revealed that pulsars were rotating neutron stars at the interior of supernova remnants. The discovery clinched the association between pulsars and supernovae and shed light on the endpoint of the evolution of massive stars.

Nearly 25 percent of the known pulsars were subsequently discovered with the 300-foot telescope.[41] It was extensively used for observations of neutral hydrogen in our galaxy and others; for radio continuum observations; for measurements of pulsars, megamasers, the interstellar medium, HII regions, and other galactic phenomena; for quasar research; for radio source variability; and for extensive

Table 5.3   Timeline of Significant Events in the History of the 300-Foot Telescope

September 21, 1962 - First observations occurred at 00:42. Dual frequency receiver operating at 1410 and 750 MHz.

By 1963 - A 20-channel, 95 kHz per channel filter bank spectrometer available.

March 1964 - Model I Autocorrelator spectrometer installed.

By early 1964 - Dual frequency receiver improved with parametric amplifier at 1410 MHz and tunnel diode amplifier at 750 MHZ.

1965 - Initiated design of solid-state Standard Receiver for continuum observations.

Summer 1966 - New focal feed support legs installed and superstructure was strengthened. Variable speed drive installed.

Winter 1966 - Surface panels removed and rolled in attempt to improve surface accuracy.

October 1967 - The 4-channel 21 cm receiver was first used.

January 1968 - Honeywell DDP-116 computer installed to control telescope motion and data acquisition.

December 1968 - Model II Autocorrelator spectrometer was first used.

July 1969 - First traveling feed was installed for receivers up to 500 MHz.

1970 - New, higher accuracy surface was installed. Cryogenic lines were installed up to the focal point.

1971 - Control building addition completed.

April 1971 - Model III Autocorrelation receiver was installed.

April 1971 - Installation of tracking focus and rotation (Sterling) receiver mount.

May 1971 - First cryogenic receiver, the cooled 18-cm, was used.

December 1971 - The 3- feed 11 cm receiver was first used.

October 1972 - The cooled 21-cm receiver was first used.

November 1974 - H-316 computer installed to position telescope and control Sterling mount and traveling feed systems.

October 1976 - The 6/25-cm receiver was first used.

Summer 1977 - Modcomp I I/25 analysis computer was installed.

January 1979 - First inductosyn was installed to replace encoders for position readout.

Fall 1980 - New, heavier traveling feed mechanism was installed for cryogenic receivers up to 1 000 MHz.

May 1981 - The 300-1000 MHz receiver was first used. First cryogenic receiver on a traveling feed was installed.

September 1983 - The 1.3-1.8 GHz receiver was first used.

August 1984 - Spoiler was installed to reduce solar interference.

August 1984 - North-south motion was added to the Sterling mount.

August 1986 - The 7-feed 4.8 GHz receiver was first used.

May 1986 - Masscomp 5500 computer was installed to replace the DDP-116 computer to control telescope positioning and data acquisition.

April 1987 - The 300-1000 MHz receiver was upgraded with FET amplifiers, and the 1.3-1.8 GHz receiver was upgraded with HEMT amplifiers.

September 21-27, 1987 - Twenty-fifth Anniversary Celebration of the 300-foot telescope was held.

November 15, 1988 - The 300-foot telescope tragically collapsed under its own weight.

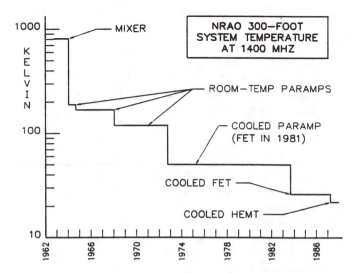

Figure 5.25 Evolution of the NRAO 300-foot telescope system temperature at 1,400 MHz from 1962 to 1988. Operating temperatures were reduced from over 800 K to 15 K in 1987 as a result of improvements in receiver technology.

sky surveys of radio sources. Over its 26-year history the 300-foot telescope greatly expanded our understanding of the universe and tremendously advanced the field of radio astronomy.

The early scientific programs involving the 300-foot telescope began to produce important results almost immediately after its completion. In 1962, Frank Drake observed Venus, Jupiter, and Saturn and greatly extended the radio spectrum of observations of these planets. The observations of Venus confirmed the surface temperature and the fact that the radio emission actually comes from the surface. The radio intensity of Jupiter was found to vary and was eventually found to be linked to the orbital period of Io, the innermost major satellite. Saturn was also found to be a radio emitter, although a weak one, with an intrinsic intensity only 0.01 that of Jupiter. No radiation was detected from Uranus, as expected.[42] The predicted intensity that a thermal radiator would emit at the distance of Uranus would be extremely low, according to the inverse square law of the dissipation of electromagnetic radiation. The value was thought to be too low to be detectable with the 300-foot telescope. Although no radiation was detected, this measurement set an upper value of the intensity of radiation emitted by Uranus, which verified the theory of the planets as primarily thermal radiators. The 300-foot telescope was used to make early high quality observations of the planets, at the closest scale in space. The 300-foot, however, because of its tremendous collecting area and sensitivity, excelled in extragalactic research.

Research in Extragalactic Astronomy

Among the earliest 300-foot telescope research programs were surveys and observations of the hydrogen in other gal-

axies. Because of the success of these early programs, surveys and mapping of extragalactic HI subsequently dominated the 300-foot research programs in extragalactic radio astronomy. The 300-foot made major contributions to the understanding of galaxies by observing the distribution, dynamics, and kinematics of atomic hydrogen within these objects. Very little was known about the gas in galaxies, and indeed, about galaxies in general in the 1960s. The existing radio telescopes used for observations of neutral hydrogen had large beams relative to the apparent size of galaxies, so little detail was revealed. The 300-foot had a very small beam which could resolve the structures within many extragalactic and galactic sources. Observations and mapping of galactic and extragalactic hydrogen therefore occupied much of the research program time.

The early research in extragalactic hydrogen produced major discoveries.[43]

1. Hydrogen distribution defines the size of galaxies.
2. Hydrogen distribution is related to the type of galaxy.
3. Galaxies are often connected by gaseous bridges.
4. Warps exist in many galaxies.

The implications of these discoveries are of major significance to the understanding of the structure and evolution of galaxies and to the science of cosmology, the study of the large-scale structure and evolution of the universe.

The hydrogen distribution of a galaxy defines its size. This understanding evolved from observations made with the 300-foot telescope by various astronomers in the early 1960s. One of the first programs using the 300-foot telescope was a group from the Department of Terrestrial Magnetism which included Merle Tuve and Bernard Burke. Tuve and Burke made high quality observations of the neutral hydrogen in the nearby galaxies M31 and M33. The DTM group used one of the earliest multifilter hydrogen line receivers that they had constructed.[44] M31, the Andromeda Galaxy, was extensively studied as it is extremely close by extragalactic standards and similar to the Milky Way. The two galaxies are now thought of as sister galaxies that anchor minor irregular galaxies in the Local Supercluster. Mort Robinson later made extensive measurements of the neutral hydrogen distribution of the Andromeda Galaxy.[45] The hydrogen is also found to be distributed differently than the visible galaxy. The data shows an absence of hydrogen gas in the region that corresponds to the optical center. This virtual absence of hydrogen gas in the center of the visible region was found to be common to most spiral galaxies. That is, most spiral galaxies were found to have holes in the center of their gas distribution. Observations with the 300-foot showed that not only does the hydrogen define the size of the galaxy, but that the hydrogen distribution among different types of galaxies varies in a systematic way. The hydrogen distribution is related to the type of galaxy. Elliptical galaxies were found to have an extremely low amount of hydrogen for their total mass whereas irregulars have a considerable amount of hydrogen and grand spirals have large amounts, primarily associated with the spiral arms. This systematic hydrogen distribution of galaxy types was a major discovery credited to research performed with the 300-foot telescope in the early 1960s.[46]

Observations of the hydrogen in other galaxies provide information not only regarding the size and shape of galaxies but also their motions, rotation rates, and interactions with other galaxies. Scientists have used the Doppler effect of extragalactic hydrogen in motion to determine galactic motions. Galaxies that are approaching the Milky Way have their 1,420 MHz hydrogen line shifted to a higher frequency while galaxies that are receding with respect to the Milky Way exhibit a shift to a lower frequency. The Doppler shift is dependent on the recession (or approach) rate. The distance to clusters of galaxies depends cosmologically on the recession rate of the cluster as the expansion of the universe takes place within inertial frameworks of clusters of galaxies. Distances to the most distant objects in the universe may therefore be estimated from the measurement of the Doppler shift of neutral hydrogen emission lines. The rotation of galaxies has also been measured by observing the Doppler shift of different parts of the galaxy. The side that is approaching is blueshifted while the receding side is redshifted. The mass of a galaxy may be estimated from its rotation rate. Observations of galactic hydrogen have provided important information regarding the structure and dynamics of galaxies.

Several surprises were revealed in the course of extragalactic research with the 300-foot telescope. The discovery of hydrogen bridges that connect galaxies and a warp in the galactic plane of the Milky Way are two classic examples. Observations of the atomic hydrogen distribution of the Milky Way have shown that the galaxy flares up on one end and down on the other. Galactic warps such as this one have subsequently been found to exist in many galaxies. Data obtained with the 300-foot telescope indicates that possibly all disk galaxies have warps in their hydrogen layer.[47] The nature and cause of galactic warps are not currently understood. In fact, a satisfactory theory of the nature of galactic warps does not yet exist. Research performed with the 300-foot telescope has also shown that disk galaxies are not isolated entities in space as all optical observations had indicated. Galaxies are often connected by bridges of gaseous hydrogen sometimes extending for thousands of light-years.[48] The physical interaction of galaxies was shown by the discovery of galactic bridges. Bridges may represent trails of matter cannibalized by large galaxies from their satellite galaxies or from other galaxies they may encounter as they gravitationally interact. Galaxies may even collide and create bridges of matter after the collision. Even a close encounter between two galaxies could create tidal forces strong enough to create bridge structures connecting the two. Bridges in local matter have also been found within the

Milky Way using the 300-foot telescope. A galactic bridge has been found connecting the clouds of neutral hydrogen of the Milky Way to the Large Magellanic Cloud, one of two satellite galaxies of the Milky Way.[49] Bridges have shown that structures of matter physically interact at the grandest cosmological scale.

Measurements of the mass of galaxies based on the rotation determined from neutral hydrogen measurements revealed an interesting problem. Galaxies seem to be far more massive, as implied by their rotation, than the visible and radio observations imply. There appears to be matter in galaxies that is invisible at optical, radio, and all other frequencies. Indeed, most of the matter in galaxies is unaccounted for in present observations, up to 90 percent by some estimates. The question of what constitutes this dark matter, which is apparently the most common form of matter in the universe, is currently a major cosmological mystery.

Quasars are another extragalactic phenomena extensively observed with the 300-foot radio telescope. Quasars and other distant radio galaxies have been found to rapidly vary in intensity. The sizes of the radiating objects can be inferred from the period of variability. Quasars and radio galaxies that produce the energy of a thousand Milky Ways have often been found to have radiating areas the size of the solar system. The 300-foot was often used as a survey instrument to investigate large numbers of sources to search for interesting phenomena such as these. Objects of interest were then investigated at high resolution by the VLA or VLBI techniques.

Census Taker of the Universe

Almost immediately after the telescope was completed, the NRAO astronomy staff and visiting astronomers began to conduct extensive sky surveys. Over the years the NRAO 300-foot telescope became known by its users as the Census Taker of the Universe as a result of its extensive survey work. The sensitivity of the 300-foot allowed it to observe sources in every direction of sky. Any area of sky without bright, obscuring sources would reveal hundreds of distant galaxies and quasars when observed with the 300-foot. Figure 5.26 shows a typical area of sky (20° declination by one hour of right ascension) in which hundreds of weak radio sources are visible.[50] When the 300-foot telescope was constructed in 1962, it could detect about 3,000 radio sources. By 1987 it could detect some 300,000 sources.[51] These technological improvements resulted in an increase of a factor of 100 in the sensitivity based on the number of radio sources detectable. Hundreds of these distant sources were observed for the first time with the 300-foot telescope.

During the first few years of the operational history of the 300-foot telescope, it became quite evident that the instrument would serve as a major survey instrument of the heavens. In 1962, Ivan Pauliny-Toth and several members of

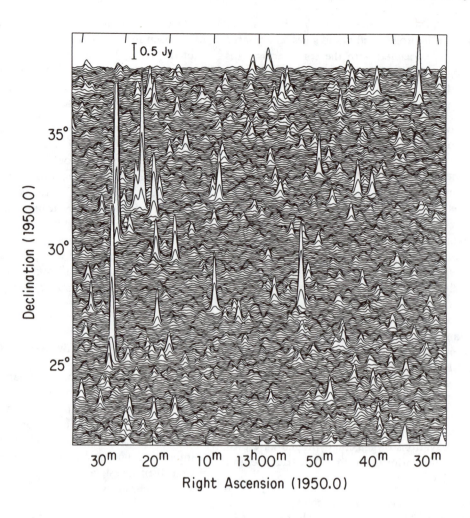

Figure 5.26 Typical area of sky (20° declination by one hour of right ascension) revealing hundreds of weak radio sources as mapped by the NRAO 300-foot radio telescope (J. J. Condon). *Courtesy J. J. Condon.*

the NRAO scientific staff observed every source in the 3C Radio Catalog. This early survey was important because the positions of many of the 3C sources were not accurately known. The 3C sources were initially observed with the Cambridge interferometer. Many of the source positions were uncertain due to lobe shifting, an uncertainty as to which interferometer lobe the source appears in. The 300-foot provided precise positions for many of these sources and went on to detect sources not listed in the 3C survey. Also in 1962, Bertil Höglund began a 1,400 MHz sky survey. The 1,400 MHz survey was repeated by various observers throughout the 300-foot history. D. S. Heeschen and C. Wade undertook a survey of normal galaxies in 1962.[52] By 1964, Heeschen and Wade had concluded the most complete radio survey of normal galaxies. The survey attempted to observe all galaxies in the Shapley-Ames catalog north of declination −20° down to a limiting photographic magnitude of 11.2. Most of the normal galaxies in the catalog were detected in this survey. The results were analyzed in terms of galactic types. Heeschen and Wade concluded that

the radio luminosities of E, S0 and Sa normal galaxies are much lower than that of other spirals and irregulars.[53]

Many important surveys were conducted with the NRAO 300-foot telescope in the 1960s, 70s, and 80s. Continuum surveys, all-sky surveys, surveys at specific frequencies, and surveys at multiple frequencies were undertaken. The GB survey, GB2 survey, several other 1,400 MHz surveys, and surveys at other frequencies provided a data base of known radio sources that has been essential to the advancement of the science of radio astronomy. Surveys have been conducted with the 300-foot at wavelengths ranging from 21 cm to 6 cm. Surveys have been conducted of specific objects such as peculiar galaxies, infrared sources, and weak galaxies; others were surveys of galactic objects such as supernova remnants and pulsars and surveys for variable sources. Many surveys were repeated yearly with the 300-foot telescope to see what changed. The 300-foot was used in this manner to patrol the Northern Hemisphere for new radio sources and variations in existing sources.

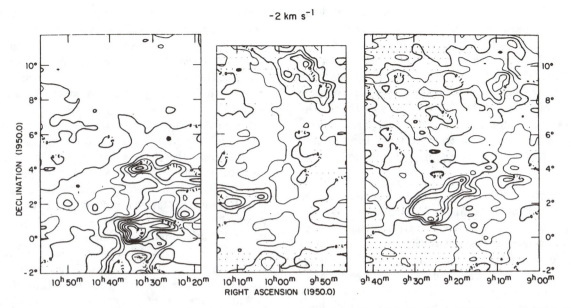

-2 km s^{-1}

Figure 5.27 HI cloud structure in the vicinity of North Polar Spur of the Milky Way (Gerrit L. Verschuur, 1973). *Courtesy G. L. Verschuur.*

A particularly significant survey was undertaken by Gerrit L. Verschuur in 1973. The HI cloud structure in the vicinity of the galaxy's North Polar Spur as well as an area roughly on the opposite side of the sky was mapped in incredible detail with the 300-foot. Elongations of the clouds were found to parallel magnetic field lines thought to be associated with the spur. The four-feed system was utilized to obtain increased efficiency in sky coverage. An enormous area was mapped, totalling over 500 square degrees of sky (the full moon by comparison subtends an angle of 1/2 degree of sky). Typical contour representations of the data are shown in Figure 5.27. The first detailed view of the HI distribution and structure of the North Polar Spur was provided by this survey. This 300-foot survey remains, to this day, the highest resolution extended HI survey ever published.[54]

During the 1980s the 300-foot continued to be used for extensive sky surveys. The increased sensitivity and multiple beams on the sky (as many as seven) allowed large areas of sky to be quickly mapped. The strategy used was to rapidly move the telescope up and down, thereby making large swaths on the sky as the earth rotates. One of the most comprehensive sky surveys ever performed at radio frequencies was a 1,400 MHz survey completed by J. J. Condon and J. J. Broderick in 1983. The 1,400 MHz survey represents a culmination of all previous surveys in the sense that most known radio sources in the northern sky were observed. The 300-foot telescope was used to survey the sky between $-5°$ and $+82°$ declination at all right ascensions with a 12 arcminute resolution. A map of the northern sky was produced from this data at 1,400 MHz (Figure 5.28).[55] The plane of the Milky Way is visible as a bright curved band stretching across the sky. The faint nebulosities represent diffuse galactic emission. Several ring-shaped supernovae

are also visible. Although the map appears similar to optical maps of the Northern Hemisphere, there is a fundamental difference. The myriad points of light dispersed over the sky do not represent stars of the Milky Way, as in optical pictures. The "stars" are actually luminous radio sources that

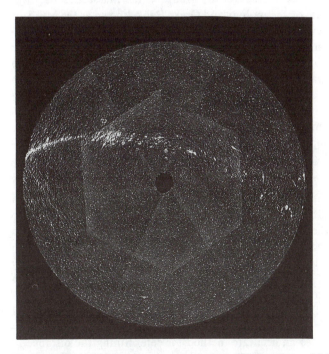

Figure 5.28 1,400 MHz northern sky survey (J. J. Condon and J. J. Broderick). The 300-foot telescope was used in 1983 to survey the entire northern sky between $-5°$ and $+82°$ declination at all right ascensions with a 12 arcminute resolution. The plane of the Milky Way is seen outlined by the faint nebulosities of diffuse galactic emission. The myriad points of light dispersed over the sky represent distant galaxies and quasars. *Courtesy J. J. Condon.*

represent extremely distant (typically billions of light-years) galaxies and quasars.[56] A similar survey of the northern sky at 5 GHz was undertaken in the late 1980s by G. Seielstad (NRAO assistant director, and director of operations at Green Bank) *et al.* The survey was underway when the telescope collapsed. Extensive sky surveys such as these conducted with the NRAO 300-foot telescope have provided invaluable information about the large-scale distribution of matter in the universe and have therefore provided insight into the cosmological structure of the universe.

Surveys with the 300-foot telescope have been conducted to search the sky for interesting or unusual cosmic phenomena. Gravitational lenses are one such phenomena for which the 300-foot telescope has extensively searched. Galaxies can act as lenses, making multiple images of objects far beyond them. Multiple images of quasars have been found to be created from a single quasar caused by a massive galaxy in between the quasar and the viewer. The gravitational field of the intervening galaxy causes refraction of the light emitted from the quasar. A dense intervening galaxy can refract the light from the quasar in such a manner that multiple images of the quasar are created. Systematic searches with the 300-foot telescope have produced candidates for gravitational lenses that are further investigated with the VLA or VLBI techniques. Discovery of 2016 + 112 in 1984 is one result of such searches for previously unknown radio sources undertaken by groups at the Massachusetts Institute of Technology and Green Bank, known as the MG Survey. 2016 + 112 consists of three images forming a right triangle. The radio and optical spectra of at least two of the objects were found to be identical, providing evidence for a gravitational lens. The redshift of two of the objects are also identical, measuring 3.7233.[57] Other less convincing candidates were also found in the MG Survey. Gravitational lenses have proven to be important phenomena in that they support Einstein's theory of general relativity at the cosmological level.

The implications of the statistical results of the 300-foot surveys to theoretical astronomy are significant although difficult to directly determine. A significant contribution to theoretical astronomy directly based on data collected with the 300-foot telescope is the discovery of Fisher-Tully effect. Rick Fisher of the NRAO and Brent Tully of the University of Hawaii conducted a survey of faint galaxies beginning in the early 1970s. They noticed a correlation between the intrinsic luminosities and the observed global galactic HI line profile width of these galaxies.[58] The correlation provides an accurate method for determining the distance to these galaxies. The relationship between the global line profile width of the HI and the intrinsic luminosities of these galaxies is a simple one, although the physics connecting the two is vague. The line width is produced by the differential Doppler shift of the hydrogen line radiation from the galaxy. One side is rotating toward the observer producing

a blueshift of the HI line, while the other side is rotating away from the observer producing a redshift in the HI line emission. The width of the line depends on the rotational velocity of the galaxy. The faster the galaxy rotates, the wider the line. The rotation rate provides a determination of the mass of the galaxy. A greater number of stars contained within a galaxy provides a higher mass and a stronger gravitational field. A high-mass galaxy therefore rotates faster to maintain itself. More stars produce a higher mass, a faster rotation, and therefore a higher luminosity. More stars therefore mean a wider line width and a higher luminosity. The line width is independent of distance, because the Doppler shift due to rotation is independent of distance. The apparent luminosity of a galaxy is equal to its distance squared. A measurement of the line width gives the absolute luminosity. If a value of the absolute luminosity is known and the apparent luminosity is measured, the distance can be determined.[59] The Fisher-Tully effect thereby provided a method of determining galactic distances. The relationship provided a distance measurement method that is completely independent of the total galactic redshift and of the velocity. The Fisher-Tully effect has since been extensively used to determine galactic distances which have provided an accurate yardstick for measuring cosmological distances. A correlation has also been found at infrared frequencies. Fisher and Tully applied the method at radio frequencies to determine the distance to the Virgo Cluster. A value for the Hubble constant was derived from this distance. The Fisher-Tully effect provided a method of determining the distance to remote galaxies as well as a new means of calculating a value of the Hubble constant, the large-scale expansion rate of the universe.

Another interesting outcome of the Fisher-Tully Survey of weak galaxies resulted in the accidental discovery of the Local Void. Fisher and Tully, at one point, made a moving picture of the survey data to give a three-dimensional picture of the local universe. The moving picture was created by rotating the dataset to allow a three-dimensional view produced by motion parallax. This form of data representation was an early attempt to show the distribution of galaxies in the local universe. Fisher plotted the data utilized in this representation on a wide format plotter. The data included positional information as well as the velocity of the galaxies. Fisher differentiated the galaxies in velocity to produce a three-dimensional representation. An interesting feature appeared in this representation, an enormous hole (in excess of ten megaparsecs) that was almost completely devoid of galaxies. Fisher was intrigued by the feature but did not pursue the research, lacking confidence in the data.[60] The feature was real and eventually became known as the Local Void. It has since been realized that voids are a fundamental cosmological structure. Tremendous voids have been discovered that possibly determine the large-scale structure of the universe. Voids currently represent an important aspect of research in radio astronomy. The Local

Void, the first to be discovered, was there in Fisher's 1974 data; or, ironically, not there.

In the 1980s many surveys were undertaken with the 300-foot which included measurements of the velocities of thousands of galaxies, as in the Fisher-Tully Survey. With this information scientists are currently mapping the distribution of matter in the local universe. For the first time, an understanding of the large-scale structure of the local universe is beginning to emerge, primarily based on extragalactic surveys to which the 300-foot telescope has been a major contributor.

Research in Galactic Astronomy and Observations of Galactic Hydrogen

The NRAO 300-foot radio telescope was used to perform important research in galactic radio astronomy. The research programs included observations of galactic hydrogen, radio source variability, megamasers, pulsars, the zone of avoidance, and the transitional lines of large atoms. Observations of galactic hydrogen represent the most extensive research in galactic astronomy performed with the 300-foot telescope. Observations of the hydrogen in the Milky Way revealed the spiral arm structure, the warp in the galactic plane, high-velocity hydrogen clouds, and hydrogen clouds associated with celestial objects. The telescope contributed to the understanding of the spiral arm structure of the galaxy through programs that mapped the hydrogen in the spiral arms. Arms moving away from the observer produce a red shift in the 21 cm hydrogen line (resulting in a shifting of the line to a slightly longer wavelength), whereas arms moving toward the observer produce a blue shift in the hydrogen line (resulting in a shifting of the line to a slightly shorter wavelength). The position and velocity of these clouds can be determined from these observations which define the spiral arms.

Observations of galactic hydrogen give insight into the large-scale structure of the Milky Way. The Milky Way is a system of 300 billion stars with at least three spiral arms and a distinct nucleus. The diameter of the galaxy is 100,000 light-years and the average thickness is around 1,000 light-years. The disk is extremely flat with a nucleic bulge and a warp at each end. The entire system is surrounded by a halo of globular clusters of stars. High-velocity hydrogen clouds were found that rain onto the disk of the Milky Way from above and below the galaxy. This hydrogen rain may partly be responsible for providing "fuel" for future stars and for the galactic nucleus. Mapping of the distribution of galactic hydrogen represents an area of research in which the 300-foot telescope has been extensively used.

One of the first research programs mapping the neutral hydrogen of the galaxy was undertaken by Gart Westerhout of the University of Maryland in 1964. The program mapped the entire galaxy visible from Green Bank with a resolution of 10 arcminutes utilizing the 100 channel autocorrelator receiver. The Maryland-Green Bank (MGB) 21-cm Line Survey, as it became known, took several years to complete. The results of the survey were compiled in a map of the distribution of neutral hydrogen in the entire galaxy visible from the Green Bank latitude.[61] The scope of the program was enormous. The quality, completeness, and uniformity of the data were exceptional. The MGB survey represented one of the most complete and highest resolution pictures of the distribution of neutral hydrogen in the Milky Way for many years.

Observations and mapping of galactic HI and HII subsequently comprised much of the 300-foot research programs in galactic astronomy. Observations over the years with the 300-foot and other telescopes provided a picture of the large-scale distribution of galactic HI and HII regions. The 300-foot telescope was used to observe galactic features such as hydrogen clouds in the galactic plane, interstellar hydrogen clouds, high-velocity hydrogen clouds, and hydrogen gas spurs, which are probably remnants of the ejecta from ancient supernova emanating from the galactic plane. The NRAO 300-foot telescope was also extensively used to observe other galactic phenomena such as emission nebula, supernova remnants, galactic variables, the flicker of radio sources caused by a medium of electrons and protons moving through magnetic fields in the intervening space, and pulsars.

Pulsars

Pulsars are rapidly rotating neutron stars that represent the leftover stellar core after a super massive star has gone supernova. Neutron stars have radii of only about 10 km, a typical mass of about 1.5 solar masses, and may rotate a thousand times a second. The density of these objects is on the order of 10^{17} kg/m^3, a density nearly that of the nucleus of atoms. Neutron stars are primarily composed of degenerate neutrons immersed in a sea of electrons and some protons. An outer solid crust composed of atomic nuclei stripped of all electrons, roughly 1 kilometer thick, surrounds the star. A 4-km thick inner crust lies below this layer that consists of a solid lattice of pure neutrons immersed in degenerate electrons and protons. Below this layer the star is composed almost exclusively of neutrons that surround a solid core that may be entirely composed of neutrons.[62] Neutron stars are bizarre cosmic objects in which equally bizarre physics is implied. Degenerate electrons, for example, at the outer layer, are disassociated with any particular nucleus and may orbit the entire star. The entire star, in this sense, acts as a single atom. As early as the 1930s, neutron stars had been theoretically predicted to exist, but they were not discovered until nearly 40 years later.

Pulsars were discovered accidentally in 1967 at Cambridge by Joscelyn Bell. Bell was, at the time, a graduate student working on a research project designed by Sir Anthony

Hewish. Hewish designed an inexpensive telescope which was essentially an array of wires stretched from over a thousand poles to observe the scintillation of radio sources. Distant radio sources that appear as points appear to scintillate as the radio waves pass through clouds of intervening electrons primarily produced by the solar wind. Bell waded through literally miles of strip chart data, in the days before powerful computers were enlisted for data reduction, to analyze the data. She noticed a "little bit of scruff" in one dataset that persisted day after day. The time of arrival of the scruff followed sidereal time and was therefore not terrestrial in origin. Further investigation of the signal revealed that the scruff was actually a pulse train whose arrival time was incredibly regular (1.3373 seconds), so regular in fact that it required the best available clocks to measure it. For months the nature of these signals was a complete mystery. The precise regularity of these pulses led the Cambridge team to speculate that they were communication signals from an extraterrestrial civilization. The source was initially designated LGM 1 by the Cambridge team for Little Green Men.[63]

The discovery of the pulsating radio source was announced in the February 1968 issue of *Nature*.[64] The announcement instigated searches for pulsating radio sources at radio astronomy observatories all around the world. In fact, several observers found pulsating sources in data they had already collected but interpreted the signals as interference.[65] It was under these circumstances that a general search for dispersed periodic signals was undertaken by David H. Staelin and Edward C. Reifenstein III, at the NRAO with the 300-foot radio telescope in October 1968. Staelin and Reifenstein monitored the band between 110 and 115 MHz with 50 channels, each 0.1 MHz wide. The time constant of the receiver was 0.05 second and all channels were sampled every 0.6 second. The output was recorded digitally on magnetic tape for computer-aided analysis. Two pulsating sources in the vicinity of the Crab Nebula were discovered.[66] One of the pulsating signals, designated PSR 0532 + 21, was shown by Staelin and Reifenstein to be coincident with the Crab Nebula, based on positional and distance information. The Crab Nebula represents the remnants of a supernova that was chronicled by Chinese astronomers in A.D. 1054. Many astronomers had begun to suspect that the pulsating sources were rapidly rotating neutron stars. The discovery of the pulsating source in the Crab Nebula confirmed the suspicions.[67] The discovery clinched the theory that pulsars are rapidly rotating neutron stars that have a hot spot on their surface or in the immediate vicinity which appears as a beacon of radiation as the star rotates, much like the beacon of a lighthouse. The nature of the hot spot is unknown and the actual details of the emission mechanism of pulsars are sketchy. The Crab Nebula pulsar provided further proof as the rotation was found to be incredibly fast, 33 times per second. No material other than the degenerate neutrons in a neutron star could

withstand such rapid rotation without self-destructing. Optical photographs of the Crab Nebula pulsar (Figure 5.29) rather impressively demonstrate the pulsations.[68] The discovery of the pulsar in the Crab Nebula by Staelin and Reifenstein is often touted as the 300-foot telescope's most famous discovery.

The NRAO 300-foot telescope was extensively used for pulsar work after the discovery of the Crab Nebula pulsar. Many surveys were conducted with the 300-foot telescope to search for pulsars. Nearly 25 percent of all known pulsars were subsequently discovered with the 300-foot telescope.[69] The distribution of pulsars in the galaxy and the major properties of pulsars were investigated with the telescope. Secular decrease in the rotation rate of certain pulsars, such as the one in the Crab Nebula, has been observed. Measurements of the spin down rate of these pulsars have led to the inference of values of the magnetic field strengths at the surfaces of neutron stars.[70] The pulses have been found to arrive in groups, similar to a city skyline. The pulse trains, as they are often called, vary in intensity from pulse to pulse. Measurements of these scintillations have provided information about the electron density of the intervening space. The pulse trains also arrive at different times at different frequencies. Longer wavelengths of the pulse signals are slowed down by the interstellar medium. The higher frequency components of the pulse, therefore, arrive first. Measurements of the differential arrival of various frequency components of the pulses lend insight into the density of the interstellar medium and into the endpoint of supermassive stars as well as the composition and density of the intervening space.

Pulsar investigators have precisely measured pulse times and numerous other pulsar characteristics such as pulse morphology, pulse-to-pulse variations, subpulses, and mode changing. The pulses have been found to range from 3.75 seconds to one thousandth of a second. The discovery of a millisecond pulsar in 1982 represented a new category of pulsars. The fastest rotating pulsar discovered to date is the millisecond pulsar PSR 1937 + 21 which rotates once every 0.0015578064488724 second.[71] The most intriguing results of pulsar measurements have shown that the pulsations of many pulsars are extremely regular, more regular than the best atomic clocks. The regularity of the pulsar PSR 1937 + 21, for example, has exceeded the present technological ability to measure it.[72] Many scientists feel that pulsars represent a new generation of extremely precise clocks. The implications of such precise timing is only now being fully investigated by groups such as the U.S. Naval Observatory.

Pulsar investigations with the 300-foot telescope represented a significant portion of the science performed with the telescope in the 1970s and 1980s. Researchers including Joe Taylor, M. Damashek, R. A. Hulse, R. M. Hjellming, and R. N. Manchester used the 300-foot and other instruments to investigate pulsar characteristics. New categories of pul-

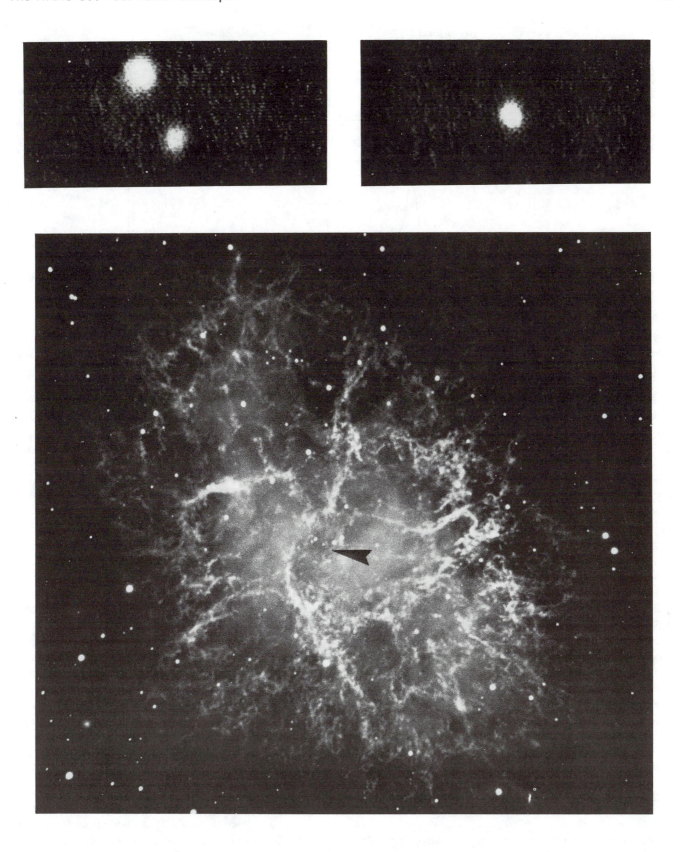

Figure 5.29 Optical photographs of the pulsar in the Crab Nebula. Photos a and b, taken at the Lick Observatory, reveal the pulsations at visible light frequencies. Photo c, taken at the Lick Observatory, shows the location of the pulsar within the Crab Nebula. The discovery of the pulsar in the Crab Nebula with the NRAO 300-foot telescope clinched the theory that pulsars are rapidly rotating neutron stars and provided the first evidence of the existence of neutron stars. *Lick Observatory photographs.*

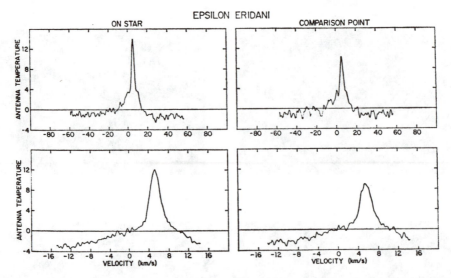

Figure 5.30 Hydrogen line profile of epsilon Eridani (Gerrit L. Verschuur, 1971). *Courtesy G. L. Verschuur.*

sars were discovered through these investigations including binary pulsars and millisecond pulsars. Joe Taylor of the University of Massachusetts discovered the first binary pulsar in 1974.[73] Taylor and Hulse were awarded the Nobel Prize in Physics in 1993 for their pioneering work in binary pulsars. Pulsars discovered in binary pairs are important in that they have allowed the first measurements of the mass of neutron stars. The orbital periods of the binary pairs indicate the mass of the two stars. The orbital period of the third binary pulsar to be discovered was measured to be 88877.7 seconds by Damashek using the 300-foot telescope. The measurements revealed a surprising result. The orbit was found to be almost perfectly circular. The implication of this finding is that the gravitational distortion of the stellar surface was sufficient to circularize the orbit in a comparably short period of time. The findings indicated that the companion star was recently a normal extended object like the sun.[74] The investigations of binary pulsars, several of which were subsequently found, provide important insight into the evolution of binary star systems, particularly binaries in which one companion is a supermassive star. Taylor and Weisberg have used investigations of binary pulsars to provide high precision verification of the theory of general relativity and possible evidence for gravitational radiation.[75] Pulsar investigations have provided great insight into many aspects of stellar evolution, the interstellar medium, and theoretical astrophysics. The contribution of the 300-foot telescope to these investigations was considerable.

SETI

Much important work in observational astrophysics was performed with the 300-foot telescope during the 1970s. Among the significant research projects of the seventies was one of a very controversial nature undertaken by Gerrit Verschuur in 1971—an extremely sensitive search for extra-

terrestrial intelligence (SETI). Although the search for radio emissions from extraterrestrial technological civilizations had begun in 1960 with the sensationalized Project OZMA, little significant SETI research had been subsequently undertaken. The philosophical appeal of detecting radio signals from exobiological life forms had tremendously excited the general public in the early 1960s, but had not incited astronomers to pursue the research in a systematic way. In 1971, despite the controversial nature of the research, Verschuur undertook a search for narrow band radio signals in the 21-cm wavelength band, in the direction of ten nearby, solar-type stars. This project was possibly the first SETI search with any hope of success. The experiment was on the order of one million times more sensitive that the Project OZMA experiment.[76] Although negative results were reported, important outcomes of the project were realized. The resulting paper published in *Icarus* was the first formal paper on SETI in which results of any significance were published. Another important consequence of the experiment was the production of 21-cm hydrogen line profiles of the ten selected stars that have been used as zero-epoch profiles for subsequent SETI searches (The profile for Epsilon Eridani is seen in Figure 5.30).[77] In many ways, the experiment represented a new era in SETI searches—systematic observation of nearby solar type stars at extremely high levels of sensitivity.

End of an Era

On November 15, 1988, a cool, windless night, the 300-foot radio telescope suddenly collapsed under its own weight. The tragic collapse dealt a severe blow to the international scientific community. The 300-foot was still one of the most productive telescopes in the world of any kind. It had performed beyond all expectations—26 years after its construction, the science of radio astronomy still had not overtaken the usefulness of the instrument. The world media was immediately flooded with speculation as to the cause of the collapse ranging from a scaled-up design from 140 feet to 300 feet that did not incorporate structural design changes corresponding to the expanded size[78] to speculation that the instrument was "zapped by space aliens because American scientists were learning too much about them."[79] These speculations were, of course, discounted. On November 18, 1988, officials at the National Science Foundation and Associated Universities Incorporated established a Technical Assessment Panel to investigate the cause of the collapse. In addition to the external assessment panel, the Observatory appointed an internal investigation panel. The Technical Assessment Panel, consisted of R. M. Matyas, G. F. Mechlin, several AUI officials, and E. Cohen, a managing partner with Amman and Whitney, a structural engineering consulting firm based in New York City, that was contracted to perform much of the technical analysis.[80]

The Technical Assessment Panel conducted a post-collapse finite element space frame stress analysis and a half-model computer analysis. The analyses were performed by Ammann and Whitney. The half-model computer analysis indicated that the stresses in a large number of circumferential and radial members were significantly higher than those permitted, in some cases by as much as 100 percent and more. The half-model computer element analysis showed that, in the zenith position, the position with the greatest structural balance and the least amount of member stress, 1.7 percent of the circumferential members were overstressed by 200 percent or more. Of the total circumferential members, 25 percent were found to be overstressed by lesser amounts. Many gusset plates, (plates found at the conjunction point of structural members used to connect the members together) were found to be overstressed by secondary forces not considered in the original design.[81] Results of the finite element analysis revealed compression indicating that many overstressed members would have eventually failed due to structural metal fatigue.

A fractured gusset plate was found in the wreckage and was submitted to a detailed finite element analysis. The gusset plate was found to be a critical connection in the diamond truss of the superstructure. The panel found that the stresses were excessive and the stress range during telescope transits indicated a limited fatigue life. Fractographic examination was conducted of the gusset plate, which indicated the metal fatigue crack propagation under cyclic loading resulted in eventual fracture of the plate. Figure 5.31 shows the failed gusset plate; a magnified view shows the small fracture that led to failure of the gusset (Figure 5.32). The panel concluded that the fracture of this gusset plate connection was the most probable cause of the telescope collapse.[82]

The gusset plate that failed was a critical connection point in the box girder. The gusset plate, designated gusset BA, was located in such a position that it could not be seen by inspection procedures or without taking major components apart, which was impossible. It was found by James F. Crews and the observatory staff among the wreckage and was sent to Lucius Pitkin, Metallurgical and Chemical Consultants and Testing Laboratories for detailed inspection. The fracture was found to be slightly oxidized indicating precollapse fracturing. The relatively flat and oxidized appearance of the fracture surface adjacent to the bolt holes indicated progressive long-term fracturing due to fatigue. Metal fatigue, which many engineers indicate is a misnomer implying that structural integrity can be regained, was indicated by this study to be the primary cause of the fracture. The fracture occurred near a hole in the gusset and fractured the gusset along a line including the hole. The holes were pre-drilled in the various fabricated components to expedite construction. In some cases where components did not line up, the hole were re-drilled in the field, occasionally through an existing hole to widen it.[83] This procedure reduced the structural integrity of the gusset plates. Further reductions in the structural integrity of the plates occurred over time due to metal fatigue until a critical fracture point was reached and failure occurred.

The Technical Assessment Panel concluded that the responsibility of the failure was solely placed on structural fatigue of the material constituting the gusset plate. The panel concluded that neither the designers of the telescope nor the observatory staff could have predicted the structural failure. In considering the designers' role, the panel concluded that engineers at the time the instrument was designed and constructed did not have the ability to perform complete analyses of large structures. The computing ability at the time prohibited finite element analysis that essentially calculates the effect that every member has on every other member throughout the cyclic transit motion. The technology was not available to calculate the forces acting on all of the components of the structure.[84] The design of the telescope was nothing short of ingenious considering that the telescope far outlived all structural expectations imposed by the original specifications and was extremely inexpensively and expeditiously constructed.

The Technical Assessment Panel also concluded that the NRAO staff could not have predicted the structural failure from the routine inspections and maintenance procedures or from the observational performance of the telescope. The inspections could not have revealed the fractured gusset plate because of its location. Periodic inspections were conducted over the entire operational life of the telescope. The successes of these inspections led to the 1964 structural improvements implemented because of structural breakdown. The observatory maintenance and inspection program of the telescope was extensive and more than sufficient to detect all structural problems physically possible. Novel maintenance programs had in fact been tested and implemented for the maintenance program. An interesting

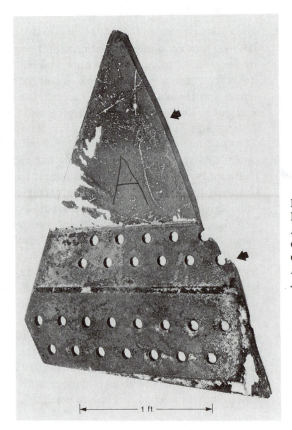

Figure 5.31 The failed gusset plate that led to the collapse of the NRAO 300-foot telescope on November 15, 1988. The Technical Assessment Panel concluded that the fracture of this gusset plate due to metal fatigue was the most probable cause of the telescope collapse. *Courtesy NRAO/AUI and Dr. Robert S. Vecchio, Lucius Pitkin, Inc., from the technical report submitted to AUI, Inc., "The 300-foot Radio Telescope—Fractured Main Box Girder Plate."*

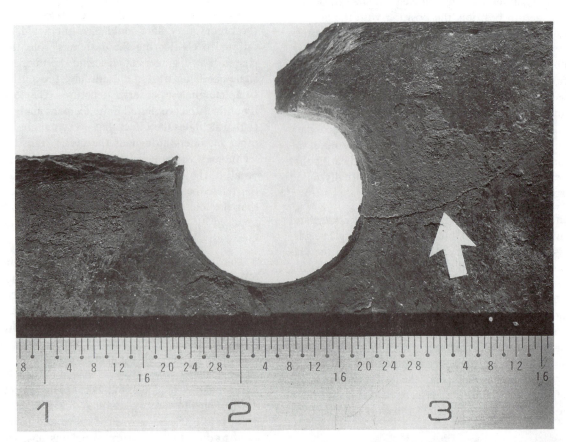

Figure 5.32 Magnified photograph of a portion of the failed gusset plate showing a secondary fatigue crack. The flat and oxidized appearance of the fracture surface at the bolt hole surface suggests progressive long-term fracturing due to fatigue. *Courtesy NRAO/AUI and Dr. Robert S. Vecchio, Lucius Pitkin, Inc., from the technical report submitted to AUI. Inc., "The 300-foot Radio Telescope—Fractured Main Box Girder Plate."*

Figure 5.33 The collapsed 300-foot telescope. These views show the collapsed superstructure and support tower of structural steel which was bent completely over by the force of the collapse. *Photographs by B. Malphrus.*

example was the use of a jet engine to melt the snow in the dish, since snow loading represented a serious danger to the structure. Although the jet engine worked well to melt the snow, the maintenance of the engine itself was not feasible.[85] The practical solution to avoid snow loads in all of the NRAO telescopes was simply to point the telescopes away from the wind and wait until conditions improved. Snow load dangers were thereby avoided with a minimal loss of observing times.

The ultimate findings of the technical assessment of the 300-foot telescope collapse revealed "no direct implication from the failure of the 300-foot telescope to other radio telescopes. There were no phenomena observed in the operation of this telescope that could not be dealt with using modern design practice."[86] There were therefore no implications regarding the present ability of engineering science to design and construct 300-plus foot aperture instruments.

The NRAO 300-foot telescope was originally designed and constructed to fill a gap in the science. The instrument outperformed all expectations and never became obsolete because of the evolution of instrumentation technology.

The versatility of the instrument was evident in the wide variety of research programs in which it was used. The 300-foot participated in programs ranging from surveys of distant galaxies to investigations of galactic hydrogen to pulsar work and even SETI searches. The loss of this instrument represented a tremendous setback to the science of radio astronomy. Figure 5.33 documents the devastation of the collapse. A promising note, however, is that, in a sense, the 300-foot telescope will still contribute to the science of radio astronomy in that it proved that there is still a tremendous need for large, filled-aperture telescopes. In this light, a replacement, the Green Bank Telescope has been proposed, funded, and designed. The GBT will be the largest and most sensitive movable telescope in the world upon its projected completion in 1996. The GBT will carry on the tradition of large single dish filled aperture telescopes.

Endnotes

1. Jay Lockman, "The History of the 300-foot Telescope or Astropolitics," Lecture given on July 20, 1988, at the National Radio Astronomy Observatory Green Bank, West Virginia.
2. *Ibid.*
3. John Findlay, "The 300-foot Transit Radio Telescope," National Radio Astronomy Observatory Internal Report, September 26, 1962, pp. 1–8.
4. James F. Crews, Interview with B. Malphrus, at the National Radio Astronomy Observatory, Green Bank, West Virginia, March 3, 1989.
5. John Findlay, "The 300-foot Transit Radio Telescope," September 26, 1962, pp. 1–8.
6. Jay Lockman, "The History of the 300-foot Telescope or Astropolitics," Lecture, July 20, 1988.
7. John Findlay, "The 300-foot Transit Radio Telescope at Green Bank," *Sky and Telescope*, vol. 25, no. 2, February 1963.
8. *Ibid.*, pp. 1–8.
9. Jay Lockman, "The History of the 300-foot Telescope or Astropolitics," Lecture, July 20, 1988.
10. John Findlay, "The 300-foot Transit Radio Telescope at Green Bank," *Sky and Telescope*, vol. 25, no. 2, February 1963.
11. *Ibid.*, pp. 1–8.
12. *Ibid.*, pp. 1–8.
13. *Ibid.*, pp. 3–8.
14. John Findlay, "The 300-foot Telescope Is Ten Years Old," *The Observer*, vol. 13, no. 6, December 1972, pp. 3–8.
15. John Findlay, "Operating Experience at the National Radio Astronomy Observatory," *Annals of the New York Academy of Science*, vol. 116, no. 1, June 26, 1964, pp. 25–37, John Findlay, "Radio Telescopes," IEEE Transactions on Military Electronics, vol. MIL-8, nos. 3 and 4, July - October, 1964, pp. 187–197.
16. John Findlay, "Operating Experience at the National Radio Astronomy Observatory," *Annals of the New York Academy of Science*, vol. 116, no. 1, June 26, 1964, pp. 25–37.
17. *Ibid.*, pp. 25–37.
18. John Findlay, "The 300-foot Transit Radio Telescope at Green Bank," *Sky and Telescope*, vol. 25, no. 2, February, 1963.
19. *Ibid.*, pp. 1–8.
20. *Ibid.*, pp. 3–8.
21. James F. Crews, Interview with B. Malphrus, March 3, 1989.
22. John Findlay, "The 300-foot Telescope Is Ten Years Old," pp. 3–8.
23. *Ibid.*, pp. 3–8.
24. D. S. Heeschen, *National Radio Astronomy Observatory Annual Report*, July 1968 - June 1969, *Bulletin of the American Astronomical Society*, vol. 2, no. 1, 1970.
25. "300-foot Telescope Technical Evolution, 1962–1987: A Quarter Century of Science on the Meridian," NRAO Electronics Division, Internal Report included in internal document for the 25 year anniversary ceremony of the 300-foot telescope, "The First 9131 Meridian Crossings, 25 Years at the Research Frontier with the 300-foot Telescope," September 1987.
26. *Ibid.*

27. *Ibid.*
28. J. Richard Fisher, Interview with B. Malphrus, at the National Radio Astronomy Observatory, Green Bank, West Virginia, March 3, 1989.
29. John Findlay, "The 300-foot Telescope Is Ten Years Old," pp. 3–8. Findlay provides figures and the curves for typical aperture efficiency of the 300-foot telescope in 1972 at frequencies of 1.4 GHz, 2.695 Ghz, and 4.695 GHz.
30. "300-foot Telescope Technical Evolution, 1962–1987: A Quarter Century of Science on the Meridian," September 1987, pp. 1–9.
31. *Ibid.*, pp. 1–9.
32. *Ibid.*, pp. 1–9.
33. *Ibid.*, pp. 1–9.
34. *Ibid.*, pp. 1–9.
35. *Ibid.*, pp. 1–9.
36. 300-foot Data from J. Condon 1986. The data represents total power observations using the NRAO 300-foot telescope and the 7-feed receiver. The data is designated 956301 SO85600W.
37. "300-foot Telescope Technical Evolution, 1962–1987: A Quarter Century of Science on the Meridian," September 1987.
38. Jay Lockman, "The History of the 300-foot Telescope or Astropolitics," Lecture, July 20, 1988.
39. "300-foot Telescope Technical Evolution, 1962–1987: A Quarter Century of Science on the Meridian," NRAO Electronics Division. The technical report summarizes the technical evolution of back-end receiver technology at the 300-foot telescope from 1962–1987.
40. *Ibid.*, pp. 1–6.
41. Jay Lockman, "The History of the 300-foot Telescope or Astropolitics," Lecture, July 20, 1988.
42. John Findlay, "The 300-foot Transit Radio Telescope at Green Bank," *Sky and Telescope*, vol. 25, no. 2, February, 1963.
43. Jay Lockman, "The History of the 300-foot Telescope or Astropolitics," Lecture, July 20, 1988.
44. John Findlay, "The 300-foot Telescope Is Ten Years Old," pp. 3–8.
45. Jay Lockman, "The History of the 300-foot Telescope or Astropolitics," Lecture, July 20, 1988.
46. *Ibid.*
47. *Ibid.*
48. *Ibid.*
49. G. L. Verschuur, "An Intermediate Velocity Cloud Showing a Velocity Bridge to Local Matter," *Astronomy and Astrophysics*, vol. 3, 1969 pp. 77–82.
50. Data showing the distribution of faint radio sources mapped by the NRAO 300-foot radio telescope, NRAO observer-J. Condon. The data field is centered around 13^h00^m right ascension and 30° declination.
51. "Science Highlights with the 300-foot Telescope: The First Quarter Century," NRAO Internal publication included in internal document for the 25 year anniversary ceremony of the 300-foot telescope, "The First 9131 Meridian Crossings, 25 Years at the Research Frontier with the 300-foot Telescope," September 1987.
52. John Findlay, "The 300-foot Telescope Is Ten Years Old," pp. 3–8.
53. David E. Hogg, "Radio Observations of the Galaxy and Extragalactic Sources," *Journal of the Royal Astronomical Society of Canada*, vol. 58, no. 5, 1964, pp. 203–216.
54. G.L. Verschuur, "Studies of Neutral Hydrogen Cloud Structure," The Astrophysicall Journal Supplement Series, no. 238, 1974, pp. 65–112.
55. The 1400 MHz survey mapped from data collected in April 1983 and December 1983 with the NRAO 300-foot radio telescope, Image description/NRAO file number: CV 83-SURV.T300R/01. Observers were J. J. Condon and J. J. Broderick. The 1400 MHz survey represents a culmination of all previous surveys in the sense that nearly every known radio source was observed. The 300-foot telescope was used to survey the sky between -5° and +82° declination at all right ascensions with a 12 arcminute resolution. A map of the Northern sky was produced from this data at 1400 MHz.
56. *National Radio Astronomy Observatory*, NRAO Brochure, NRAO/AUI, 1988, operated by Associated Universities Inc., under contract with the National Science Foundation.
57. Dietrick E. Thomsen, "Gravitational Refractions," *Science News*, vol. 125, pp. 154–155,
58. R. Brent Tully and J. Richard Fisher, "A New Method of Determining Distances to Galaxies," Astronomy and Astrophysics, vol. 54, 1977, pp. 661–673.
59. J. Richard Fisher, Interview with B. Malphrus, March 3, 1989.
60. J. Richard Fisher, Interview with B. Malphrus, March 3, 1989. Fisher's discovery was never announced or published.
61. John Findlay, "The 300-foot Telescope Is Ten Years Old," pp. 3–8.
62. William K. Hartmann, *Astronomy, the Cosmic Journey*, 3rd edition, Wadsworth Publishing Company, Belmont, California, 1985, pp. 318–322.
63. Gerrit L. Verschuur, *The Invisible Universe Revealed: The Story of Radio Astronomy*, Springer-Verlag, New York, New York, 1987, pp. 142–148.
64. A. Hewish, Bell, S. J., Pilkington, J. D., Scott, P. F., Collins, R. A., "Observation of a Rapidly Pulsating Radio Source," *Nature*, vol. 217, February 1968, pp. 709–713.
65. Jay Lockman, "The History of the 300-foot Telescope or Astropolitics," Lecture, July 20, 1988.
66. David H. Staelin and Edward C. Reifenstein III, "Pulsating Radio Sources Near the Crab Nebula," *Science*, vol. 162, December 1968, pp. 1481–1483.
67. Mark Damashek, "Discovery of the Third Binary Pulsar," *The Observer*, vol. 21, no. 2, July 1980, pp. 6–9.
68. Optical photograph of the pulsar in the Crab Nebula included in *The National Radio Astronomy Observatory*, NRAO Brochure, NRAO/AUI, 1988, operated by Associated Universities Inc., under contract with the National Science Foundation.
69. Jay Lockman, "The History of the 300-foot Telescope or Astropolitics," Lecture, July 20, 1988.
70. David J. Helfand, J. H. Taylor, P. R. Backus, and J. M. Cordes, "Pulsar Timing: Observations From 1970 to 1978," *The Astrophysical Journal*, vol. 273, April 1, 1980, pp. 206–215.
71. Gerrit L. Verschuur, *The Invisible Universe Revealed: The Story of Radio Astronomy*, pp. 142–148.
72. *Ibid.*
73. Mark Damashek, "Discovery of the Third Binary Pulsar," pp. 6–9.
74. *Ibid.*, pp. 6–9.
75. Gerrit L. Verschuur, *The Invisible Universe Revealed: The Story of Radio Astronomy*, pp. 142–148.
76. Gerrit L. Verschuur, E-mail message to Benjamin K. Malphrus, January 10, 1994.
77. Gerrit L. Verschuur, "A Search for Narrow Bank 21-cm Wavelength Signals from Ten Nearby Stars," *Icarus*, no. 19, 1973, pp. 329–340.
78. Associated Press Release, "Radio Telescope Collapses in a Major Blow to Research," *Orlando Sentinel*, Thursday, November 17, 1988.
79. Ragan Dunn, "America's Most Powerful Radio Telescope Is Zapped by Hostile Space Aliens," *Weekly World News*, January 31, 1988.
80. Report of the Technical Assessment Panel, "Collapse of the 300-foot Radio Telescope," including "National Radio Astronomy Observatory 300-foot Telescope Finite Element Analysis, Volume I," prepared by Henry Ayvazyan, Joel Stahmer, and Joseph Vellozzi, of Amman and Whitney, Consulting Engineers, New York, New York, February 1989.
81. *Ibid.*
82. James F. Crews, "Collapse of the 300-foot Telescope," Lecture given at the National Radio Astronomy Observatory, Green Bank, West Virginia on June 19, 1989.
83. James F. Crews, Interview with B. Malphrus, March 3, 1989.
84. Report of the Technical Assessment Panel, "Collapse of the 300-foot Radio Telescope," February 1989.
85. James F. Crews, Interview with B. Malphrus, March 3, 1989.
86. Report of the Technical Assessment Panel, "Collapse of the 300-foot Radio Telescope," February 1989.

Chapter 6

The Green Bank Interferometer

In 1964, a second 85-foot (25.9-meter) telescope was acquired from the Blaw-Knox Corporation and incorporated into an interferometer system. 85-Two, as the second 85-foot telescope is called, is a twin to the original Tatel telescope. 85-Two was designed to be moved by an 80-wheel dolly along a track and to be placed on one of six foundations to operate in conjunction with the Tatel telescope (often subsequently called 85-One). Spacings ranging from 1,200 to 2,700 meters from 85-One that remained fixed were provided. The axis on which the foundations are arranged is at 242° azimuth (west-southwest). The two elements were physically connected by cable supported above ground along the axis. Operations with the two-element interferometer began in October 1964.[1]

A third 85-foot telescope was added in 1967 to facilitate earth rotation aperture synthesis observations. Additional foundations were added to provide more spacings which were required to perform aperture synthesis. The maximum baseline spacing of the system (85-One to 85-Three) remained 2,700 meters. Figure 6.1 shows the Deer Creek Valley in which the interferometer is visible along its baseline at the right. In Figure 6.2, 85-One and Two are shown being serviced as they sit on foundations located along the baseline axis. Figure 6.3 shows the dolly on which the two mobile dishes are moved to new spacings. The 1967 three-element interferometer gave the same resolution as the two-element interferometer because the maximum baseline distance remained the same. A 2 arcsecond resolution was achieved.[2]

A 42-foot (12.8-meter) remote telescope was incorporated into the Green Bank interferometer system in 1969. The 42-foot portable polar mounted telescope was located on Spencer Mountain, a site roughly 35 kilometers from Green Bank. The 42-foot was designed to work with one of the 85-foot telescopes as a microwave-linked interferometer pair.[3] The 42-foot was replaced by a new 45-foot portable elevation over azimuth-turret mounted telescope. The basic design of the 45-foot portable telescope is seen in Figure 6.4. Electronic Space Systems Incorporated designed

and fabricated the instrument which was brought on-line in 1973. In that year, the 45-foot telescope was incorporated with all three on-site 85-foot telescopes into a four-element interferometer system.[4]

The initial concept of the Green Bank interferometer included more than the immediate research it was to perform. The Green Bank instrument was also designed as a prototype for the VLA, whose preliminary design studies were begun as early as 1961.[5] The NRAO preliminary designs for the VLA envisioned the most ambitious aperture synthesis instrument in the world. The Green Bank interferometer was, in part, designed to provide experience in interferometric systems before the far more ambitious VLA project was undertaken. The 35-kilometer baseline provided by the remote component of the Green Bank interferometer is the same as the largest separation between telescopes of the proposed VLA in its widest configuration.[6] In 1982, a portable 14.2-meter telescope was added at a site on Point Mountain, 32 kilometers from Green Bank, to add a west arm to the GB interferometer.[7] The two remote telescopes used in conjunction with 85-One and 85-Two used as a four-element interferometer produced a Y-shaped configuration of elements to permit earth-rotation aperture synthesis observations. The Y-shaped configuration was decided upon early in the design stages of the VLA. Image synthesis processes could therefore be developed for the Green Bank interferometer that would provide insight into image synthesizing with the VLA.

The Green Bank interferometer receiver electronics underwent many stages of improvements as receiver technology evolved at the NRAO. The new developments in receiver technology were incorporated into the interferometer electronics in the same manner as the NRAO 140-foot and 300-foot telescopes. The main prime-focus front-end receiver system contains 3.7 cm, 6 cm, and 11.1 cm receivers that operate simultaneously and that each simultaneously receive right- and left-handed circularly polarized signals. The selection of the three operating wavelengths was not arbitrary. The proposed VLA was designed to have front-

Figure 6.1 The NRAO telescopes in the Deer Creek Valley. Three antennas of the Green Bank interferometer are seen at the right aligned along a west-southwest axis. *Courtesy NRAO/AUI.*

Figure 6.2 85-Two, one of the two movable 85-foot (25.9-meter) antennas of the Green Bank interferometer, is shown as it is prepared to be moved along the baseline axis. *Courtesy NRAO/AUI.*

Figure 6.3 The 80-wheel dolly with which the two mobile 85-foot antennas of the interferometer are moved to new stations for aperture synthesis observations. *Courtesy NRAO/AUI.*

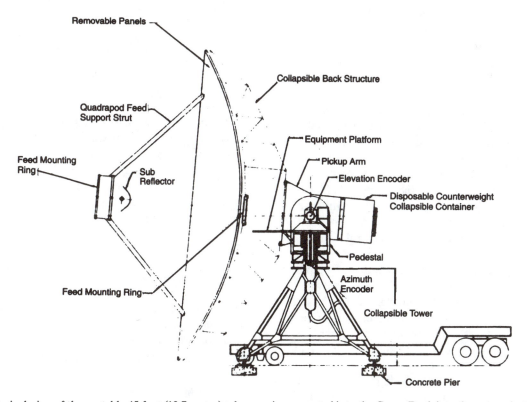

Figure 6.4 Basic design of the portable 45-foot (13.7-meter) telescope incorporated into the Green Bank interferometer in 1973. The portable 45-foot elevation over azimuth-turret mounted telescope was connected to the on-site interferometer antennas by a microwave link.

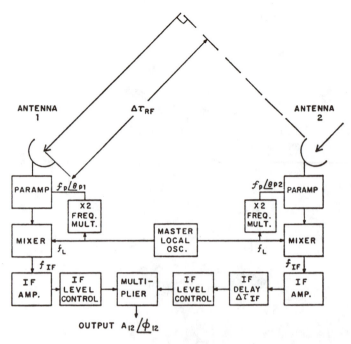

Figure 6.5 Basic schematic design of the Green Bank two-element interferometer system. The three-element system has the same basic design. Two remote antennas are also incorporated into the system by a microwave link.

end receivers operating at 3.7 cm, 6 cm, and 11.1 cm.[8] All elements of the interferometer were equipped with identical three-receiver front-end systems.

The basic interferometer design of the two-element system is shown in a block diagram in Figure 6.5. The three-element system had the same basic design. Interferometric techniques in radio astronomy and basic instrumentation are described in Chapter 2. Microwave lengths and relays provided the connection between the two-element system and the two remote elements to produce a four-element interferometer. At each physically connected antenna, the radio signals collected by the feeds are amplified and converted to a lower, intermediate frequency (IF) by mixing the incoming signal with a signal generated by the master local oscillator. The IF signals are then transmitted to a central location where the delays required to obtain correlation of the signals are inserted. The power is kept constant at each antenna by an automatic level control system. The power signals from each antenna are then fed to a multiplier. The multiplier produces the interferometer output by detecting the portion of the IF signals received from each of the elements. The Green Bank interferometer is basically a double sideband type because both of the frequency bands above and below the local oscillator frequency are received and processed. The entire system, including all electronics, delay lines, antenna positioning, and data collection, integration time, and data recording, was handled by a Honeywell DDP-116 computer.[9]

The interferometer system produces extremely high reso-

lution. The two- and three-element versions produced a resolution of 2 arcseconds. The addition of the remote antenna produced a four-element interferometer with a fringe separation (resolution) of 0.2 arcsecond at 3.7 cm wavelength, and 100,000 wavelengths separation at 11.1 cm.[10] The 0.2-arcsecond resolution is more than 10 times greater than the three-element interferometer and roughly 1,000 times greater than the 140-foot or 300-foot telescopes.[11] The tremendous resolution afforded with the interferometer system is ideal for precise source positioning, and therefore for identification of radio sources, as well as other astrometric measurements. The advances in receiver electronics incorporated into the Green Bank interferometer system combined with the remote baselines produced a tremendously powerful instrument.

Green Bank Interferometer Research Programs

The Green Bank interferometer system, in its various stages of technical evolution, has carried out a wide variety of excellent research and made significant discoveries. The discoveries of radio stars, radio novae, and the highest accuracy proof of general relativity via observations of gravitational deflections from a star at radio frequencies are credited to the Green Bank interferometer. The interferometer has provided detailed mapping of radio sources, quasar measurements, pulsar timing and measurements of other pulsar characteristics, and precision position measurements. Much work in astrometry has been performed because of its high resolution. It has also been used to monitor the earth's rotation and polar motion, and was, for a while, the first and only radio telescope devoted to radio astrometry.[12]

A wide range of astronomical research is possible with the interferometer design. Research programs have ranged from statistical surveys of large numbers of objects to monitoring one or a few sources for long periods of time for measurements of weak or variable sources. One of the first programs performed with the interferometer was a statistical survey of celestial objects in order to determine which objects had small apparent diameters and which had large enough apparent diameters to be investigated using aperture synthesis techniques.[13]

Earth-Rotation Aperture Synthesis

Earth-rotation aperture synthesis investigations have represented a major research program conducted with the interferometer. Hydrogen-line aperture synthesis with the three-element version of the interferometer has provided rather detailed maps of extended galactic and extragalactic radio sources. Although not enough data sampling for extremely detailed maps like the VLA is produced by the instrument, good mapping is possible by moving the antennas several times. Four configurations of the movable telescopes produce interferometer separations every 300 meters out to

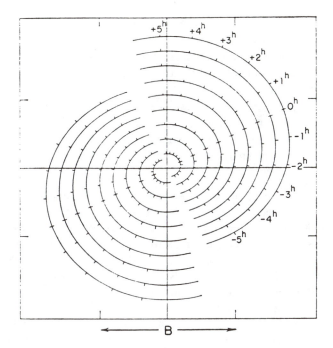

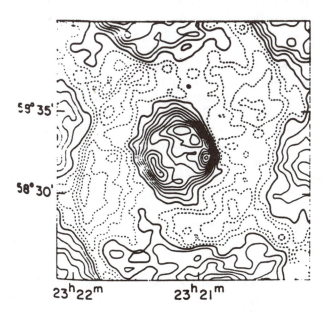

the coordinates u and v) is greatest at 68° declination and decreases with declination. Figure 6.6 shows the earth-rotation aperture synthesis coverage of the u-v plane with nine configurations of the three-element interferometer at 68°. Sufficient two dimensional coverage to synthesize images can be obtained at northern declinations greater than 20°. This synthesis technique has been used to produce two dimensional maps at high declinations and fan-beam surveys of sources at low declinations.[14] Hydrogen-line earth rotation aperture synthesis techniques with the interferometer have been used to make detailed maps of galactic sources, such as emission nebula and supernova remnants, and extragalactic sources such as radio galaxies. Figure 6.7 shows a hydrogen-line aperture synthesis map of Cass A, a supernova remnant. Figure 6.8 shows a hydrogen-line aperture synthesis map of Cygnus A, a distant radio galaxy.[15] Earth-rotation aperture synthesis mapping of galactic and extragalactic sources provided significant structural detail that lends insight into the nature of the cosmic source. Aperture synthesis techniques used with the Green Bank interferometer produced important research while providing the NRAO staff with valuable experience in interferometry techniques.

Figure 6.6 Earth-rotation aperture synthesis coverage of the *u-v* plane combining nine configurations of the three-element interferometer at 68° declination, which represents the highest spatial coverage. Sufficient coverage to synthesize images can be obtained at declinations greater than 20°. *Courtesy B. Clark and M. C. H. Wright.*

Radio Stars

Investigations with the Green Bank interferometer opened the door to stellar radio astronomy, which has subsequently evolved into an important research area. For the first 40 years of the science of radio astronomy, many radio-emitting cosmic objects were observed. Ordinary stars, however, had not been detected at radio frequencies. Normal stars (ones that are not protostellar nebulae or highly

the 2,700 meter maximum baseline. A 15-arcsecond synthesized beam with a grating sidelobe repetition at 2.4 arc minute radius is produced at 21 cm. Additional spacings can provide a total of nine configurations that can be obtained in nine telescope moves. The spatial coverage (as defined by

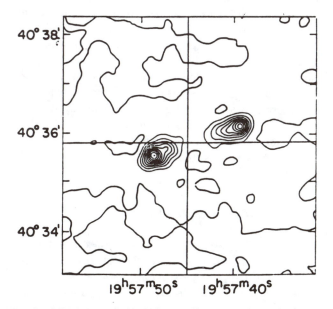

Figure 6.7 High resolution hydrogen-line aperture synthesis map of Cassiopeia A, supernova remnant made with the Green Bank three-element interferometer (M. C. H. Wright, B. G. Clark, C. H. Moore, and J. Coe). *Courtesy B. Clark and M. C. H. Wright.*

Figure 6.8 High resolution hydrogen-line aperture synthesis map of Cygnus A, distant radio galaxy made with the Green Bank three-element interferometer (M. C. H. Wright, B. G. Clark, C. H. Moore, and J. Coe). *Courtesy B. Clark and M. C. H. Wright.*

evolved exotic objects) had not been observed by radio telescopes because they emit only a very small percentage of
their total radiative output at radio frequencies. Normal,
main-sequence stars are thermal emitters with effective
temperatures ranging from 2,500 K to 20,000 K.[16] Following
thermal radiator curves, these objects emit only about one
millionth of their total output at radio frequencies. Radio-
emitting stars, therefore, had not been detected due to their
extremely weak radio emission. The discovery of radio-
emitting stars in 1970 is credited to R. M. Hjellming and C.
M. Wade using the Green Bank interferometer.[17]

Wade and Hjellming discovered radio stars in June of
1970 with the three-element interferometer shortly after it
was equipped to simultaneously operate at three wavelengths. Wade and Hjellming had theorized that the interferometer could detect stellar radio emission and that the
best candidates would be red supergiants. The first three
weeks of time with the newly improved interferometer were
awarded to Wade and Hjellming to search for stellar emissions. They put together a list of candidates of red supergiants whose emissions might be detectable. A few "wild
cards" including a nova were also included on the observing
list. A certain amount of serendipity was involved in the discovery as observations of one of these "wild cards," objects
that did not fit into the class of objects under investigation,
produced a discovery. Radio emission from one of the "wild
cards," Nova Herculis 1934, was detected in June 1970. Radio emission from two other nova, Nova Delphini 1967 and
Nova Serpentis 1970, was also detected at 3.7 cm and 11.1
cm wavelengths in June.[18]

During the same observing program, radio emission
from a red supergiant, Antares, was detected. Subsequent
observations of Antares with the Green Bank interferometer showed that the emission was actually coming from a
blue dwarf companion to the star.[19] Stellar radio emission,
nonetheless, had been detected from two distinct classes of
stars. Six classes of radio stars have subsequently been observed including: flare stars, red supergiants, blue dwarf
companions to a red supergiants, novae, X-ray stars, and
pulsars. Of these, the first detection of a nova, a blue dwarf
companion star, and an X-ray star not previously observed
at radio frequencies is credited to the Green Bank interferometer.[20]

Experimental Relativity

A classic experiment performed with the Green Bank interferometer in 1974 resulted in the highest precision verification of general relativity at radio frequencies. Einstein's
theory of general relativity had been experimentally verified by various methods beginning with the measurement
of the deflection of starlight by Sir Frank Dyson in 1919.[21]
Experimental relativity has since produced verifications of
general relativity through a variety of experiments including:[22]

1. measuring radar echoes from the planets to show
time-delay effects
2. measuring the parameters of binary pulsars and the
orbital deflections thereby verifying gravitational radiation
3. various laboratory measurements of the gravitational
redshifts, such as the slowing of the time of atomic
clocks accelerated on aircraft and rockets experimentally compared to the time of similar ground-based
atomic clocks.

Interferometric measurements of the bending of radio
waves as they passed near the sun were made beginning in
1969 that experimentally verified the theory of general relativity at radio frequencies. By the early 1970s, groups using
interferometers—at Caltech, Jet Propulsion Laboratory,
Westerbork, Cambridge, NRAO, and groups using VLBI
techniques between MIT and NRAO—had made successful measurements of the gravitational deflection of radio
source emission due to the sun's gravity well. The gravity
well of the sun is produced as the mass of the sun warps the
space-time continuum, in accordance with the theory of
general relativity.

E. B. Fomalont and R. A. Sramek devised an innovative
experiment using the Green Bank interferometer in 1974 to
measure the gravitational deflection of microwave radiation
with great accuracy to undeniably confirm the theory of
general relativity at radio frequencies. Fomalont and Sramek observed the position of a radio source $0116 + 08$, with
respect to two other sources $0119 + 11$ and $0111 + 02$, that
were farther away from the sun. The sources were observed
over the duration of a month while the sun moved through
their region of the sky. Fomalont and Sramek measured the
position of the middle source relative to the position of the
two sources farther out. The distance between the positions
of the two farthest sources was measured for calibration.
The three sources were alternately observed in the experiment which is the equivalent of stretching a ruler between
the sources. An illustration of the experiment is seen in Figure 6.9 which shows the apparent size and path of the sun
and the three radio sources observed.[23] Alternate observations of the three sources showed that the position of $0116 +
08$ increasingly shifted away from the sun as the sun ap-
proached. Microwave link interferometric observations that
included the remote instrument combined with the differential measurement strategy afforded a resolution high
enough to measure the deflection to an accuracy of
$1/1,000,000$ of the size of the apparent solar disk, providing
an accuracy of 1 percent.[24]

The precision of the Fomalont and Sramek measurements is compared to that of previous measurements in Figure 6.10.[25] The Fomalont and Sramek measurements provided the highest accuracy measurement of gravitational
deflection at radio frequencies which helped to verify Ein-
steinian general relativity over the Braun-Dickey theory
which had received much attention at the time. The gravita-

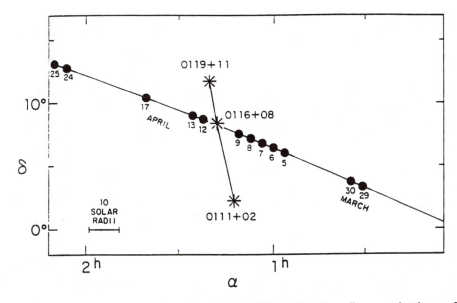

Figure 6.9 The Fomalont/Sramek experiment measuring the gravitational deflection of a radio source by the sun. The path and apparent size of the sun is indicated. The radio source 0116 + 08 exhibited gravitational deflection as the sun approached, providing the highest-accuracy confirmation of Einstein's theory of general relativity at radio frequencies. Measurements of the deflection were made with respect to the separational distance between the two outer sources to increase the accuracy. *Courtesy E. Fomalont and R. Sramek.*

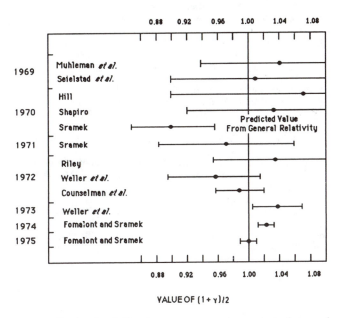

Figure 6.10 Radio deflection measurements as a test of general relativity.

tional deflection of radio waves experiment conducted by Fomalont and Sramek confirmed the theory of general relativity and demonstrated that radio astronomy instrumentation could provide an integral tool in the science of experimental relativity.

Pulsars, Quasars, and Precise Measurement

The Green Bank interferometer was and is currently used for research programs including pulsar measurements, observations of quasars, and precise measurements of other radio sources. The precise positions of quasars have been measured with the interferometer to an accuracy better than most optical telescopes. Quasars are extremely small in terms of their apparent size (due to their extreme distance) and are therefore primarily investigated using interferometers. Many pulsar measurement programs are undertaken using the interferometer by groups such as Princeton and the U.S. Naval Observatory. Binary pulsars have been extensively observed with the interferometer as well as characteristics of single pulsars such as intensity, pulse variation, and pulse dispersion. The U.S. Navy has a particular interest in pulsar timing as many scientists believe that pulsars represent a new generation of extremely accurate galaxy-wide clocks.

Astrometry, Geophysics, Timekeeping, and Operation for the U.S. Naval Observatory

The Green Bank interferometer began full-time operation for the U.S. Naval Observatory in 1978. The NRAO staff, for a time, operated the instrument for the U.S. Naval Observatory which exclusively devoted the instrument to astrometric measurements. The program continuously observes 15 extragalactic and several galactic radio sources that subtend an apparent angle of less than 0.01 arcsecond and therefore appear unresolved by the interferometer. Total sky coverage is obtained by choosing sources widely distributed about the northern hemisphere. The precise positions of these sources can be unambiguously measured as they appear unresolved by the instrument. The positions of the sources on the sky with respect to the axis of rotation of the earth are measured to determine the rate of rotation of the earth and the precession of its axis. The earth is found to rotate at various rates about an axis that periodically precesses. Annual oscillations in the position of the north celestial pole (NCP) arise because of the earth's spherical shape and from other variables such as geophysical and meteorological effects.[26] The variations in the rotation rate and the position of the NCP must be precisely known for navigational purposes. Navigation depends on fixed sources in the sky, which are affected by the earth's rotation. The U.S. Naval Observatory uses information obtained with the interferometer to provide daily and weekly corrections for navigational fix points.

Important astrometric research has been performed with the interferometer since 1978. Observations of quasars and radio galaxies have provided accurate coordinates for source position surveys and data bases. Recently, time on the interferometer has been devoted, once again, to a variety of research programs such as pulsar timing, quasar investigations, astrometry, and polar motion measurements. The Green Bank interferometer has proven to be an important, if not well-recognized, instrument in the science of radio astronomy. It has provided important information about a vast array of phenomena ranging from radio stars and quasars to information about general relativity and planet earth.

Endnotes

1. "The Interferometer System," *The Observer*, Vol. 3, no. 12, October 30, 1964, p. 4.
2. W. C. Tyler, D. E. Hogg, and C. M. Wade, "First Results with the National Radio Astronomy Observatory Interferometer," *American Institute of Physics for the American Astronomical Society Abstracts*, paper presented at the 118th meeting of the American Astronomical Society held 14–17 March 1965, at the University of Kentucky, Lexington, Kentucky.
3. John Ralston, "The 45-foot Portable Antenna," *The Observer*, vol. 13, no. 5, October 1972, p. 8.
4. "New 45-foot Portable Antenna," *The Observer*, vol. 13, no. 1, January 1972, p. 5.
5. John W. Findlay, "The National Radio Astronomy Observatory," *Sky and Telescope*, December 1974, pp. 353–356.
6. *Ibid.*, p. 353.
7. "National Radio Astronomy Observatory Timeline," Internal Publication for Open House 1983.
8. John W. Findlay, "The National Radio Astronomy Observatory," p. 353.
9. James R. Coe, "NRAO Interferometer Electronics," *Proceedings of the Institute of Electrical and Electronics Engineers*, vol. 61, no. 9, September 1973, pp. 1335–1339.
10. J. P. Basart, B. G. Clark, and J. S. Kramer, "A Phase Stable Interferometer of 100,000 Wavelengths Baseline," *Publications of the Astronomical Society of the Pacific*, vol. 80, no. 474, June 1968, pp. 273–279.
11. Bruce Balick, "First Fringes form Huntersville," *The Observer*, vol. 14, no. 8, August 1973, p. 8.
12. *National Radio Astronomy Observatory*, NRAO Brochure, NRAO/AUI, 1988, operated by Associated Universities Inc., under contract with the National Science Foundation.
13. Tony Distasio, "From the Interferometer," *The Observer*, vol. 4, no. 2 May 28, 1965, p. 5.
14. M. C. H. Wright, B. G. Clark, C. H. Moore, and J. Coe, "Hydrogen-line Aperture Synthesis at the National Radio Astronomy Observatory: Techniques and Data Reduction," *Radio Science*, vol. 8 nos. 8 and 9, August–September 1973, pp. 763–773.
15. *Ibid.*, pp. 763–773.
16. William K. Hartmann, *Astronomy, the Cosmic Journey*, 3rd edition, Wadsworth Publishing Company, Belmont, California, pp. 286–286.
17. C. M. Wade, "The Discovery of Radio Novae," *Serendipitous Discoveries in Radio Astronomy*, Proceedings of a workshop held at the National Radio Astronomy Observatory, Green Bank, West Virginia, on May 4, 5, and 6, 1983, K. Kellerman, ed., pp. 291–293.
18. R. M. Hjellming and C. M. Wade, "Radio Novae," *The Astrophysical Journal*, vol. 162, October 1970, pp. L1–L4.
19. R. M. Hjellming and C. M. Wade, "Detection of Radio Emission From Antares," *The Astrophysical Journal*, vol. 163, February 1971, pp. L105–106.
20. R. M. Hjellming and C. M. Wade, "Radio Stars," *Science*, vol. 173, no. 4002, September 1971, p. 1087–1092.
21. Clifford M. Will, "Testing General Relativity: 20 Years of Progress," *Sky and Telescope*, October 1983, pp. 294–299.
22. *Ibid.*, pp. 294–299.
23. E. B. Fomalont, "Testing the Theory of Relativity," *The Observer*, vol. 15, no. 6, December 1974, pp. 3–5.
24. E. B. Fomalont and R. A. Sramek, "A Confirmation of Einstein's General Theory of Relativity by Measuring the Bending of Microwave Radiation in the Gravitational Field of the Sun," *The Astrophysical Journal*, vol. 199, August 1975, pp. 749–755.
25. Clifford M. Will, "Testing General Relativity: 20 Years of Progress," pp. 294–299.
26. *National Radio Astronomy Observatory*, NRAO Brochure, NRAO/AUI, 1988.

Chapter 7

The NRAO and the Evolution of Radio Astronomy

The science of radio astronomy revolutionized astronomy. In the 50 years since its accidental beginning, it has answered many astronomical questions, revealed new, previously unsuspected phenomena, and raised new cosmological questions. The National Radio Astronomy Observatory has contributed significantly to this revolution since the early 1960s. In 1957 when the NRAO began in earnest, the science of radio astronomy was well established but still in the formative stages. Several factors drove the evolution of radio astronomy into "big" science. First, the rapid development of receiver and instrumentation technologies during and after World War II led to significant discoveries. Second, the detection of Sputnik by the British with the Jodrell Bank telescope provided impetus to the science. The capability to detect signals from earth-orbiting satellites appealed to developers of radar and radio astronomy technology. The important discoveries being made in Great Britain, Australia, and the Netherlands had a direct impact on the establishment of the NRAO. International competition became a major sociopolitical factor in the establishment of a U.S. observatory for radio astronomers. The establishment of the National Science Foundation in 1950 showed that the federal government accepted a major role in the funding of large-scale scientific endeavors. The establishment of the NRAO represented a commitment of the NSF to support large-scale research. The scale of research at the NRAO, from the planning to the construction and development of the major research instruments, reflects the evolution of radio astronomy into "big" science in the United States.

The revolution in radio astronomy was (and still is) reflected in the growth of the NRAO. The observatory and the science developed a symbiotic relationship in the sense that many major contributions were made at the NRAO that advanced the science which, in turn, increased the needs of the nation's radio astronomers. Growth outward from Green Bank began in 1965 and continues to occur. In December 1965, the NRAO headquarters, including most of the scientific staff, the computer division, the main library, and administrative and some electronics personnel moved to the Charlottesville laboratory.[1] In January 1968, the NRAO 36-foot millimeter-wavelength telescope began operations at the Kitt Peak site near Tucson, Arizona.[2] The 36-foot telescope made many contributions to the field of molecular radio astronomy. The most ambitious interferometer in the world, the National Radio Astronomy Observatory VLA, came on-line in 1980.[3] The VLA, in many aspects, represents the world's premier facility for research in radio astronomy. Continuing in the spirit of the national observatory, the NRAO is currently planning and constructing two major, state-of-the-art instruments for the 1990s and beyond. The VLBA, a recently completed continental array of permanent 25-meter antennas and the GBT, a 100-meter high-precision telescope designed to replace the 300-foot, represent the next generation of major NRAO instruments. The NRAO will continue to provide state-of-the art instrumentation for the future with the VLA, VLBA, and the GBT.

The NRAO 36-Foot Telescope

The NRAO 36-foot (11-meter) telescope had an enormous impact on astronomy. Its legacy includes the discovery of a large number of molecules, more than any other telescope in the world (refer to Table 4.2). Discoveries made with it changed the way astronomers understand the processes of star formation. Star formation was almost exclusively a theoretical science before the 36-foot telescope. Observations of molecular clouds and the interstellar medium led astronomers to revise theories about how stars are formed from interstellar gas and dust. The discovery of complex molecules was unexpected because of the environment of interstellar space. The discovery of organic molecules was even more bizarre. Astronomers have since found that organic molecules are abundant in the interstellar medium. Complex organic molecules are involved in all life on earth. The discovery of the same molecular structures that are essential to the origin and evolution of life has had a tremendous impact on astronomy. The NRAO 36-foot, serendipitously facilitated the inception and development of new astronomical sciences including astrochemistry, observational star formation, new techniques in observations of galactic structure, and star formation in other galaxies.

The 36-foot radio telescope is historically important for another reason. It is possibly one of the last risky ventures funded by the NSF. Prior to 1968 only a few molecules had been detected at microwave or any other frequencies. There was little evidence, either theoretical or observational, that millimeter wavelength (mmwave) observations would be useful. The fact that the project was funded reflects the level of federal support for scientific research at the time.

The 36-foot had its beginning in the early 1960s at Green Bank. In the early days of the NRAO, Frank Drake believed that the observatory should offer an instrument suitable for mmwave observations. After learning about the development of a new germanium bolometer that could detect mmwaves, Drake brought its inventor, Frank Low, to Green Bank in 1960 to develop a sensitive mmwave receiver. High atmospheric water vapor at Green Bank made it difficult for Drake and Low to make observations. Drake, Low, and Heeschen, the NRAO director, discussed the value of developing a large mmwave telescope to be located somewhere in the Southwest where the drier air is more suited to mmwave astronomy. With the consent of the AUI board, they appended the NSF annual request, asking for $1 million to build the instrument. Drake based the cost on the 85-foot telescope, using the Diameter$^{3.2}$ Frequency$^{0.5}$ scaling law in vogue at the time. The NSF, with no guarantee that the instrument would ever be useful, granted the funds, and the project got underway.[4]

The story of the design is intriguing. The telescope began as a risky venture with no guarantees of success and although it never met its design specifications, it became one of the most productive telescopes in the world. During the early design stage, Frank Low learned that a naval shipyard in Norfolk, Virginia, had a precision mill capable of cutting highly precise figures as large as 36 feet in diameter. Low verified that the shipyard could cut a 36-foot diameter, segmented surface telescope to an rms of 0.001 inch (0.002 centimeter), exactly that required to reflect radiation efficiently at 300 GHz, the desired operating frequency of the instrument. John Findlay, the observatory deputy director, believed a high-tech company would be more capable of machining such a highly precise paraboloid and sought out an aerospace company to design and build the new telescope. The Rohr Corporation of San Diego was awarded the contract. Findlay and the Rohr Corporation completed the design and work began in 1966. Rohr Corporation tried to machine the reflecting surface but could not meet the specifications.[5] The result was a completed "mmwave telescope" in place at Kitt Peak, Arizona, that could not meet the design goal of 300 GHz. The first observations, nonetheless, took place in October 1968.

By this time most of the principal players had abandoned interest in the project. A few NRAO scientists used the tele-

Figure 7.1   The NRAO 36-foot (11-meter) millimeter wavelength telescope at Kitt Peak, Arizona. *Courtesy NRAO/AUI.*

scope to observe the planets but generally there was little interest among the astronomical community in the use of the telescope. Frank Low, frustrated with delays in the instrument's completion, left and joined the staff of the University of Arizona where he almost singlehandedly developed modern infrared astronomy.[6] Then, in 1970, Penzias and Wilson detected the carbon monoxide molecule with the instrument. This detection, along with the nearly simultaneous detections of ammonia and water molecules by astronomers at the University of California, Berkeley, revived interest in the telescope. For the next decade, it became one of the most desired of all telescopes. Astronomers through the 1970s used the instrument to discover an impressive list of molecules, far more than any other telescope in the world.

The telescope consists of an altitude-azimuth mounted solid aluminum reflector enclosed in a 29-meter (95-foot) diameter weather-protecting dome (Figure 7.1). Observations are made through a 12-meter opening which may be covered with a protective shutter. The dome rotates with the telescope and permits unobstructed views through the opening. The original solid surface reflector, machined to a parabolic form in one piece, was precise to 0.1 mm. The surface was replaced by a slightly larger, higher precision paraboloid in 1983 and became the NRAO 12-meter telescope. Actually, everything above the elevation axis was replaced. The superstructure was replaced by a truss of steel tubes cantilevered from a central hub made of steel plate based on the VLA design. The design proved far more rigid than the original instrument. An rms of 100 microns was achieved.[7] This improved surface precision combined with new bolometric receivers provided an extremely powerful continuum system at the 1-, 2- and 3-mm atmospheric bands.

The contributions of the 36-foot telescope to astronomy and technology are among the NRAO's greatest achievements. Even with its original limitations the instrument influenced the development of mmwave telescopes in other countries, including China. Technological advances fostered with the 36-foot include the design of highly sensitive mmwave receivers and the creation of a new computer language, FORTH, that is now used extensively by major industries and corporations including Federal Express.[8] The history of the instrument represents a bold venture in scientific instrumentation that paid off. Many astronomers believe its contributions rival that of the VLA.

The VLA

The Very Large Array (VLA) is undisputedly the world's most powerful and advanced aperture synthesis radio telescope. Planning for an ambitious high-resolution earth-rotation aperture synthesis array began at the NRAO in 1961. Preliminary design studies for the VLA continued through the 1960s. Many of the technical concepts developed for the system were incorporated into the Green Bank interferometer, the VLA prototype. In 1972 Congress funded the NRAO VLA project and the preliminary detailed designs and site acquisition began. The design consisted of 25-meter (82-foot) azimuth-elevation mounted antennas—27 of them—arranged in a Y-configuration spread over a 32-kilometer (20-mile) area. Site construction on the plains of San Augustin, near Socorro, New Mexico, began in 1974. The site was chosen because it is the center of a high, flat plain (7,000 feet in altitude) with a desert climate, ideal for aperture synthesis at higher frequencies because of low atmospheric water vapor. The first antenna was assembled on-site in 1975. Six antennas were assembled by 1977 when the first astronomical observations began. The assembly of all 27 antennas was complete and dedication ceremonies were held on October 10, 1980. The VLA was completed on schedule and essentially within the original budget at a total cost of $77.6 million.[9] Several of the 27 antennas of the VLA are seen in Figure 7.2.

The 27 identical antennas are located on a track system consisting of dual railroad tracks arranged in a Y-configuration. Each antenna can be moved along the system for aperture synthesis imaging. Four telescope configurations, each

Figure 7.2 The 25-meter antennas of the NRAO Very Large Array on the plains of San Agustin, New Mexico. Antennas that comprise the VLA are visible pointing toward the zenith. *Photograph by B. Malphrus.*

with 9 telescopes on each 13-mile arm, are possible. Hybrid configurations that represent various combinations of these four arrays are also available. The configurations range from the tightest in which all antennas are placed within 0.4 mile (0.65 kilometer) of the center to the widest configuration utilizing the entire 13-mile (21-kilometer) arm lengths. Increasing resolution is provided as the configuration widens, and increased sensitivity is gained as the array tightens. The highest attainable resolution at 6 cm is 0.5 arcsecond.[10] The antennas are physically connected by tubular waveguides which direct the signals from each antenna to a common back-end autocorrelation receiver.

The 28 telescopes (including one extra that is cycled as repairs are needed) consist of azimuth-elevation mounted solid-surface parabolic dishes. The telescope design is a Cassegranian reflector, with a subreflector and a vertex cabin which contains the front-end receiver. Each antenna weighs 235 tons, 100 tons of which comprises the moving weight. Tracking within an accuracy of 15 arcseconds is achieved. The aluminum panels are accurate to 0.5 millimeter which allows operation down to 1 centimeter. Standard NRAO receivers operating at primary wavelengths of 20 cm, 6 cm, 2 cm, and 1.3 cm were initially installed. Receivers operating at 90 cm and 3.5 cm were later added. All receivers are cryogenically cooled to 18 K.[11]

The autocorrelator adds time delays to account for the differing travel times of signals from each antenna due to the varying distances. The autocorrelator system operates in either a continuum or spectral line mode. The correlated data is sent to a computer for calibration. The signals are combined to produce a data base. Long integration times generate copious amounts of data, up to 10 million data points every minute. Standard sources are typically observed every 30 minutes for calibration. An image processing system is then used to edit and reduce the data and to create images via a Fourier transform and other processes. The images are processed and stored as pixel information with each pixel representing location and radio intensity; specifically each data point in the *u-v* plane is produced by

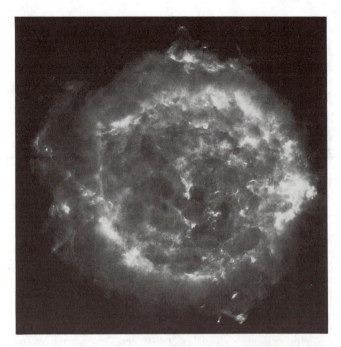

Figure 7.3 Cassiopeia A SNR, VLA observations made in 1983 by P. E. Angerhofer, R. Braun, S. F. Gull, R. A. Perley, and R. J. Tuffs reveal detailed structure in the expanding shell of this SNR that went supernova in A.D. 1680. This radio image of Cass A at 6 cm with a resolution of 0.2 × 0.2 arcsecond, required four million picture elements to construct, making it the most detailed radio image ever constructed. *Courtesy R. Braun, R. A. Perley, and NRAO/AUI.*

one antenna pair for one integration. Images are enhanced to compensate for atmospheric fluctuation for array distortion, an effect caused by the geometry of the combined configurations used to synthesize the image. Computer algorithms to enhance the images by other processes including deconvolution of the beam pattern, self-calibration, normalization of antenna performance, and numerous other routines have recently been developed. These algorithms and processes ultimately evolved into a complex and powerful image processing system called AIPS—Astronomical Image Processing System.[12] The AIPS program is an effective and versatile image processing system utilized with VLA, VLBA, and VLBI observations. The system has greatly enhanced the performance and versatility of the VLA far beyond the original expectations.

Positioning the array and choice of center frequency, bandwidth, integration time, and other parameters are selected through the on-line computer that controls the array. The VLA was initially operated with five Modcomp II minicomputers working in tandem, referred to as the synchronous computer system. Additions were made to produce a specialized distributed processing network. A DEC-10 acted as the interactive device with peripheral processor systems used for specific functions. A PDP-11/70-FPS AP120B system, for example, was installed to perform map making and related operations.[13] Continual upgrades have

been incorporated into the computing systems and into the receiver electronics as well. These techniques and developments have allowed astronomers to produce the highest quality, most detailed radio images of cosmic phenomena ever obtained.

The VLA has been used to investigate a wide range of cosmic phenomena ranging from the sun and planets, to emission nebula, supernova remnants, and the nucleus of the Milky Way, to radio galaxies, quasars, and gravitational lenses. Tremendous structural detail has been revealed in these objects. VLA observations of Cass A made in 1983 by P. E. Angerhofer, R. Braun, S. F. Gull, R. A. Perley, and R. J. Tuffs reveal detailed structure in the expanding shell of this SNR that went supernova in A.D. 1680 (Figure 7.3). This radio image of Cass A at 6 cm with a resolution of 0.2 × 0.2 arcsecond required four million picture elements, making it the most detailed radio image ever constructed.[14] Observations of extragalactic objects such as Cygnus A, radio galaxy (Figure 7.4) made by R. A. Perley, J. W. Dreher, and J. J Cowan, have revealed equivalent structural detail.[15] The bipolar jet structure of Cygnus A is a common structure of intense radio galaxies. The central engine in the galactic nucleus occupies a space roughly the size of the solar system while the jets extend outward for millions of light-years. An unprecedented view of the galactic nucleus revealing unknown structures is another major contribution of the VLA. The radio-emitting region, Sagittarius A (Figure 7.5), has been found to correspond to the galactic nucleus. The image reveals filamentary arcs and a ring-like core which contains a spiral structure which corresponds to the highest intensity radio-emitting region.[16] The structure unambiguously corresponds to the galactic nucleus although the nature of its engine as well as the nature of the arcs is unknown.

Investigations with the VLA have revealed many bizaare phenomena, which have collectively become known as the "extragalactic zoo" because of their unusual and unimaginably energetic natures. The compact nuclei of radio galaxies, for example, are found to radiate as much as 10^{39} watts, or 100 million times the output of a normal galaxy from areas only a few light-years across.[17] Among the most unusual of these energetic phenomena is MG1131 + 0456, an Einstein ring, a projection of a quasar showing a double quasar image immersed in a ring structure caused by gravitational lensing.[18] The VLA has been used for a variety of scientific programs from investigation of distant cosmic phenomena to spacecraft tracking and telemetry of the NASA Voyager 2 space probe during the 1989 Neptune encounter. Observations with the VLA have revealed some of the most beautiful and most complex images of the universe yet obtained.

The VLBA

The needs for increasingly higher resolution led to the development of very long baseline interferometry (VLBI) in 1967. Many extragalactic phenomena remained unre-

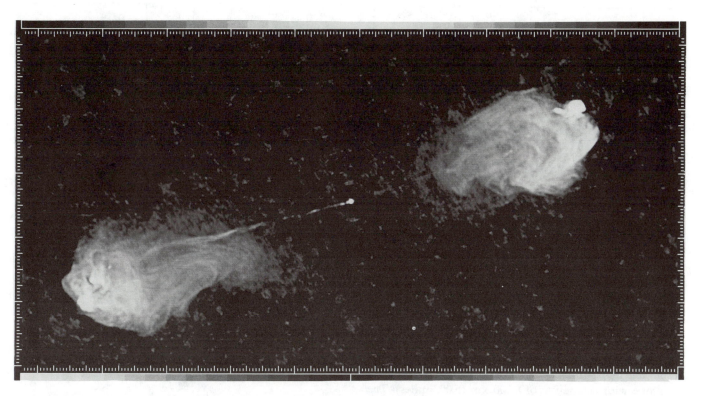

Figure 7.4 Cygnus A, radio galaxy, VLA observations made in 1983 by P. A. G. Scheuer, R. A. Laing, and R. A. Perley. The image of Cygnus A was made at 20 cm with a resolution of 1.25 × 1.25 arcseconds. The field of view is approximately 2 × 1 arcminute.. Detailed source structure in the bipolar jets is visible. *Courtesy P. A. G. Scheuer, R. A. Perley, and NRAO/AUI.*

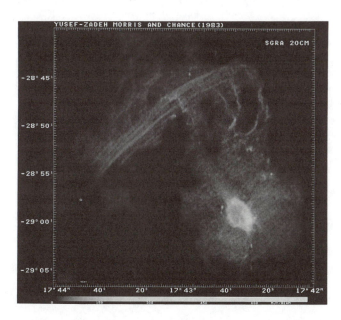

Figure 7.5 Sagittarius A, the galactic nucleus, VLA observations made in 1982–1983 by F. Yusef-Zadeh, M. R. Morris, and D. R. Chance. The image of the galactic nucleus was made at 20 cm with a resolution of 5 × 8 arcseconds. The continuum arc near the galactic center, filamentary structure, and an 8-arcminute halo surrounding the Sagittarius A East shell are visible. *Courtesy NRAO/AUI.*

solved with physically connected interferometer systems. The nucleus of radio galaxies, for example, typically subtend an angle of 0.001 arcsecond, a factor of 100 to 1,000 beyond the resolution limit of the VLA.[19] Intercontinental VLBI measurements have been extremely successful in achieving resolutions on this order of magnitude. In 1975 the NRAO began planning a permanent array of 10 telescopes distributed across the United States that would be exclusively devoted to VLBI measurements. VLBI techniques were often difficult because of individual differences between the various antennas used. The Very Long Baseline Array (VLBA) high resolution synthesis radio telescope was funded in 1982. Construction of the antennas began in 1985.[20] The Array Operations Center (AOC), located on the campus of New Mexico Technical College in Socorro, was dedicated and occupied in 1987. The VLA and VLBA now share essentially the same computing facilities. The completed VLBA went on-line in May 1993.

The VLBA consists of 10 permanent VLBI stations that utilize 25-meter antennas located throughout the United States from Hawaii to the U.S. Virgin Islands. Each antenna is equipped with low-noise receivers operating throughout the wavelength range from 1 cm to 1 m. Each station has a hydrogen-maser frequency standard for time and frequency referencing. The signals are recorded on broadband digital tape at each station and simultaneously replayed and corre-

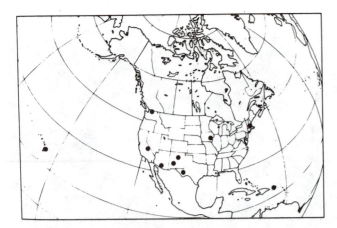

Figure 7.6 Geographic distribution of the 10 25-meter antennas of the VLBA, showing a baseline of 8,000 km (5,000 miles), nearly equal to the diameter of earth. *With permission of Krieger Publishing.*

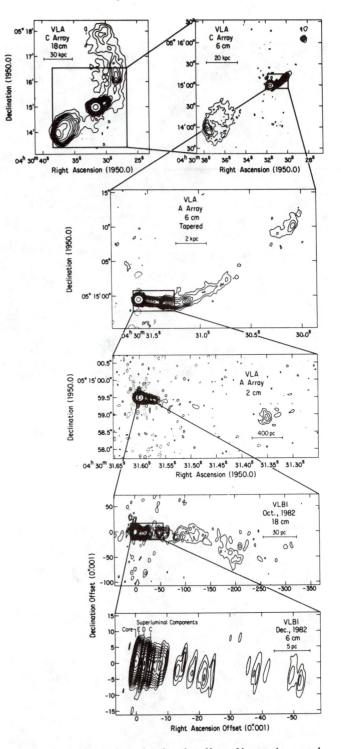

Figure 7.7 Observations showing the effect of increasing angular resolution made with the VLA and VLBI techniques providing resolution similar to that of the NRAO, VLBA, (R. C. Walker, J. M. Benson, and S. C. Unwin). *Courtesy NRAO/AUI.*

lated by a digital correlator at the AOC. Images are constructed from the correlated signals via the Fourier transformation. The VLBA can produce images with an angular resolution of 0.001 arcsecond.[21] The VLBA represents a telescope with a baseline of 8,000 km (5,000 miles). The geographical distribution of the 10 antennas is shown in Figure 7.6. This baseline is nearly equal to the diameter of earth and therefore represents the largest baseline and the highest possible resolution achievable on earth.

The VLBA provides extremely high resolution images of galactic and extragalactic objects that remained unresolved in all previous investigations. The VLBA also provides unrivaled resolution of detail and motion in distant radio sources. Figure 7.7 shows the effect of increasing angular resolution in observations of an extragalactic source 3C 120.[22] The image is zoomed-in on using resolutions provided by the VLA in the first four images and by VLBI techniques in the last two images that are comparable to the VLBA resolution. The extreme resolution of the VLBA will allow precise astrometric measurements that will provide direct measurement of continental drift, the earth's rate of rotation, the wobble of its axis, and even the effect of wind on the rotation of the earth. The range of frequencies and resolution of the various NRAO instruments is tremendous. The VLBA and VLA together will provide coverage of angular scales that range over a factor of more than 100,000.[23] The VLA and VLBA represent a new generation of earth-rotation aperture synthesis arrays with a new level of resolution and sensitivity. Even so, many cosmic objects will still remain unresolved.

The GBT

The need for a large, filled-aperture telescope in the United States is critical due to the collapse of the NRAO 300-foot telescope. Interferometers and arrays afford high resolution imaging and sensitivity, but there are many research programs for which a large, filled-aperture telescope is far more desirable. Large, filled-aperture telescopes tend to be more versatile than arrays which are primarily used to provide high-resolution views of objects or areas of small

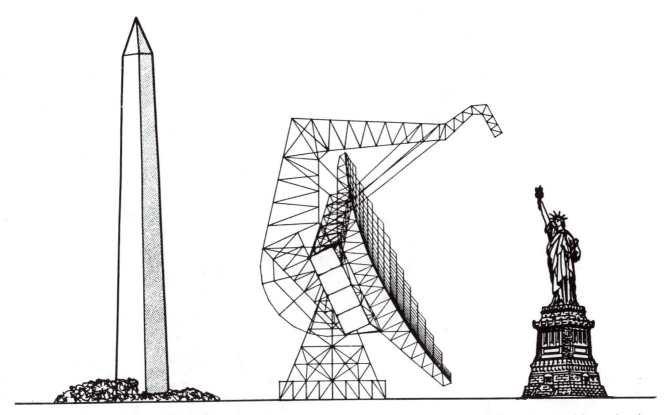

Figure 7.8 The GBT, scheduled for completion in 1996, will have an aperture of 100 meters and provide surface precision and optics systems to allow high-precision observations over a broad wavelength range of 1 meter to 3 millimeters. Its enormous scale is pictured in comparison with other national landmarks.

angular extent. A large, fully steerable filled-aperture telescope is ideal for surveys, mapping (of extended HI, HII, and other regions), molecular spectroscopy, pulsar timing and other variable phenomena measurements, and observations of weak stellar and extragalactic sources. A single, filled-aperture telescope can respond more quickly to new scientific demands than an array.[24]

A replacement instrument for the NRAO 300-foot telescope has been designed and planned by the NRAO scientific, engineering, and electronics staffs. In July 1989, Congress agreed to provide $75 million to fund the design and construction of the Green Bank Telescope (GBT), which will be the largest fully steerable filled-aperture telescope in the world.[25] The GBT, scheduled for completion in 1996, will have an aperture of 100 meters and provide surface precision and optics systems to allow high-precision observations over a broad wavelength range of 1 meter to 3 millimeters. Its enormous scale is pictured in comparison with other national landmarks in Figure 7.8. The instrument will stand approximately 500 feet tall; its reflecting surface is enormous (2.3 acres or 5,780 square meters). The GBT is a radical unblocked aperture design, which consists of a portion of a large paraboloid alt-azimuth mounted with a long cantilever arm extending from underneath to support the subreflector and prime focus feed (Figure 7.9). The cantilever arm is supported by two secondary feed support structures. The radio waves are focused to the side so that aperture blockage will be minimized allowing high beam efficiency and low sidelobe levels. The offset geometry of the dish has a much shallower curvature than a symmetrical antenna. The reflector geometry is actually a portion of a much larger parent parabola, designed to move the focus to the side of the dish (Figure 7.10). Sidelobes and extraneous radiation patterns resulting from aperture blockage by feed legs and other support structures are greatly reduced by this design (Figure 7.11). The aperture efficiency—a performance characteristic that partly determines the sensitivity of a radio telescope—is inherently high in an unblocked aperture. A focal ratio of 0.6 allows satisfactory polarization performance and illumination at the prime focus as well as the Cassegrain focus. Both types of observing will be made possible by incorporating a removable hyperbolic subreflector that will be deformable and nutating. The instrument will incorporate 2,620 actively controlled surface panels to compensate for gravitational deformations and thereby support high-frequency observations (Figure 7.12). These panels will be individually adjusted by actuators to conform to a predetermined shape (Figure 7.13). An elaborate laser ranging system will be employed to measure the deformations of the panels for adjustment (Figure 7.14). Ground-based laser range finders will precisely measure

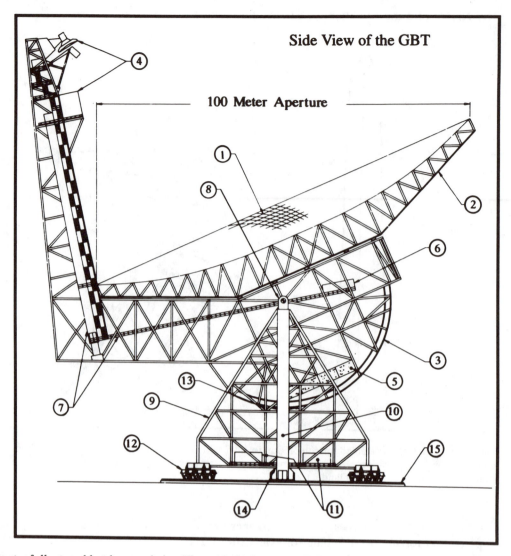

1. Primary Reflector Surface
2. Reflector Support Structure
3. Elevation Wheel
4. Secondary Reflector Receiver Room
5. Counterweight
6. Active Surface Control Room
7. Access Way to Focal Point
8. Elevation Bearing
9. Alidade
10. Elevator
11. Equipment Rooms
12. Azimuth Trucks and Drives
13. Elevation Drives
14. Pintle Bearing
15. Azimuth Track

Side View of the GBT

100 Meter Aperture

Figure 7.9 The NRAO GBT 100-meter fully steerable telescope design. The unblocked aperture design will greatly improve the aperture efficiency. The GBT will be the largest fully steerable telescope in the world and the only large unblocked aperture telescope ever constructed.

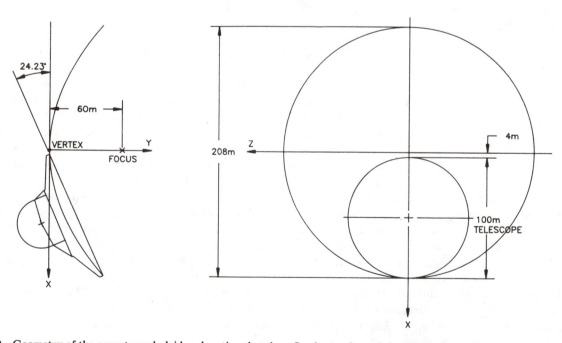

Figure 7.10 Geometry of the parent paraboloid and section that the reflecting surface of the GBT will replicate.

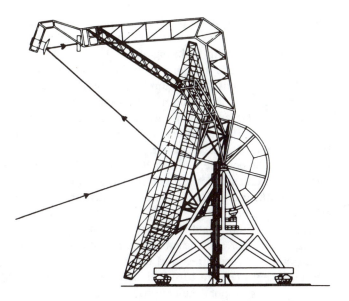

Figure 7.11 Radio frequency ray diagram showing the focusing of radio waves by the reflecting surface of the GBT.

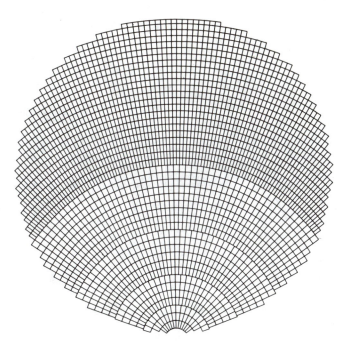

Figure 7.12 Surface panel configuration of the 2,620 panels comprising the reflecting surface.

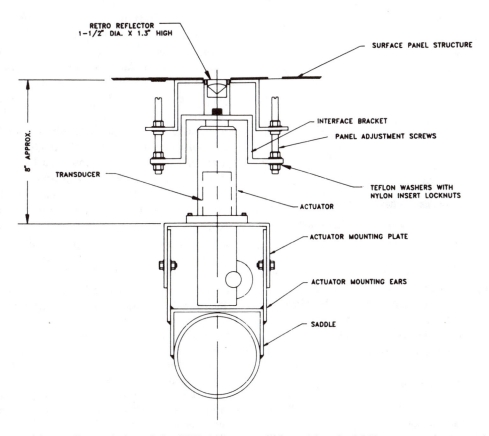

Figure 7.13 Acutator with retroflector design of the GBT. Actuators will be used at the 2,844 support points to actively reposition the surface panels to conformed to the predetermined shape.

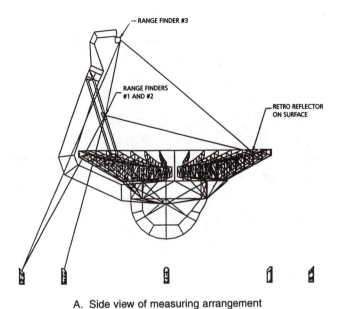

A. Side view of measuring arrangement

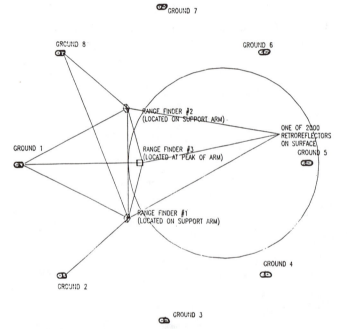

B. Top view of measuring arrangement (telescope structure omitted)

Figure 7.14 Laser ranging system employed to measure the deformations of the panels for adjustment.

distances to pick points on the surface in various positions to determine the gravitational and thermal deformations. The necessary surface panel adjustments will be based on these measurements. The superstructure will use conventional hoop-and-rib geometry, with the center oriented at the middle of the reflector. The hoops will be elliptical, rather than circular as in symmetrical antennas, because of the unusual shape of the reflector.[26]

The GBT performance characteristics are impressive. The geometric aperture blockage will be less than 3 percent, less than half that of conventional designs. The high

surface accuracy will be provided by panels adjusted to within 0.1 mm, allowing operations down to wavelengths of 3 mm. The beamwidth will range from 9 arcminutes at 21 cm to 18 arcseconds at 7 mm. Pointing accuracy will be roughly 2 arcseconds at optimum frequencies. System noise temperatures of 15 K should be achievable at a frequency range of 1–10 GHz. The sky coverage will include 85 percent of the celestial sphere including two-thirds of the galactic lane and the galactic center. Tracking from horizon to horizon will be possible, allowing long integration times for weak and distant sources.[27] Rapid instrumentation changes will add to the versatility of the instrument. Many of the receivers will be permanently mounted in a receiver housing affixed to a large turntable that may be rotated into the focal plane (Figure 7.15). The subreflector mounted above the focal point will then direct and focus the radiation to the selected receiver (Figure 7.16). This configuration will allow astronomers to respond quickly to weather conditions favorable to short wavelength observations and to fast-breaking astronomical discoveries like supernova. The large, high-precision surface, increased aperture efficiency, low-noise receivers, and rotating receiver design of the GBT will produce an enormously powerful and versatile instrument.

The GBT research programs will include investigations similar to the 300-foot telescope and other large, filled-aperture instruments. The GBT will be used for continuum and line emission surveys, mapping of the spatial distribution of extended HI, HII, and other regions, measurements of magnetic fields in interstellar clouds, molecular spectroscopy, masers, pulsar timing and variable phenomena measurements, and observations of stellar and extragalactic sources including radio galaxies and quasars. The GBT will also be used in conjunction with the VLBA to improve the system performance of the VLBA. The GBT will enhance the sensitivity and dynamic range in high resolution VLBA images of sources of small angular extent. Interestingly, the GBT will have more collecting area than the combined area of all 10 25-meter VLBA antennas. Depending on the type of observation, the VLBA will have two to four times the sensitivity when the GBT is included into the system.[28] The GBT will also allow high-quality millimeter VLBI observations. Uses for the GBT will likely include planetary radar investigations and spacecraft tracking. As is historically the case with cutting edge instruments in radio astronomy, the scientific uses of the GBT can only partially be foreseen. This notion is well stated in the introduction to the scientific programs section of the GBT proposal:

The history of radio astronomy suggests that we can anticipate only a fraction of of the most exciting science that will emerge from the Green Bank Telescope—most recent programs on the 140-foot and 300-foot telescopes at Green Bank observed astronomical objects (pulsars, quasars) or phenomena (radio recombination lines, astrophysical masers) unknown when those telescopes were proposed and required equipment (low-noise solid-state amplifiers, autocorrelation spectrometers, fast data-taking computers,

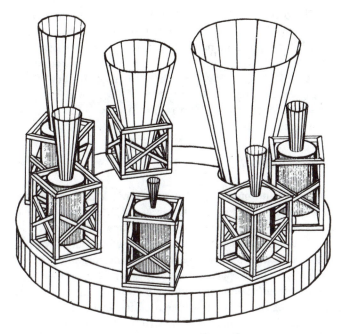

Figure 7.15 Feed room showing permanently mounted feed horns and turntable that may be rotated into the focal plane.

VLBI recorders, maser time and frequency standards) unavailable when those telescopes were constructed.[29]

Construction of the GBT began in the spring of 1991. During that year, 4,500 cubic yards of reinforced concrete were poured for the foundation of the telescope, indicating the extent of the scale. When the telescope is completed in 1996 it will be unique in the world. The design features and expected performance characteristics will make the GBT extremely accurate, sensitive, and adaptable. The GBT should prove to be the world's leading filled-aperture telescope well into the 21st century.

Evolution of Technology, Instrumentation, and Understanding

The continual development of technology and instrumentation in radio astronomy has played a major role in the evolution of the science. The development of more sensitive and versatile instruments has allowed astronomers to detect weaker signals, and therefore see farther out into space and farther back into time. The NRAO has made major contributions to the technology development. The resulting increases in sensitivity and resolution of the instruments as well as in their aperture efficiency and other characteristics have produced cutting edge instrumentation for radio astronomy investigations. The increasing resolution of instrumentation at the NRAO is given in Figure 7.17.[30] Increases in receiver sensitivity follow the evolution of noise temperature in NRAO receivers given in Chapter 5, Figure 5.24. The tremendous gains in angular resolution and receiver sensitivity have been the most critical factors that

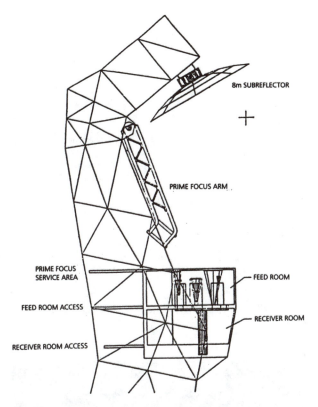

Figure 7.16 Subreflector and receiver room assemblies.

have advanced the evolution of radio astronomy instrumentation at the NRAO.

The cutting edge instrumentation has allowed frontier research from which major contributions logically followed. The major contributions have been outlined in the preceding chapters. The most significant contributions of NRAO research programs include:

1. investigations of the HI distribution, size, and kinematics of galaxies
2. implications of the missing matter
3. the discovery of galactic bridges and interactions
4. the discovery of radio stars
5. the pulsar association and pulsar timing
6. the discovery of numerous molecules in the interstellar medium
7. high-resolution images of galactic and extragalactic sources
8. surveys
9. astrometric measurements
10. measurements that verify general relativity theory
11. measurements of the earth's polar motion, rotation, and continental drift
12. a new view of our own galaxy's structure and nucleus
13. investigations of quasars and radio galaxies

Although it is the nature of contributions of research per-

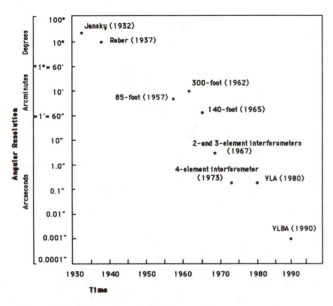

Figure 7.17 Increasing resolution of instrumentation at the NRAO.

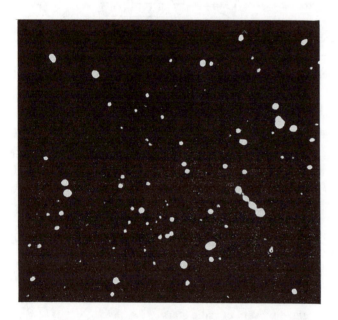

Figure 7.18 VLA image of a typical area of sky containing completely unidentified faint extragalactic radio sources (J.J. Condon). The vertical size of the VLA image is 20 arcseconds. The potential to discover faint radio sources is demonstrated. *Courtesy J. J. Condon.*

formed at the NRAO to be primarily statistical, fundamental discoveries of new phenomena have also occurred. A VLA image of faint extragalactic radio sources in a small area of sky demonstrates the ability of the new generation instruments to detect new sources and discover new phenomena (Figure 7.18). Most of the sources in this image were previously undetected and therefore previously unidentified.

The primary goal of the NRAO is to providing cutting edge instrumentation for research in radio astronomy with which these phenomena may be investigated. The contributions of the NRAO to the science of radio astronomy are historically reflected by the achievement of this goal. Technological developments at the NRAO have essentially defined the state of the art in radio astronomy instrumentation since the early 1960s.

Speculation

The resolution achieved with current VLBI techniques represents the highest resolution attainable on earth. Baselines between intercontinental telescopes have been routinely used that are effectively the size of the earth's diameter. The next step in radio astronomy instrumentation will logically be into space. Space arrays will connect orbiting antennas with telescopes on earth such as the GBT, VLA, and VLBA into extremely long baseline space arrays. A tremendous increase in resolution will be provided with earth-space baselines. The NRAO scientific staff has exhibited great insight in foreseeing this possibility. The VLBA elements are designed to be compatible with orbiting antennas for incorporation into an earth-orbital space array.[31]

Eventually, even larger baselines will be desired that are greater than earth-orbital distances. The next generation of space arrays may eventually achieve a baseline distance equivalent to the diameter of the earth's orbit around the sun. Such an interorbital baseline may be achieved by leaving two stationed antennas at diametric points in the earth's orbit and controlling them via radio link similar to the NASA space probes. A baseline of 300 million kilometers (186 million miles), twice the distance of the earth's average separation from the sun, could be achieved that would provide an incredible resolution of 8.4×10^{-13} arcseconds at 21 cm (based on the resolution equation given in Chapter 2). Such an interorbital baseline would provide resolution more than a billion times greater than the highest resolution attainable with earthbound VLBI systems.[32] The NRAO is currently participating in an international venture that will place a radio telescope into earth orbit to work with the GBT in creating such an earth-space interferometer. Russia plans to launch an orbiting instrument in 1996. The resulting interferometer, called RADIOASTRON, will simulate a telescope with a diameter equal to the orbital radius of the space based radio telescope, achieving unprecedented resolution. Japanese astronomers are working on a similar project, VSOP (VLBI Space Observing Program) which is scheduled for launch in September 1996. It may also be possible to incorporate multiple antennas along the ecliptic plane to synthesize an instrument at the periphery of the earth's orbital path. The potential for space arrays is tremendous considering that the technology essentially exists now to produce such an instrument.

Figure 7.19 Parallel universes—View of the radio sky as seen with the NRAO 300-foot radio telescope (J. J. Condon). This image was produced from data collected with the 300-foot telescope. The power of the image in revealing an invisible yet parallel universe is a posthumous tribute to the science performed with the instrument. *Courtesy J. J. Condon*. See Frontispiece.

The VLA, VLBA, and VLBI systems currently represent the largest operating scientific instruments. Only particle accelerators for high energy physics approach the scale of these instruments. These systems are also among the most sensitive of scientific instruments. Space arrays represent the largest scale conceived for scientific instruments. A new level of resolution and therefore insight will inevitably be achieved with space array radio telescopes.

Toward a Cosmic Perspective

The science of radio astronomy asks the big questions of how the universe is structured and how it has evolved. SETI searches and the presumed existence of extraterrestrials raise another of the big questions. Project OZMA stimulated subsequent SETI programs often conducted in the midst of controversy. Research in radio astronomy involves questions about the universe that are at once scientific, philosophical, and even religious in nature. Although it is the nature of cutting edge science to be shrouded in a degree of mystery, radio astronomy provides a tool for seeking the answers.

Astronomy, geometry, and history are inseparable aspects of the cosmic scheme. This fact has been revealed as the large-scale structure of the universe has been investigated using radio astronomy techniques. Observations in radio astronomy are investigations into the past. Light (and therefore radio waves) travels at a finite speed. The time it takes light to transverse the immense distances of intergalactic space is enormous. The closest galaxy, for example,

outside the Milky Way, is the Andromeda galaxy which is 2.2 million light-years distant. Observations of Andromeda see the galaxy as it was 2.2 million years ago because the photons currently detected have taken that long to traverse the distance. The farther away the object under investigation, the farther back into time the astronomer sees. It is interesting to realize that many phenomena being investigated represent events that occurred before the earth existed. Observations of extremely distant objects such as quasars provide clues into the formative stages of the universe. Investigations of phenomena at various cosmological distances allow observations of different stages of the evolution of cosmic structures from the beginning of the universe to the present. A picture of the sequence of events in the evolution of the universe is made possible by this basic premise of astrophysics.

Parallel Universes

In making the invisible universe visible, radio astronomy has revealed a parallel universe of bizarre phenomena that coexists with our own conventionally viewed universe. The optical and radio skies occupy the same spatial coordinates in the sky but the views are of totally different phenomena and dimensions. The optical and radio "parallel universe" is dramatically illustrated in a composite image of a portion of the radio sky compiled by J. J. Condon from data collected with the NRAO 300-foot telescope (Figure 7.19). The image itself is a posthumous tribute to an instrument that played an important role in the evolution of a modern science. The picture displays a 45° wide view of the radio sky superimposed on an optical photograph of the Deer Creek Valley. The NRAO 300-foot, 140-foot, and three 85-foot dishes of the interferometer are visible. The image is strangely similar to the optical sky in that there is a fairly random distribution of objects. Close inspection, however, reveals a totally different universe at these low frequencies. The familiar stars that speckle our sky are totally missing in this image. In fact, very few of the objects in the picture are stars. Most of the objects in the image are extragalactic sources, and fewer than 1% of these are near enough to coincide with the relatively nearby galaxies detected by optical telescopes. The Milky Way is visible as a concentration of extended sources that extend in a band from the lower left to the top right of the image. The objects that define the plane of the galaxy are actually fairly distant objects in the outer reaches of the Milky Way. The objects revealed in this image outside of the galactic plane are primarily energetic beasts with powerful engines that represent some of the most luminous and distant objects in the universe. By contrast, out optical view of the same sky reveals a very local universe—stars within a local corner of our own galaxy. Most stars in the visible sky lie within a few thousand light-years of our sun. Distant radio galaxies and quasars in the radio view represent objects from 400,000,000 to 12,000,000,000 light-years distant.[33] These objects also necessarily represent galaxies in formative evolutionary stages, seen as they appeared billions of years ago. In fact, a picture of the evolution of the large scale structures in the universe is revealed by looking at various distances into space and therefore back in time. The universe itself has evolved since the time of these ancient cosmic phenomena. The radio sky offers a dramatic view through the local sky of our conventional perspective to reveal a distant view of an unfamiliar, parallel universe.

Radio astronomy has incited a revolution in astronomy. The National Radio Astronomy Observatory has greatly contributed to this revolution. The amazing discoveries of radio astronomy are providing clues to our past, insight into the present, and indications of the future evolution of cosmic phenomena. With new insight into the nature of cosmic objects ranging from planet earth to the most distant quasars and the largest scale structures in the universe, advances in radio astronomy have allowed us to understand our place in the cosmos and to achieve a truly cosmic perspective.[34] The positions of the earth, solar system, Milky Way, and Local Supercluster have been shown to be non-preferential. This notion has historically been difficult for Western cultures to accept, as it is the nature of elements within these cultures to imagine humankind to be the center of the universe. Investigations in radio astronomy have unambiguously revealed the large-scale structure of the universe and our place within its geometry. The science of radio astronomy provides insight into our place in the cosmic scheme, which is perhaps the greatest enrichment of all.

Endnotes

1. *National Radio Astronomy Annual Report* 1966, July 1965–June 1966, *The Astronomical Journal*, vol. 71, no. 9, November, 1966, pp. 798–802.
2. *National Radio Astronomy Annual Report* 1968, July 1967–June 1968, *American Astronomical Society*, vol. 1, no. 1, January 1969, pp. 86–92.
3. Jack Lancaster, "VLA Dedicated: October 10, 1980," *The Observer*, vol. 22 no. 1, January, 1981, pp. 4–7.
4. Mark A. Gordon, Letter to Benjamin K. Malphrus, January 31, 1991.
5. *Ibid.*
6. *Ibid.*
7. Dewey E. Ross, "History of the 36-Foot", *The Observer,* vol. 20, no. 4, December 1979, p. 3.
8. Mark A. Gordon, Letter to Benjamin K. Malphrus, January 31, 1991.
9. *The National Radio Astronomy Observatory Very Large Array*, operated by Associated Universities Inc., under contract with the National Science Foundation, Brochure, 1989 NRAO/AUI, pp. 11–13.
10. A. R. Thompson, B. G. Clark, C. M. Wade, and P. J. Napier, "The Very Large Array," *The Astrophysical Journal Supplement Series*, vol. 44, October, 1980, pp. 151–167.
11. *The National Radio Astronomy Observatory Very Large Array*, Brochure.
12. Ronald D. Ekers, "The Almost Serendipitous Discovery of Self-Calibration," *Serendipitous Discoveries in Radio Astronomy*, Proceedings

of a workshop held at the National Radio Astronomy Observatory at Green Bank, West Virginia on May 4, 5, and 6, 1983, K. Kellerman and B. Sheets, ed.
13. A. R. Thompson, B. G. Clark, C. M. Wade, and P. J. Napier, "The Very Large Array," pp. 151–167.
14. Cass A SNR, VLA observations made in 1983 by P. E. Angerhofer, R. Braun, S. F. Gull, R. A. Perley, and R. J. Tuffs.
15. Cygnus A, Radio Galaxy, VLA observations made in 1979–1983 by R. A. Perley, J. W. Dreher, and J. J. Cowan.
16. Sagittarius A, The Galactic Nucleus, VLA observations made in 1982–1983 by F. Yusef-Zadeh, M. R. Morris, and D. R. Chance.
17. K. I. Kellerman and A. R. Thompson, "The Very Long Baseline Array," *Science*, July 12, 1985, vol. 229, no. 4709, pp. 123–130.
18. *The National Radio Astronomy Observatory Very Large Array*, Brochure, Ronald A. Schorn, "The Extragalactic Zoo," *Sky and Telescope*, January 1988, pp. 23–27.
19. K. I. Kellerman and A. R. Thompson, "The Very Long Baseline Array," *Science*, July 12, 1985, vol. 229, no. 4709, pp. 123–130.
20. *Ibid.*, pp. 123–130.
21. *Ibid.*, pp. 123–130.
22. *The National Radio Astronomy Observatory Very Large Array*, Brochure.
23. K. I. Kellerman and A. R. Thompson, "The Very Long Baseline Array," pp. 123–130.
24. *The Green Bank Telescope: A Radio Telescope for the Twenty-First Century*, Final Proposal to the National Science Foundation, June 1989, submitted by Associated Universities, Inc.
25. "Telescope to Be Built," *The Los Angeles Times*, Science Watch, July 30, 1989.
26. *The Green Bank Telescope: A Radio Telescope for the Twenty-First Century*, Final Proposal to the National Science Foundation, June 1989.
27. *Ibid.*, pp. 2–23.
28. *Ibid.*, pp. 2–23.
29. *Ibid.*, pp. 2–23.
30. The increasing resolution of instrumentation at the NRAO is given in Figure 7.17. The datapoints are based on information provided in the following references: J. D. Kraus, *Radio Astronomy*, Cygnus-Quasar Books, Powell, Ohio, 1986, p. 6–97; W. C. Tyler, D. E. Hogg, and C. M. Wade, "First Results with the National Radio Astronomy Observatory Interferometer," *American Institute of Physics for the American Astronomical Society Abstracts*, paper presented at the 118th meeting of the American Astronomical Society held on 14–17 March 1965 at the University of Kentucky, Lexington, Kentucky; W. M. Baars and P. G. Mezger, "First Observations at Short Wavelengths with the 140-foot Radio Telescope," *Sky and Telescope*, vol. 29, no. 11, November 1965, pp. 272–273; John Findlay, "The 300-foot Transit Radio Telescope at Green Bank," *Sky and Telescope*, vol. 25, no. 2, February 1963; A. R. Thompson, B. G. Clark, C. M. Wade, and P. J. Napier, "The Very Large Array," *The Astrophysical Journal Supplement Series*, vol. 44, October, 1980, pp. 151–167; *The National Radio Astronomy Observatory Very Large Array*, operated by Associated Universities Inc., under contract with the National Science Foundation, Brochure, 1989 NRAO/AUI, pp. 34–26.
31. K. I. Kellerman and A. R. Thompson, "The Very Long Baseline Array," pp. 123–130.
32. A baseline of 300 million kilometers (186 million miles), twice the distance of the earth's average separation from the sun, could be achieved that would provide an incredible resolution of 8.4×10^{-13} arcseconds at 21 cm (based on the resolution equation given in Chapter 2). Such an interorbital baseline would provide resolution more than a billion times greater than the highest resolution attainable with earth bound VLBI systems based on VLBI resolutions of 0.001 arcsecond resolution.
33. J. J. Condon, "Parallel Universes," Image Caption, National Radio Astronomy Observatory, 1994.
34. George A. Seielstad, *Cosmic Ecology, The View from the Outside In*, University of California Press, Berkeley, California, 1983.

Glossary

absolute magnitude. A measure of the brightness a star would have if it were placed at a standard distance of 10 pc from the sun; the intrinsic brightness of a star.

absorption (dark) lines. Colors (specific frequencies of optical radiation) missing in a continuous spectrum because of the absorption of those colors by atoms.

absorption-line spectrum. Dark lines superimposed on a continuous spectrum.

accretion. The colliding and collection of small particles into larger masses.

accretion disk. A disk made by infalling material around a massive object; the conservation of angular momentum results in the disk shape.

active galaxies. Galaxies characterized by nonthermal spectra and a large energy output compared to normal galaxies.

active galaxy nucleus (AGN). The tiny central engine (possibly a supermassive black hole) that drives the active galaxy phenomenon.

albedo. A measure of an object's reflectivity; the ratio of reflected light to incoming light for a solid surface, where complete reflection gives an albedo of 1, total absorption gives 0.

Alpha Centauri. The closest star to the sun, a triple star system; component A happens to have almost the same luminosity and surface temperature as the sun.

Andromeda galaxy (M31). The closest spiral galaxy to the Milky Way galaxy at a distance of 680 kpc; it has a diameter of about 50 kpc. M31 is considered to be a sister galaxy to the Milky Way.

angular diameter. The apparent diameter of an object; the angular separation of two points on opposite sides of the object.

angular distance. The apparent angular spacing between two objects in the sky.

angular separation. The observed angular distance between two celestial objects, measured in degrees, minutes, and seconds of angular measure.

angular speed. The rate of change of angular position of a celestial object viewed in the sky.

antiparticles. Subatomic particles that annihilate with normal particles when they collide; an example is the positron, which is the antiparticle of an electron.

aperture synthesis. Interferometric techniques in which an array of radio antennas are used to attain high resolution images of cosmic phenomena at radio frequencies.

aphelion. For a body orbiting the sun, the point on its orbit that is farthest from the sun.

apogee. The point in its orbit where an earth satellite is farthest from the earth.

apparent magnitude. The visual brightness of a star on the magnitude scale as seen from earth; an astronomical measure of the object's flux.

astronomical unit (AU). The semimajor axis of the earth's orbit and average distance between the earth and the sun; 149.6 million km or 8.3 light-minutes.

astrophysical jets. Collimnated beams of material (usually ions and electrons) expelled from astrophysical objects, such as the nuclei of active galaxies.

atmosphere. A gaseous envelope surrounding a planet, or the visible layers of a star; also a unit of pressure (atm) equal to the pressure of air at sea level on the earth's surface.

atmospheric escape. The process by which particles at the exosphere with greater than escape velocity and unhindered by collisions leave a planet.

atom. The smallest particle of an element that exhibits the chemical properties of the element.

AU. Abbreviation of astronomical unit.

aurora. Visible light emission from atmospheric atoms and molecules excited by collisions with energetic, charged particles from the magnetosphere, driven by the solar wind.

Balmer series. The set of transitions of electrons in a hydrogen atom between the second energy level and higher levels; also the set of absorption or emission lines corresponding to these transitions that lies in the visible part of the spectrum, the first of which is the H-alpha line.

bandwidth. The range of frequencies which a radio receiver (or another electromagnetic sensor) detects at the same time.

barred spirals. A subclass of spiral galaxies that have a structural bar across the nuclear region.

big bang model. A model of the evolution of the universe that postulates its origin and evolution, in an event called the big bang, from a hot, dense state that rapidly expanded to cooler, less dense states.

binary-accretion model. A model for the origin of the moon in which the moon and earth form by accretion of material from the same cloud of gas and dust.

binary galaxies. Two galaxies bound by gravity that orbit a common center of mass.

bipolar outflows. High-speed outflows of gas in opposite directions from a young stellar object; possibly the result of a magnetized accretion disk around the object.

blackbody. A (hypothetical) perfect radiator of light that absorbs and reemits all radiation incident upon it; its flux depends only on its temperature and is described by Planck's law.

black dwarf. The theoretical cold remains of a white dwarf after all its thermal energy is exhausted.

black hole. A massive star that has collapsed to such a degree that the escape velocity from its surface is greater than the speed of light, so that light is trapped by the intense gravitational field.

blue shift. A decrease in the wavelength of the radiation emitted by an approaching celestial body as a consequence of the Doppler effect; a shift toward the short wavelength (blue) end of the spectrum.

Bohr model of the atom. A simple picture of atomic structure in which electrons have well-defined orbits about the nucleus of the atom.

Boltzmann's constant. The number that relates pressure and temperature, or kinetic energy and temperature in a gas; the gas constant per molecule.

bound-bound transition. A transition of an electron between two bound energy state of an atom or ion.

bound-free transition. A transition of an electron between a bound and an unbound (free) state.

bright-line spectrum. *See* emission-line spectrum.

broad absorption line (BAL) quasars. High red-shift quasars in which very broad troughs of absorption are found.

carbon-nitrogen-oxygen (CNO) cycle. A series of thermonuclear reactions taking place in a star's core, in which carbon, nitrogen, and oxygen aid the fusion of hydrogen into helium; it is a secondary energy-production process in the sun, but the major process in high-mass, main-sequence stars.

Cassegrain reflector. A design of a reflecting telescope where a secondary mirror directs the beam to a focus through a hole in the center of the primary mirror.

cD galaxies. *See* supergiant elliptical galaxies.

celestial equator. An imaginary projection of the earth's pole onto the celestial sphere; a point about which the apparent daily rotation of the stars takes place.

celestial sphere. An imaginary sphere of very large radius centered on the earth on which the celestial bodies appear and against which their motions occur.

center of mass. The balance point of a set of interacting or gravitationally connected bodies; the single point of an object on which gravity appears to act.

cepheid variables (cepheids). Stars that vary in brightness as a result of a regular variation in size and temperature; a class of variable stars for which the star Delta Cephei is the prototype.

Chandrasekhar limit. The maximum amount of mass for a white dwarf star, about 1.4 solar masses; this amount leads to the highest density and smallest radius for a star made of a degenerate electron gas.

chemical condensation sequence. The sequence of chemical reactions and condensation of solids that occurs in a low-density gas as it cools at specific pressures and temperatures.

circumpolar stars. For an observer north of the equator, those stars that are continually above the northern horizon and never set; for a southern observer, those stars that never set below the southern horizon.

CNO cycle. *See* carbon-nitrogen-oxygen cycle.

collisional deexcitation. Loss of energy by an electron of an atom in a collision in which the electron drops to a lower energy level.

collisional excitation. Forcing an electron of an atom to a higher energy level by a collision.

color index. The difference in the magnitudes of an object measured at two different wavelengths; a measure of the color and hence the temperature of a star.

color temperature. Temperature inferred from color, usually by fitting a Planck function to the continuous spectrum of a star at two wavelengths.

coma. The bright, visible head of a comet.

comets. Bodies of small mass that orbit the sun, usually in highly elliptical orbits, and consist, in the dirty snowball model, of small, solid particles of rocky material imbedded in frozen gases. The nuclei of comets contain a high percentage of organic materials.

compact radio galaxies. Active galaxies that have a small, strong radio source in their nuclei.

compound. A substance composed of the atoms of two or more elements bound together by chemical forces.

condensation. The growth of small particles by sticking together of atoms and molecules.

conduction. Transfer of thermal energy by particles colliding into one another.

conservation of angular momentum. The principle stating that, with no torques, the total angular momentum of an isolated system is constant.

conservation of energy. A fundamental principle in physics that states that the total energy of an isolated system remains constant regardless of whatever internal changes may occur.

conservation of magnetic flux. The physical principle stating that, under certain circumstances, the number of magnetic field lines passing through an area remains constant.

conservation of momentum. The physical principle stating that, with no outside net forces, the total momentum of an isolated system is constant.

constellation. An apparent arrangement of stars on the celestial sphere, usually named after ancient gods, heroes, animals, or mythological beings; now an agreed-upon region of the sky containing a group of stars.

continental drift. A geophysical model in which the present continents were at one time a joined landmass (Pangea) that fragmented and drifted apart.

continuous spectrum. A spectrum showing emission at all wavelengths, unbroken by either absorption lines or emission lines.

contour map. A diagram showing how the intensity of some kind of radiation varies over a region of the sky; lines on such a map connect points of equal intensity; closely spaced lines indicate that the intensity changes rapidly over a small distance, widely spaced lines indicate it changes more slowly.

convection. The transfer of energy by the moving currents of a fluid.

core (of the earth). The central region of the earth; it has a high density, is in part liquid, and is believed to be composed of iron and iron alloys.

core (of the sun). The inner 25 percent of the sun's radius, where the temperature and pressure are great enough for thermonuclear reactions to take place.

core (of the galaxy). The inner few parsecs of the galactic nucleus, which contains a small, nonthermal radio source and fast-moving clouds of ionized gas.

core-mantle grains. Interstellar dust particles with cores of dense materials (such as silicates) surrounded by a mantle of icy materials (such as water).

Cosmic Background Explorer (COBE). A satellite launched in 1989 to measure the spectrum and intensity distribution of the microwave background radiation.

cosmic blackbody microwave radiation. Radiation with a blackbody spectrum at a temperature of about 2.7 K permeating the universe; believed to be the remains of the primeval fireball in which the universe was created; often referred to as the 3 K radiation.

cosmic nucleosynthesis. The production of the lightest few nuclei during the first 100 seconds of the big bang.

cosmic rays. Charged atomic particles moving in space with very high energies (the particles travel at a velocity near the speed of light); most originate beyond the solar system, but some of low energy are produced in solar flares.

cosmological principle. The statement that the universe, averaged over a large enough volume, appears the same from any location.

cosmology. The study of the nature and evolution of the physical universe.

cosmos. The universe considered as an orderly and harmonious system.

Crab Nebula. A supernova remnant, located in the constellation Taurus, produced by the supernova explosion visible from earth in A.D. 1054; a pulsar in the nebula marks the neutron-star corpse of the exploded star.

critical density. In cosmology, the density that marks the transition from an open to a closed universe; the density that provides enough gravity to just bring the expansion to a stop after infinite time.

Cygnus arm. A segment of one of the spiral arms in the outer part of our galaxy, about 14 kpc from the center.

Cygnus X-1. Binary X-ray source in the constellation Cygnus; it contains a probable black hole orbiting a blue supergiant star.

dark cloud. An interstellar cloud of gas and dust that contains enough dust to blot out the light of stars behind it (as seen from the earth).

dark-line spectrum. *See* absorption-line spectrum.

dark matter. The probable dominant form of matter in the universe; it may be non-baryonic and does not form stars or galaxies—hence is "dark." Potential candidates for dark matter include neutrinos, brown dwarfs, and stellar black holes.

declination. Celestial latitude, measured in degrees north or south of the celestial equator.

decoupling. The time in the universe's history when the density became low enough that matter and light became independant phenomena.

degenerate electron gas. An ionized gas in which nuclei and electrons are packed together as much as possible, filling all possible low energy states, so that the perfect gas law relating pressure, temperature, and density no longer applies.

degenerate gas pressure. Gas pressure caused by matter made up of neutrons packed together as tightly as possible.

density-wave model. A model for the generation of spiral structure in galaxies that pictures density waves (similar to sound waves) plowing through the interstellar matter and sparking star formation.

differential rotation. The tendency of a fluid, spherical body to rotate faster at the equator than at the poles.

diffuse (bright) nebula. A cloud of ionized gas, mostly hydrogen, with an emission-line spectrum.

dirty-snowball comet model. A model for comets that pictures the nucleus as a compact solid body of frozen materials, mixed with pieces of rocky matter, that turns into gases as a comet nears the sun, creating the coma and tail.

disk (of a galaxy). The flattened wheel of stars, gas, and dust outside the nucleus of a galaxy.

dispersion. The effect that causes pulses of radiation at different frequencies emitted simultaneously to arrive at different times after traversing the interstellar medium.

Doppler shift. A change in the wavelength of waves from a source reaching an observer when either the source or the observer is moving with respect to each other along the line of sight; the wavelength increases (red shift) or decreases (blue shift) according to whether the motion is away from or toward the observer

double quasar. Two images on the sky of a single quasar, produced by a gravitational lens.

dust tail. The part of a comet's tail containing dust particles, pushed out by radiation pressure from the sun.

dwarf. A star of relatively low light output and relatively small size; a main-sequence star of luminosity class V.

dynamo model. A model for the generation of a planet's (or star's) magnetic field by the organized circulation of conducting fluids in its core.

eccentric. An ancient geometric device used to account for nonuniform planetary motion; a point offset from the center of circular motion.

eccentricity. The ratio of the distance of a focus from the center of an ellipse to its semimajor axis.

eclipse. The phenomenon of one body passing in front of another, cutting off its light.

eclipsing binary system. Two stars that revolve around a common center of mass, the orbits lying edge-on to the line of sight, so that each star periodically passes in front of the other.

ecliptic. From the earth, the apparent yearly path on the celestial sphere of the sun with respect to the stars; also, the plane of the earth's orbit.

effective temperature. The temperature a body would have if it were a blackbody of the same size radiating the same luminosity.

electromagnetic force. One of the four fundamental forces of nature; particles with electromagnetic charge either attract or repel each other depending upon whether the two charges are opposite or identical. The electromagnetic force gives rise to the phenomena of electricity and magnetism.

electromagnetic radiation. A self-propagating electric and magnetic field, such as light, radio, ultraviolet, X-ray, gamma or infrared radiation; all types travel at the same speed and can be differentiated by wavelength or frequency.

electromagnetic spectrum. The range of all wavelengths of electromagnetic radiation that are naturally produced by objects in the universe. The electromagnetic spectrum is divided into gamma ray, X-ray, ultraviolet, visible light, infrared, microwave and radio wave bands.

electron. A lightweight, negatively charged subatomic particle, considered to be one of the fundamental particles in nature.

element. A substance that is made of atoms with the same chemical properties and cannot be decomposed chemically into simpler substances.

ellipse. A plane curve drawn so that the sum of the distances from a point on the curve to two fixed points is constant.

elliptical galaxy. A gravitationally bound system of stars that has rotational symmetry but no spiral structure and that contains mainly old stars and little gas or dust.

emission (bright) lines. Light of specific wavelengths or colors emitted by atoms; sharp energy peaks in a spectrum caused by downward electron transitions from one discrete quantum state to another discrete state.

emission-line spectrum. A spectrum containing only emission lines.

emission nebula. A hot cloud of hydrogen gas whose visible spectrum is dominated by emission lines. *See* also diffuse nebula.

energy. The ability to do work.

energy level. One of the possible quantum states of an atom, with a specific value of energy.

equilibrium. A state of a physical system in which there is no overall change.

escape velocity. The speed a body must achieve to break away from the gravity of another body and never return to it.

Euclidean (flat) geometry. Geometry in which only one

parallel line can be drawn through a point near another line; the sum of the angles in a triangle drawn on a flat surface is always 180°.

event. A point in four-dimensional spacetime.

evolutionary track. On a temperature-luminosity diagram, the path made by the points that describe how the temperature and luminosity of a star change with time.

excitation. The process of raising an atom to a higher energy level.

expanding (3 kpc) arm. The segment of a spiral arm structure encircling the center of our galaxy at a distance of about 3 kpc; it appears to be moving toward us and away from the galaxy's center.

extended radio galaxies. Active galaxies that show extended radio emission, usually in the form of two lobes on either side of the nucleus.

extinction. The dimming of light when it passes through some medium, such as the earth's atmosphere or interstellar material.

extragalactic. Outside of the Milky Way galaxy.

flux. The amount of energy flowing through a given area in a given time.

focal length. The distance from a lens (or mirror) to the point where it brings light to a focus for a distant object.

focus. The point at which light is gathered in a telescope.

forbidden line. An emission line from an atom produced by a transition with a low probability of occurrence.

frame of reference. A set of axes with respect to which the position or motion of an object can be described or physical laws can be formulated.

free-bound transition. A transition of an electron between a free energy state and one bound to an atom; results in the atom adding an electron with the emission of a photon; the reverse process is a bound-free transition.

free-free transition. A transition of an electron between two different unbound states; if energy is lost by the electron, the process is free-free emission; if gained, it is free-free absorption.

frequency. The number of waves that pass a particular point in some time interval (usually a second); usually given in units of hertz, one cycle per second.

galactic cannibalism. A model for galaxy interaction in which more massive galaxies strip material, by tidal forces, from less massive galaxies. The model suggests that a massive galaxy can completely incorporate the stars and mass of a less massive galaxy into its own structure.

galactic center (core). The innermost part of the galaxy's nuclear bulge.

galactic (open) cluster. A small group, about ten to a few hundred, of gravitationally bound stars of Population I, found in or near the plane of the galaxy.

galactic equator. The great circle along the line of the Milky Way, indicating the central plane of the galaxy.

galactic latitude. The angular distance north or south of the galactic equator.

galactic longitude. The angular distance along the galactic equator from a zero point in the direction of the galactic center.

galactic rotation curve. A description of how fast an object some distance from the center of a galaxy revolves around it.

galaxy. A huge assembly of stars (typically between 10^6 and 10^{12}), plus gas and dust, that is held together by gravity; our own galaxy, containing the sun.

gamma ray. A very high energy photon of electromagnetic radiation with a wavelength shorter than that of X-rays.

gas tail. The part of a comet's tail that consists of ions and molecules; it is shaped by its interaction with the solar wind.

gauss. A physical unit measuring magnetic field strength (not the SI unit, which is the tesla, but commonly used by astronomers).

general theory of relativity. The idea developed by Albert Einstein that mass and energy determine the geometry of spacetime and that any curvature of this spacetime by massive objects shows itself by what we commonly call gravitational forces; Einstein's theory of gravity.

geomagnetic axis. The axis that connects the earth's magnetic poles; it is inclined about 12° from the geographic spin axis and does not pass through the earth's center.

giant molecular clouds (GMCs). Large interstellar clouds, with sizes up to tens of parsecs and containing up to 100,000 solar masses of material; found in the spiral arms of the galaxy, giant molecular clouds are the sites of massive star formation.

globular cluster. A gravitationally bound group of about 10^5 to 10^6 Population II stars (of roughly solar mass), symmetrically shaped, found in the halo of the galaxy.

grand design spirals. Spiral galaxies that have a well-defined spiral arm structure.

grand unification theories (GUTs). Physical theories that attempt to unite the elementary particles and the four forces in nature as the actions of one particle and one force.

gravitation. In classical Newtonian terms, a force between masses that is characterized by their acceleration toward each other; the size of the force depends directly on the product of the masses and inversely on the square of the distance between the masses; in relativistic terms, the curvature of spacetime.

gravitational bending of light. The effect of gravity on the usually straight path of a photon.

gravitational collapse. The unhindered contraction of any mass from its own gravity.

gravitational field. The property of space having the potential for producing gravitational force on objects within it; characterized by the acceleration of free masses.

gravitational focusing. The directing of the paths of small masses by a larger one so that their paths cross, which enhances their accretion onto the larger mass.

gravitational force. The weakest of the four forces of nature; all particles with nonzero mass attract each other.

gravitational instability. The tendency for a disturbed region in a gas to undergo gravitational collapse.

gravitational lens. The bending effect of a large mass on light rays so that they form an image of the source of light.

gravitational mass. The mass of an object as determined by the gravitational force it exerts on another object.

greenhouse effect. The effect producing a greatly increased equilibrium temperature at the surface of a planet due to the opacity of its atmosphere in the infrared, trapping outgoing heat radiation.

ground state. The lowest energy level of an atom.

HI region. A region of neutral hydrogen in interstellar space.

HII region. A region of ionized hydrogen in interstellar space; it usually forms a bright nebula around a hot, young star or cluster of hot stars.

Halley's Comet. The periodic comet (orbital period approximately 76 years) whose orbit was first worked out by Edmund Halley from Newton's laws; has a small nucleus (about 10 km diameter), dark, irregular in shape and emitting jets of gas and dust.

halo (of a galaxy). The spherical region around a galaxy, not including the disk or the nucleus, containing globular clusters, some gas, and probably a few stray stars.

H-alpha line. The first line of the Balmer series, the set of transitions in a hydrogen atom between the second energy level and levels with higher energy; it lies in the red part of the visible spectrum.

head-tail radio galaxy. An active galaxy whose radio lobes have been swept back to form a tail because of interaction with a surrounding medium.

helium burning. Fusion of helium into carbon by the triple-alpha process.

hertz. A physical unit of frequency equal to 1 cycle/s.

Hertzsprung-Russel (H-R) diagram. A graphic representation of the classification of stars according to their spectral class (or color or surface temperature) and luminosity (or absolute magnitude or flux density); the physical properties of a star are correlated with its position on the diagram, so a star's evolution can be described by its change of position on the diagram with time. *(See also* evolutionary track).

high velocity clouds. Clouds of gas associated with the galaxy, moving at speeds of hundreds of kilometers per second.

high velocity stars. Stars in the galaxy with velocities greater than 60 km/s relative to the sun; they have orbits with high eccentricities, often at large angles with respect to the galactic plane.

homogeneous. Having a consistent and even distribution of matter, the same in all parts.

horizon. The intersection with the sky of a plane tangent to the earth at the location of the observer.

HPBW. Half power beamwidth; performance characteristic of a radio telescope that is analogous to the angular resolution of an optical telescope. HPBW is usually determined by measuring the point at which the intensity of far field radiation pattern drops to one-half its maximum value, sometimes described as FWHM-full width at half maximum.

Hubble constant. The proportionality constant relating velocity and distance in the Hubble law; the value, now believed to be around 75 km/s/Mpc, changes with time as the universe expands.

Hubble law. A description of the expansion of the universe, such that the more distant a galaxy lies from us, the faster it is moving away; the relation, $v = Hd$, between the expansion velocity (v) and distance (d) of a galaxy, where H is the Hubble constant.

hydrogen burning. Any fusion reaction that converts hydrogen (protons) to heavier elements.

hyperbola. A curve produced by the intersection of a plane with a cone; the shape of the orbit of a body with more than escape velocity.

hyperbolic geometry. An alternative to Euclidean geometry, constructed by N. I. Lobachevski on the premise that more than one parallel line can be drawn through a point near a straight line; the sum of the angles of a triangle drawn on a hyperbolic surface is always less than 180°.

image processing. The computer manipulation of digitized images to enhance specific aspects of these images.

inertia. The resistance of an object to a force acting on it because of its mass; an inherent quality of matter to resist a change in its dynamical state.

inertial frame. Any system of reference in special relativity that is not accelerated.

inertial mass. Mass determined by subjecting an object to a known force (not gravity) and measuring the acceleration that results.

inflation theory. A modification of the big bang model in which the universe undergoes a brief interval of rapid expansion.

Infrared Astronomy Satellite (IRAS). A satellite that surveyed the sky at wavelengths of 12, 25, 60, and 100 μm.

infrared telescope. A telescope, optimized for use in the infrared part of the electromagnetic spectrum, utilizing an infrared detector.

intergalactic medium. The gas and dust found between the galaxies.

interferometer. *See* radio interferometer.

interstellar dust. Small (micrometers in diameter) solid particles in the interstellar medium.

interstellar extinction curve. The amount of extinction from interstellar dust as a function of wavelength.

interstellar gas. Atoms, molecules, and ions in the interstellar medium.

interstellar medium. All the gas and dust found between stars.

inverse-square law for light. The decrease of the flux of light with the inverse square of the distance from the source.

ion. An atom that has become electrically charged by the gain or loss of one or more electrons.

ionization The process by which an atom loses or gains electrons.

ionized gas. A gas that has been ionized so that it contains free electrons and charged ions; a plasma if it is electrically neutral overall.

ionosphere. A layer of the earth's atmosphere ranging from about 100 to 700 kilometers above the surface where oxygen and nitrogen are ionized by sunlight, producing free electrons.

irregular galaxy. A galaxy without spiral structure or rotational symmetry, containing mostly Population I stars and abundant gas and dust.

isophote. A contour of constant intensity in a digital image.

isotope. Atoms with the same number of protons but different numbers of neutorns.

isotropic. Having no preferred direction in space.

joule. A physical unit of work and energy.

Kepler's laws. Kepler's three laws of planetary motion that describe the properties of elliptical orbits with an inverse-square force law.

kiloparsec (kpc). One thousand parsecs.

kinetic energy. The ability to do work due to motion.

Large Magellanic Cloud (LMC). A small galaxy, irregular in shape, about 50 kpc from the Milky Way.

light curve. A graph of a star's changing brightness with time.

lighthouse model. Model developed to explain pulsars, rapidly rotating neutron stars with strong magnetic fields; the rotation provides the pulse period and the magnetic field generates the electromagnetic radiation.

light rays. Imaginary lines in the direction of propagation of a light wave.

light-year. The distance light travels in a year, approximately 3.09×10^{13} kilometers.

Local Group. A gravitationally bound group of about 20 galaxies to which our Milky Way galaxy belongs.

Local Supercluster. The supercluster of galaxies in which the Local Group is located; spread over 10^7 pc, it contains the Virgo and Coma clusters.

low velocity stars. Stars with close to circular orbits in the plane of the galaxy; they travel at less than 60 km/s with respect to the sun.

luminosity. The total rate at which radiative energy is given off by a celestial body, at all wavelengths; the sun's luminosity is about 4×10^{26} watts.

Lyman series. All transitions in a hydrogen atom to and from the lowest energy level; they involve large energy changes, corresponding to wavelengths in the ultraviolet part of the spectrum; also, the set of absorption or emission lines corresponding to these transitions.

Magellanic Clouds. Two satellite galaxies of the Milky Way, the Large Magellanic Cloud (LMC) and the Small Magellanic Cloud (SMC), visible in the Southern Hemisphere to the unaided eye; companions to our galaxy.

magnetic field. The property of space having the potential of exerting magnetic forces on bodies within it.

magnetic field lines. A graphic representation of a magnetic field showing its direction and, by the density of the lines, its intensity.

magnetic flux. The number of magnetic field lines passing through an area.

magnetosphere. The region around a planet in which particles from the solar wind are trapped by the planet's magnetic field.

magnitude. An astronomical measurement of an object's brightness; larger magnitudes represent fainter objects.

main sequence. The principal series of stars in the Hertzsprung-Russell diagram; such stars are converting hydrogen to helium in their cores by the proton-proton process or by the carbon-nitrogen-oxygen cycle; this is the longest stage of a star's active life.

major axis. The larger of the two axes of an ellipse.

mass. A measure of an object's resistance to change in its motion (inertial mass); a measure of the strength of gravitational force an object can produce (gravitational mass).

mass loss. The rate at which a star loses mass, usually by a stellar wind, per year.

mass-luminosity ratio. For galaxies, the ratio of the total mass to the luminosity; a rough measure of the kind of stars in a galaxy.

mass-luminosity relation. An empirical relation, for main-sequence stars, between a star's mass and its luminosity, roughly proportional to the third power of the mass.

mechanics. A branch of physics that deals with forces and their effects on bodies.

megaparsec. 1 million parsecs.

meteor. The bright streak of light that occurs when a solid particle (a meteoroid) from space enters the earth's atmosphere and is heated by friction with atmospheric particles; sometimes erroneously called a falling star.

meteorite. A solid body from space that survives a passage through the earth's atmosphere and falls to the ground.

meteoroid. A very small solid body moving through space in orbit around the sun.

meteor shower. A rapid influx of meteors that appear to come out of a small region of the sky, called the *radiant*.

microwave background radiation. A universal bath of low energy photons having a blackbody spectrum with a temperature of about 2.7 K., which is a remnant of the highly energetic beginning of the universe according to the big bang model.

Milky Way. The galaxy to which the sun belongs; the Milky Way is a spiral galaxy containing approximately 300,000,000,000 stars.

millisecond pulsar. Generic name given to any pulsar with a pulse period on the order of milliseconds.

minute of arc. 1/60 of a degree.

mmwave. Millimeter wave; referring to that portion of the electromagnetic spectrum whose wavelengths are measured in millimeters.

molecular cloud. Large, dense, massive clouds in the plane of a spiral galaxy; they contain dust and a large fraction of gas in molecular form.

molecular maser. Microwave amplification by stimulated emission of radiation from a molecule.

molecule. A combination of two or more atoms bound together electrically; the smallest part of a compound that has properties of that substance.

momentum. The product of an object's mass and velocity.

nanometer. 10^{-9} of a meter; common unit of wavelength measurement for light.

narrow-tailed radio galaxies. Radio galaxies that show a U-shaped tail behind the nucleus; they are fast-moving galaxies in a cluster of galaxies.

nebula. (Latin for "cloud") A cloud of interstellar gas and dust.

nebular model. A model for the origin of the solar system, in which an interstellar cloud of gas and dust collapsed gravitationally to form a flattened disk out of which the planets formed by accretion.

neutrino. An elementary particle (lepton) with no (or very little) mass and no electric charge produced in certain nuclear reactions. Neutrinos only weakly interact with matter particles.

neutron. One of the main constituents of an atomic nucleus; the union of a proton and an electron.

neutron star. A star of extremely high density and small size that is composed mainly of very tightly packed neutrons; cannot have a mass greater than about 3 solar masses.

Newton. The SI unit of force.

nonthermal radiation. Emitted energy that is not characterized by a blackbody spectrum; usually used to refer to synchrotron radiation.

Norma arm. The segment of a spiral arm of our galaxy, about 4000 pc from the Sun toward the center of the galaxy in the direction of the constellation Norma.

nova. (Latin for "new") A star that has a sudden outburst of energy, temporarily increasing its brightness by hundreds to thousands of times; now believed to be the outburst of a degenerate star in a binary system; also used in the past to refer to some stellar outbursts that modern astronomers now call supernovas.

nuclear bulge. The central region of a spiral galaxy, containing old Population I stars.

nuclear fission. A process that releases energy from matter; in this process, a heavy nucleus hit by a high-energy particle splits into two or more lighter nuclei whose combined mass is less than the original, the missing mass being converted into energy.

nuclear fusion. A process that releases energy from matter by the joining of nuclei of lighter elements to make heavier ones; the combined mass is less than that of the constituents, the difference appearing as energy.

nucleosynthesis. The chain of thermonuclear fusion processes by which hydrogen is converted to helium, helium to carbon, and so on through all the elements of the periodic table up to Iron-57.

nucleus (of an atom). The massive central part of an atom, containing neutrons and protons, about which the electrons orbit.

nucleus (of a comet). Small, bright, starlike point in the head of a comet; a solid, compact (diameter of a few tens of kilometers) mass of frozen gases with some rocky material as dust embedded in it.

nucleus (of a galaxy). The central portion of a galaxy, composed of old Population I stars, some gas and dust, and, for many galaxies, a concentrated source of nonthermal radiation.

objective. The main light-gathering lens or mirror of a telescope.

observable universe. The parts of the universe that can be detected by the light they emit.

occultation. The eclipse of a star or planet by the moon or another planet.

Oort cloud. A cloud of comet nuclei in orbit around the solar system, formed at the time the solar system formed; the reservoir for new comets.

opacity. The property of a substance that hinders (by absorption or scattering) light passing through it; opposite of transparency.

open cluster. Same as galactic cluster.

optics. The manipulation of light by reflection or refraction.

orbital inclination. The angle between the orbital plane of a body and some reference plane; in the case of a planet in the solar system, the reference plane is that of the earth's orbit; in the case of a satellite, the reference is usually the equatorial plane of the planet; for a double star; it is the plane perpendicular to the line of sight.

Orion arm. A segment of the galaxy's larger spiral arm structure; the solar system lies within it.

Orion Nebula. A hot cloud of ionized gas including an extensive GMC that is a nearby region of recent star formation, visibly located in the sword of the constellation of Orion; also called Messier 42 (M42).

ozone layer (ozonosphere). A layer of the earth's atmosphere about 40 to 60 kilometers above the surface, characterized by a high content of ozone, O_3.

parabola. A geometric figure that describes the shape of an escape-velocity orbit.

parallax. The change in an object's apparent position when viewed from two different locations; specifically, half the angular shift of a star's apparent position as seen from opposite ends of the earth's orbit.

parsec (pc). The distance an object would have to be from the earth so that its heliocentric parallax would be 1 second of arc; equal to 3.26 light-years; a kiloparsec is 1,000 parsecs, and a megaparsec is 10^6 parsecs.

perihelion. The point at which a body orbiting the sun is nearest to it.

period. The time interval for some regular even to take place; for example, the time required for one complete revolution of a body around another.

periodic (regular) variables. Stars whose light varies with time in a regular fashion from various causes.

period-luminosity relationship. For cepheid variables, a relation between the average luminosity and the time period over which the luminosity varies; the greater the luminosity, the longer the period.

Perseus arm. The segment of a spiral arm that lies about 3 kpc from the sun in the direction of the constellation Perseus.

photodissociation. The breakup of a molecule by the absorption of light with enough energy to break the molecular bonds.

photometer. A light-sensitive detector placed at the focus of a telescope; it is used to make accurate measurements of small photon fluxes.

photon. A discrete amount of light energy; the energy of a photon is related to the frequency f of the light by the relation $E = hf$, where h is Planck's constant.

photon excitation. Raising an electron of an atom to a higher energy level by the absorption of a photon.

photosphere. The visible surface of the sun; the region of the solar atmosphere from which visible light escapes into space.

physical universe. The parts of the universe that can be seen directly plus those that can be inferred from the laws of physics.

pixel. The smallest picture element in a two-dimensional detector.

Planck curve. The continuous spectrum of a blackbody radiator.

Planck constant. The number that relates the energy and frequency of light; it has a value of $h = 6.63 \times 10^{-34}$ J·s.

planet. From the Greek word for "wanderer"; any of the nine major bodies that orbit the sun.

planetary nebula. A thick shell of gas ejected from and moving out from an extremely hot star; thought to be the outer layers of a red giant star thrown out into space, the core of which eventually becomes a white dwarf.

planetesimals. Asteroid-sized bodies that, in the formation of the solar system, combined with each other to form the protoplanets.

plasma. A degenerate gas consisting of equal numbers of ionized atoms and electrons.

plate tectonics. A model for the evolution of the earth's surface that pictures the interaction of crustal plates driven by convection currents in the mantle.

Polaris. The present north pole star; the outermost star in the handle of the Little Dipper.

polarization. A lining up of the planes of vibration of light waves.

polarized light. Light waves whose planes of oscillation are all the same.

Population I stars. Stars found in the disk of a spiral galaxy, especially in the spiral arms, including the most luminous, hot, and young stars, with a heavy element abundance similar to that of the sun (about 2 percent of the total); an old Population I is found in the nucleus of spiral galaxies and in elliptical galaxies.

Population II stars. Stars found in globular clusters and the halo of a galaxy; may be older than any Population I stars, and contain a smaller abundance of heavy elements.

potential energy. The ability to do work because of position; it is storable and can later be converted into other forms of energy.

precession of the equinoxes. The slow westward motion of the equinox points on the sky relative to the stars of the zodiac caused by the wobbling of the earth's spin axis.

primary. The brighter of the two stars in a binary system.

proper motion. The angular displacement of a star on the sky determined from its motion through space.

protogalaxies. Clouds with enough mass that they are destined to collapse gravitationally into galaxies.

proton. A massive, positively charged elementary particle; one of the main constituents of the nucleus of an atom.

proton-proton (PP) chain. A series of thermonuclear reactions that occur in the interiors of stars, by which four hydrogen nuclei are fused into helium; this process is believed to be the primary mode of energy production in the sun.

protoplanet. A large mass formed by the accretion of planetesimals; the final stage of formation of the planets from the solar nebula.

protostar. A collapsing mass of gas and dust out of which a star will be born (when thermonuclear reactions turn on) whose energy comes from gravitational contraction.

pulsar. A radio source that emits signals in very short, regular bursts; thought to be a highly magnetic, rotating neutron star.

quantum. A discrete packet of energy.

quantum number. In quantum theory, one of the four special numbers that determine the energy structure and quantum state of atoms.

quark. An elementary particle with third-integral charge that makes up others, such as protons and neutrons.

quasar or quasi-stellar object (QSO). An intense, point-like source of light and radio waves that is characterized by large red shifts of the emission lines in its visible spectrum.

radar mapping. The topographic surveying of the geographic features of a planet's surface by the reflection of radio waves from its surface.

radial velocity. The component of relative velocity that lies along the line of sight.

radian (rad). A unit of angular measurement; 1 rad equals $57.3°$; 2π radians equals $360°$.

radiant. The point in the sky from which a meteor shower appears to come.

radiation. Electromagnetic waves, including gamma rays, X-rays, ultraviolet, visible light, infrared, microwave, and radio also sometimes used to refer to atomic particles of high energy, such as electrons (beta-radiation), and helium nuclei (alpha-radiation).

radiation belts. In a planet's magnetosphere, regions with a high density of trapped solar wind particles.

radiation era. In the big bang model, the time in the universe's early history in which the energy in the universe was dominated by radiation.

radiative energy. The capacity to do work that is carried by electromagnetic waves.

radio galaxies. Galaxies that emit large amounts of radio energy by the synchrotron process, generally characterized by two giant lobes of emission situated on opposite ends of a line drawn through the nucleus; they are divided into two types, compact and extended.

radio interferometer. A radio telescope that achieves high angular resolution by combining signals from at least two widely separated antennas.

radio recombination line emission. Sharp energy peaks at radio wavelengths caused by low energy transitions in atoms from one very high energy level to another nearby level following recombination of an electron with an ion.

radio telescope. A telescope designed to collect and detect radio emissions from cosmic objects and phenomena.

recombination. The joining of an electron to an ion; the reverse of ionization.

reddening. The preferential scattering or absorption of blue light by small particles, allowing more red light to pass directly through.

red giant. A large, cool star with a high luminosity and a low surface temperature (about 2,500 K), which is largely convective and has fusion reactions going on in shells.

red shift. An increase in the wavelength of the radiation received from a receding celestial body as a consequence of the Doppler effect; a shift toward the long wavelength (red) end of the spectrum.

red variables. A class of cool stars variable in light output.

reference frame. A set of coordinates in which position and motion may be specific.

reflection. The return of a light wave at the interface between two media.

reflection nebula. A bright cloud of gas and dust that is visible because of the reflection of starlight by the dust.

refraction. Bending of the direction of a light wave at the interface between two media, such as air and glass.

relativistic Doppler shift. Wavelength shift from the radial velocity of a source as calculated in special relativity, so that very large red shifts do not imply that the source moves faster than light.

relativistic jet. A beam of particles moving at speeds close to that of light.

relativity. Two theories proposed by Albert Einstein; the special theory describes the motion of nonaccelerated objects, and general relativity is a theory of gravitation.

resolving power. The ability of a telescope to separate close stars or to pick out fine details of celestial objects.

retrograde motion. The apparent anomalous westward (backward) motion of a planet with respect to the stars, which occurs near the time of opposition (for an outer planet) or inferior conjunction (for an inner planet).

retrograde rotation. Rotation from east to west.

revolution. The motion of a body in orbit around another body or a common center of mass.

right ascension. Celestial longitude, measured eastward along the celestial equator in hours of time from the vernal equinox.

Roche lobe. In a binary star system, the region in the space around the stars where their gravitational fields provide a path from one star to another.

rotation. The turning of a body, such as a planet, on an internal axis.

rotation curve. The relation between rotational velocity of objects in a galaxy and their distance from its center.

RR Lyrae stars. A class of giant, pulsating variable stars with periods of less than one day; they are Population II objects and commonly found in globular clusters.

S0 galaxy. A type of galaxy intermediate between ellipticals and spirals; they have a disk but no spiral arms; also known as linticular galaxies.

Sagittarius A (Sgr A). A collection of radio sources at the center of the Milky Way galaxy; Sgr A West is a thermal radio source (H II region), Sgr A East a nonthermal source, and Sgr A* a pointlike source that may mark the galaxy's core and gravitational center.

Sagittarius arm. A portion of spiral-arm structure of the Milky Way galaxy that lies about 2000 pc from the center of the galaxy in the direction of the constellation Sagittarius.

Schwarzschild radius. The critical size that a mass must reach to be dense enough to trap light by its gravity, that is, to become a black hole.

scientific model. A mental image of how the natural world works, based on physical, mathematical, and aesthetic principles.

second of arc. 1/3600 of a degree, or 1/60 of a minute of arc.

secular parallax. A method of determining the average distance of a group of stars by examining the components of their proper motions produced by the straight-line motion of the sun through space.

seismometer. An instrument used to detect earthquakes and moonquakes.

semimajor axis. Half of the major axis of an ellipse; distance from the center of an ellipse to its farthest point.

Seyfert galaxies. A type of AGN galaxy; the nuclear spectrum shows intense emission lines with either narrow (Type 2) or both broad and narrow (Type 1) components; the host galaxy is usually a spiral.

sidereal period. The time interval needed by a celestial body to complete one revolution around another with respect to the background stars.

signs of the zodiac. The twelve equal angular divisions of 30° each into which the ecliptic is divided; each corresponds to a zodiacal constellation.

singularity. A theoretical point of zero volume and infinite density to which any mass that becomes a black hole must collapse, according to the general theory of relativity.

Small Magellanic Cloud (SMC). The smaller of the two companion galaxies to the Milky Way; it is an irregular galaxy containing about 2×10^9 solar masses.

solar cosmic rays. Low energy cosmic rays generated in solar flares.

solar flare. Sudden burst of electromagnetic energy and particles from a magnetic loop in an active region.

solar mass. The amount of mass in the sun, about 2×10^{30} kg.

solar nebula. The disk of gas and dust, around the young sun, out of which the planets formed.

solar wind. A steam of charged particles, mostly protons and electrons, that escapes into the sun's outer atmosphere at high speeds and streams out into the solar system.

solstice. The time at which the day or the night is the longest; in the Northern Hemisphere, the summer solstice (around June 21), the time of the longest day;. and the winter solstice (around December 21), the time of the shortest day; the dates are opposite in the Southern Hemisphere.

space. A three-dimensional region in which objects move

and events occur and have relative direction and position.

spacetime. A four dimensional universe with space and time unified; a continuous system of one time coordinate and three space coordinates by which events can be located and described.

spacetime curvature. The bending of a region of spacetime due to the presence of mass and energy.

space velocity. The total velocity of an object through space, combining the components of radial and transverse velocities.

special theory of relativity. Einstein's theory describing the relations between measurements of physical phenomena as viewed by observers who are in relative motion at constant velocities.

spectral line. A particular wavelength of light corresponding to an energy transition in an atom.

spectral sequence. A classification scheme for stars based on the strength of various lines in their spectra; the sequence runs O–B–A–F–G–K–M–R–N–S, from hottest to coolest.

spectral type (or class). The designation of the type of a star based on the relative strengths of various spectral lines.

spectroscope. An instrument for examining spectra; also a spectrometer or spectrograph if the spectrum is recorded and measured.

spectroscopic binary. Two stars revolving around a common center of mass that can be identified by periodic variations in the Doppler shift of the lines of their spectra.

spectroscopic parallax. A technique for measuring distance by comparing the brightnesses of stars with their actual luminosities, as determined by their spectra.

spectroscopy. The analysis of light by separating it by wavelengths (colors).

spectrum (pl., spectra). The array of colors or wavelengths obtained when light is dispersed, as by a prism; the amount of energy given off by an object at every different wavelength.

speed. The rate of change of position with time.

spherical (closed) geometry. An alternative to Euclidean geometry, constructed by G. F. B. Riemann on the premise that no parallel lines can be drawn through a point near a straight line; the sum of the angles of a triangle drawn on a spherical surface is always greater than 180°.

spin angular momentum. The angular momentum of a rotating body; the product of a body's mass distribution, rotational velocity, and radius.

spiral arm. A structure, part of a spiral pattern in a galaxy, composed of gas, dust, and young stars, that winds out from near the galaxy's center.

spiral galaxy. A galaxy with spiral arms; the presumed shape of our Milky Way galaxy.

spiral tracers. Objects that are commonly found in spiral arms and so are used to trace spiral structure; for example, Population I Cepheids, H II regions, and OB-stars.

spontaneous emission. The emission of a photon by an excited atom in which an electron falls to a lower energy level.

stellar interior model. A table of values of the physical characteristics (such as temperature, density, and pressure) as a function of position within a star for a specified mass, chemical composition, and age, calculated from theoretical ideas of the basic physics of stars.

stellar nucleosynthesis. A process in which nuclear fusion builds up continually heavier nuclei while supplying the energy by which stars shine.

stimulated emission. Radiation produced by the effect of a photon stimulating an atom in an excited state to emit another photon of the same wavelength.

strong nuclear force. One of the four forces of nature; the strong force acts over short distances to keep the nuclei of atoms together.

superclusters. A system containing multiple clusters of galaxies.

supergiant. A massive star of large size and high luminosity.

supergiant elliptical (cD) galaxies. Largest and most massive elliptical galaxies, sometimes with more than one nucleus; usually found at the core of a rich cluster of galaxies.

superluminal motion. Motion apparently faster than the speed of light that is exhibited by some distant objects in the universe.

supermassive black hole. A black hole with a mass of 10^6 solar masses or more; such black holes probably power active galaxies and quasars.

supernova. A stupendous explosion of a massive star, which increases its brightness hundreds of millions of times in a few days.

supernova remnant. Expanding gas cloud from the outer layers of a star blown off in a supernova explosion; supernova remnants are detectable at radio wavelengths and move through the interstellar medium at high speeds.

synchrotron radiation. Radiation from an accelerating charged particle (usually an electron) in a magnetic field; the wavelength of the emitted radiation depends on the strength of the magnetic field and the energy of the charged particles.

temperature. A measure of the average random speeds of the microscopic particles in a substance.

temperature gradient. The change in temperature over a unit change in distance.

terrestrial planets. Inner planets of the solar system that are similar in composition and size to the earth; Mercury, Venus, Mars, and Earth.

thermal equilibrium. Steady-state situation characterized by no large-scale temperature changes.

three-degree cosmic blackbody microwave radiation. The relic radiation from the big bang that permeates all space in the universe; has a black body spectrum and a current temperature near 3 K (2.7 K).

time. A measure of the flow of events.

ton (metric). 1000 kilograms.

transition (in an atom). A change in the electron arrangements in an atom, which involves a change in energy.

Trapezium cluster. A small cluster of young massive stars located in the Orion Nebula.

trigonometric parallax. A method of determining distances by measuring the angular position of an object as seen from the ends of a baseline having a known length.

triple-alpha reaction. A thermonuclear process in which three helium atoms (alpha particles) are fused into one carbon nucleus.

T-Tauri stars. Newly formed stars of about 1 solar mass; usually associated with dark clouds; some show evidence of flares and starspots, of which T-Tauri is the prototype.

Tully-Fisher relation. The relation between the luminosity of a galaxy and the width of its 21-cm emission line; an important tool in determining the absolute luminosities of galaxies and therefore their distances.

21-cm line. The emission line, at a wavelength of 21.11 cm, from neutral hydrogen gas; it is produced by atoms in which the direction of spin of their proton and electron changes from parallel to opposed.

Type I, Type II supernovae. Classification of supernovae by their light curves and spectral characteristics; Type I show a sharp maximum and slow decline with no hydrogen lines; Type II have a broader peak and a very sharp decline after 100 days with strong hydrogen lines in the spectrum.

universal law of gravitation. Newton's law of gravitation. *See* also gravitation.

universe. The totality of all space and time; all that is, has been, and will be.

variable star. Any star whose luminosity changes over a short period of time.

velocity. The rage and direction in which distance is covered over some interval of time.

vernal equinox. The spring equinox; the point on the celestial sphere where the sun crosses the celestial equator passing from south to north; corresponds to around March 21.

Very Large Array (VLA). A radio interferometer located on the plains of San Augustin, New Mexico; it consists of 27 antennas spread in a Y-shaped pattern.

Very Large Baseline Array (VLBA). A radio interferometer with antennas spread across the United States; the processing and control center is in New Mexico.

Virgo cluster of galaxies. The nearest large cluster of galaxies; it appears to lie in the direction of the constellation of Virgo.

visual binary. Two stars that revolve around a common center mass, both of which can be seen through a telescope so that their orbits can be plotted.

void (cosmic void). A large region of space empty of visible galaxies.

watt. A unit of power; 1 joule of energy expended per second.

wavelength. The distance between two successive peaks or troughs of a wave.

weak nuclear force. A short-range force that operates in radioactive decay and governs the behavior of leptons.

white dwarf. A small, dense star that has exhausted its nuclear fuel and shines from residual heat; such stars have an upper mass limit of 1.4 solar masses, and their interior is composed of degenerate electron gas.

Wien's law. The relation between the wavelength of maximum emission in a blackbody's spectrum and its temperature; the higher the temperature, the shorter the wavelength at which the peak occurs.

ZAMS. *See* zero-age main sequence.

zenith. The point on the celestial sphere that is located directly above the observer at 90° angular distance from the horizon.

zero-age main sequence (ZAMS). The position on the H-R diagram reached by a protostar once it derives most of its energy from thermonuclear reactions rather than from gravitational contraction.

zodiac. The twelve constellations through which the sun apparently travels in its yearly motion, as seen from the earth.

zone of avoidance. A region near the plane of the Milky Way galaxy where very few other galaxies are visible because of obscuration by dust.

Bibliography

Books

Bamford, James, *The Puzzle Palace: A Report on America's Most Secret Agency*, (Penguin Books, New York, New York, 1983).

Blitz, Leo, and Kutner, Marc, eds., *Extragalactic Molecules*, (Proceedings of a workshop held at National Radio Astronomy Observatory operated by Associated Universities Inc., Green Bank, West Virginia, 1981).

Cardwell, D. S. L., *Technology, Science and History*, (Heinemann Educational Books Ltd., Cambridge, England, 1971).

Christiansen, W. N., and Hogbom, J. A., *Radiotelescopes*, (Cambridge University Press, Cambridge, England, 1969).

Edge, D., and Mulkay, M. J., *Astronomy Transformed: The Emergence of Radio Astronomy in Britain,* (Wiley Interscience, New York, 1976).

Gingerich, O., ed., *Astrophysics and 20th Century Astronomy to 1950: Part A*, General History of Astronomy Vol. 4, (Cambridge University Press, Cambridge, England, 1984).

Hartmann, William K., *Astronomy, the Cosmic Journey*, 3rd edition, (Wadsworth Publishing Company, Belmont, California, 1985).

Harwitt, M., *Cosmic Discovery*, (Basic Books, New York, 1981).

Henbest, Nigel, and Marten, Michael, *The New Astronomy*, (Cambridge University Press, London, 1983).

Hey, J. S. *The Evolution of Radio Astronomy*, (Science History Publications, New York, 1973).

Jena, P., Rao, B. K., and Khana, S. N., Ed., *Physics and Chemistry of Small Clusters*, (Plenum Publishing Corporation, 1987).

Kellerman, K., and Seielstad, G. A., *The Search for Extraterrestrials*, (Proceedings of a workshop held at National Radio Astronomy Observatory operated by Associated Universities Inc., Green Bank, West Virginia, 1985).

Kellerman, K., and Sheets, Beaty, *Serendipitous Discoveries in Radio Astronomy*, (Proceedings of a workshop held at National Radio Astronomy Observatory operated by Associated Universities Inc., Green Bank, West Virginia, 1983).

Kraus, J. D., *The Big Ear*, (Cygnus-Quasar, Powell, Ohio, 1976).

Kraus, J. D., *Radio Astronomy*, (Cygnus-Quasar, Powell, Ohio, 1986).

Kuhn, Thomas, *The Structure of Scientific Revolutions*, (Chicago University Press, Chicago, Illinois. 1970).

Lovell, Bernard, *The Story of Jodrell Bank*, (Oxford University Press, Oxford, 1968).

Lovell, Bernard, *The Jodrell Bank Telescopes*, (Oxford University Press, Oxford, 1985).

Lovell, Bernard, and Brown, R. H., *Exploration of Space by Radio*, (Wiley Press, New York, 1958).

Mar, James W., and Liebowitz, Harold, *Struture Technology for Large Radio and Radar Telescope Systems*, (The Massachusetts Institute of Technology Press, Cambridge, Massachusetts, 1969).

McDonough, Thomas, *The Search for Extraterrestrial Intelligence: Listening for Life in the Cosmos*, (Wiley Interscience, New York, 1987).

Pacholczyk, A. G., *Radio Astrophysics Nonthermal Processes in Galactic and Extragalactic Sources*, (W. H. Freeman and Company, 1982).

Seielstad, G. A., *A Cosmic Ecology, The View From the Outside In*, (University of California Press, Berkeley, California, 1983).

Skobel'tsyn, D. V., *Radio Astronomy- Instruments and Observations*, (Consultants Bureau, London, 1971).

Smith, F. Graham, *Radio Astronomy*, (Penguin Books, Baltimore, Maryland, 1960).

Sullivan, W. T., *Classics in Radio Astronomy*, (D. Reidel Press, New York, 1982).

Sullivan, W. T., *The Early Years of Radio Astronomy: Fifty*

Years after Jansky's Discovery, (Cambridge University Press, Cambridge, Mass., 1986).

Tucker, Wallace and Karen, *The Cosmic inquirers: Modern Telescopes and Their Makers*, (Harvard University Press, Cambridge, Mass., 1986).

Verschuur, Gerrit L., *The Invisible Universe Revealed: The Story of Radio Astronomy*, (Springer-Verlag, London, 1987).

Verschuur, Gerrit L. and Kellerman, Kenneth I., eds., *Galactic and Extragalactic Radio Astronomy*, (Springer-Verlag, New York, 1974).

Periodicals

Altenhoff, W. J., Downes, D., Goad, L., Maxwell, A., and Rinehart, R., "Surveys of the Galactic Plane at 1.414, 2.695, and 5.000 GHz," (*Astronomy and Astrophysics*, Supplement 1, 1969).

Appleton, E. V., and Hey, J. S., "Circular Polarization of Solar Radio Noise," (*Nature*, vol. 158, 1946).

Baade, W., and Minkowski, R., "Identification of the Radio Sources in Cassiopeia, Cygnus A, and Puppis A," (*Astrophysical Journal*, vol. 119, 1954).

Baars, J., "The Synthesis Radio Telescope at Westerbork," (*Proceedings of the IEEE*, vol. 61, no. 9, September 1973).

Baars, J., and Mezger, P. G., "First Observations at Short Wavelengths with the 140-foot Radio Telescope," (*Sky and Telescope*, vol. 31, no. 1, 1966).

Backer, D. C., Kulkarni, S. R., Heiles, C., Davis, M. M., and Goss, W. M., "A Millisecond Pulsar," (*Nature*, vol. 300, 1982).

Balick, B., "Radio Observations of Early-Type Stars," (*Astrophysical Letters*, vol. 12, 1972).

Balick, B., "First Fringes from Huntersville," (*The Observer*, vol. 14, no. 8, August 1973).

Bare, C., and Clark, B. G., "Interferometer Experiments with Independent Local Oscillators," (*Science*, vol. 157, no. 3785, 1967).

Basart, J. P., Clark, B. G., and Kramer, J. S., "A Phase Stable Interferometer of 100,000 Wavelengths Baseline," (*Publications of the Astronomical Society of the Pacific*, vol. 80, no. 474, June 1968).

Berkner, Lloyd, "NRAO First Annual Report," (*Bulletin of the American Astronomical Society*, 1969).

Bijleveld, Willem, and Burton, W. Butler, "Leiden Observatory: 350 Years of Astronomy," (*Sky and Telescope*, vol. 69, no. 6, 119, 1985).

Blum, E. J., Denisse, J. F., and Steinberg, J. L., "Radio Astronomy at the Meudon Observatory," (*Proceedings of the IRE*, vol. 46, 1958).

Bok, B. J., "New Science of Radio Astronomy," (*Scientific Monthly*, vol. 80, 1955).

Bok, B. J., "A National Radio Astronomy Observatory," (*Scientific American*, vol. 195, 1956).

Bolton, J. G., "Discrete Sources of Galactic Radio Frequency Noise," (*Nature*, vol. 162, 1948).

Bolton, J. G., and Stanley, G. J., "The Position and Probable Identification of the Source of Galactic Radio-Frequency Radiation Taurus A," (*Australian Journal of Scientific Research*, vol. 75, 1949).

Bolton, J. G., Stanley, G. J., and Slee, O. B., "Positions of Three Discrete Sources of Galactic Radio-Frequency Radiation," (*Nature*, vol. 164, 1949).

Bracewell, Ronald N., "The New Cambridge Radio Telescope," (*The Proceedings of the IEEE*, vol. 61, no. 9, September 1973).

Briggs, F. H., "The Microwave Properties of Saturn's Rings," (*The Astrophysical Journal*, vol. 189, 1974).

Broderick, J. J., *et al.*, "Interferometric Observations on the Green Bank-Crimea Baseline- Summary," (*Radio Science*, vol. 5, no. 10, 1970).

Brown, R. H., and Hazard, C., "Radio-Frequency Radiation from Tycho Brahe's Supernova (A. D. 1572)," (*Nature*, vol. 170, 1952).

Brown, R. H., and Lovell, A. C. B., "Large Radio Telescopes and Their Use in Radio Astronomy," (*Vistas in Astronomy*, vol. 1, 1955).

Brown, R. H., Palmer, H. P., and Thompson, A. R., "Polarization Measurements of three Intense Radio Sources," (*Monthly Notices of the Royal Astronomical Society*, vol. 115, 1955).

Buhl, D., "Chemical Constituents of Interstellar Clouds," (*Nature*, vol. 234, no. 5328, 1971).

Buhl, D., "Molecules and Evolution in the Galaxy," (NRAO Reprint Series A, 1972).

Buhl, D., and Snyder, L. E., "Unidentified Interstellar Microwave Line," (*Nature*, vol. 228, no. 5268, 1970).

Buhl, D., and Snyder, L. E., "Molecules in the Interstellar Medium," (*Sky and Telescope*, vol. 40, nos. 5 and 6, 1970).

Buhl, D., and Snyder, L. E., "From Radio Astronomy Towards Astrochemistry," (*Technology Review*, April 1971).

Buhl, D., and Snyder, L. E., "Microwave Receivers for Molecular Line Radio Astronomy," (*Nature Physical Science*, vol. 232, no. 34, 1971).

Buhl, D., Snyder, L., Zuckerman, B., and Palmer, P., "Microwave Detection of Interstellar Formaldehyde," (*Physical Review Letters*, vol. 22, no. 13, March 31, 1969).

Burke, B. F., and Franklin, K. L., "Observations of a Variable

Radio Source Associated with the Planet Jupiter," (*Journal of Geophysical Research,* vol. 60, 1955).

Burns, Jack O., "Dark Matter in the Universe," (*Sky and Telescope,* vol. 68, no. 5, 1984).

Burns, W. R., and Ewing, M. S., "Radio Astronomical Synthesis Arrays-Real Time Processing Needs," (*Society of Photo-Optical Instrumentations Engineers,* vol. 431, 1983).

Burns, W. R., and Yao, Stanton S., "A New Approach to Aperture Synthesis Processing," (*Astronomy and Astrophysics,* vol. 6, 1970).

Burton, W. B., "The Kinematics of Galactic Spiral Structure," (*Publications of the Astronomical Society of the Pacific,* vol. 85, no. 508, 1973).

Clark, T. A., et al., "Meter-Wavelength VLBI. The Observations" (*The Astronomical Journal,* vol. 80, no. 11, 1975).

Cocconi, Guiseppe, and Morrison, Philip, "Searching for Interstellar Communications," (*Nature,* vol. 184, no. 4690, September 19, 1959).

Coe, James R., "NRAO Interferometer Electronics," (*Proceedings of the IEEE,* vol. 61, no. 9, September 1973).

Cogdell, John R., and Hvatum, Hein, et al., "High Resolution Millimeter Reflector Antennas," (*Transactions on Antennas and Propagation,* vol. AP-18, no. 4, 1970).

Cohen, M. H., Jauncey, D. L., Kellerman, K. I., and Clark, B. G., "Radio Interferometry at One-Thousandth Second of Arc," (*NRAO Reprint Series A,* 1967).

Cohen, M. H., "Introduction to Very Long Baseline Interferometry," (*Proceedings of the IEE,* vol. 61, no. 9, September 1973).

Cohen, M. H., Moffet, A. T. Romney, J. D., Schilizzi, R. T., Kellerman, K. I., Purcell, G. H., Grove, G. Yen, J. L., Pauliny-Toth I. I. K., Preuss, E., Witzel, A., and Graham, D., "Observations with a VLB Array," (*The Astrophysical Journal,* vol. 201, 1975).

Colander, Valerie, "Other Stars and Planets," (*West Virginia Times,* Sunday, September 4, 1983.

Cooley, R. C., and Roberts, M. S., "Observations of the Andromeda Galaxy at 11-Centimeter Wavelength," (*Science,* vol. 156, no. 3778, 1967).

Counselman, C. C. III, Kent, S. M., Knight, C. A., Shapiro, I. I., Clark, T. A., Hinteregger, Rogers, A. E. E., and Whitney, A. R., "Solar Gravitational Deflection of Radio Waves Measured by Very-Long-Baseline Interferometry," (*Physical Review Letters,* vol. 33, no. 27, 1974).

Covington, Arthur E., "Origins of Canadian Radio Astronomy," (*The Journal of the Royal Astronomical Society of Canada,* vol. 82, no. 4, 1988).

Crews, James F., "The Tatel 85-foot Radio Telescope," (*The Observer,* vol. 1, March 31, 1964).

Crutcher, R. M., Kazes, I., and Troland, T. H., "Magnetic Field Strengths in Molecular Clouds," (*Astronomy and Astrophysics,* vol. 181, 1987).

Damashek, Mark, "Discovery of the Third Binary Pulsar," (*The Observer (NRAO),* vol. 21, no. 2, 1980).

Dick, Stephen J., "Book Review: *The Jodrell Bank Telescopes,* A. C. B. Lovell," (*ISIS,* vol. 73, no. 2, 1986).

Dicke, R. H., "The Measurement of Thermal Radiation at Microwave Frequencies," (*Review of Scientific Instrumentation,* vol. 17, 1946).

Dickey, John, "A New Interferometer for Green Bank?" (*The Observer (NRAO),* vol. 21, no. 3, 1980).

Dickinson, Dale F., et al., "Observations of Interstellar Silicon Monoxide," (*The Astrophysical Journal,* vol. 206, 1976).

Distasio, T., "From the Interferometer," (*The Observer,* vol. 4, no. 2, May 28).

Drake, F. D. "Radio Resolution of the Galactic Nucleus," (*Sky and Telescope,* 1959).

Drake, F. D., "The Position-Determination Program of the National Radio Astronomy Observatory," (*Publications of the Astronomical Society of the Pacific,* vol. 72, no. 429, 1960).

Drake, F. D., "How Can We Detect Radio Transmissions from Distant Planetary Systems?" (*Sky and Telescope,* vol. 19, no. 3, 1960).

Drake, F. D., "Radio Emission from the Planets," (*Physics Today,* April 1961).

Drake, F. D., "Project OZMA," (*Physics Today,* April 1961).

Drake, F. D., "Microwave Spectrum of Saturn," (*Nature,* vol. 195, no. 4844, 1962).

Dunn, Raga, "America's Most Powerful Telescope is Zapped By Space Aliens," (*Weekly World News,* January 31, 1988).

Emberson, Richard M., "National Radio Astronomy Observatory: The Early History and Development of the Observatory at Green Bank, West Virginia," (*Science,* vol. 130, no. 3385:, 1956).

Emberson, Richard M., "The Telescope Program for the National Radio Astronomy Observatory at Green Bank, West Virginia," (*Proceedings of the IRE,* vol. 46, no. 1, 1958).

Erickson, William C., "The Clark Lake Array," (*Proceedings of the IRE,* vol. 61, no. 9, September 1973).

Erkes, Joseph W., and Dickel, John R., "Radio Observations of the Supernova Remnant HB 21," (*Astronomical Journal,* vol. 74, no. 6, 1969).

Ewen, H. I., and Purcell, E. M., "Observation of a Line in the Galactic Radio Spectrum- Radiation from Galactic Hydrogen at 1,420 Mcs," (*Nature,* vol. 165, 1950).

Feld, Jacob, "Design Study for the Construction of a 600-

foot Radio Telescope," (*Annals of the New York Academy of Science*, vol. 70, 1955).

Findlay, John W., "Noise Levels at the National Radio Astronomy Observatory," (*Proceedings of the IRE*, vol. 46, no. 1, 1958).

Findlay, John W., "The 300-ft Transit Radio Telescope," (*NRAO Internal Document*, September 13, 1962).

Findlay, John W., "Operating Experience at the National Radio Astronomy Observatory," (*Annals of the New York Academy of Science*, vol. 116, no.1, June 26, 1964).

Findlay, John W., "The 300-Foot Radio Telescope at Green Bank," (*Sky and Telescope*, vol. 25, no. 2, 1972).

Findlay, John W., "Radio and Radar Astronomy," Special Issue, ed. by J. F. Findlay, (*Proceedings of the IEE*, vol. 61, 1973).

Findlay, John W., "The National Radio Astronomy Observatory," (*Sky and Telescope*, vol. 48, 1974).

Findlay, John W., "Book Review: *Astronomy Transformed: The Emergence of Radio Astronomy in Britain*, D. Edge and M. J. Mulkay," (*Sky and Telescope*, June 1977).

Findlay, John W., "The 300-ft Telescope Is Ten Years Old," (*The Observer (NRAO)*, vol. 13, no. 6, 1980).

Findlay, John W., and Hvatum, H., "An Absolute Flux-Density Measurement of Cassiopeia A at 1440 MHz," (*Astrophysical Journal*, vol. 141, no. 3, April 1, 1965).

Findlay, John W., and Payne, John M., "Upgrading the 11-Meter Antenna of the National Radio Astronomy Observatory," (*Proceedings of the Third International Conference on Antennas and Propagation*, vol. 219, 1983).

Fomalont, E. B., "Testing the Theory of Relativity," (*The Observer*, vol. 15, no. 6, 1974).

Fomalont, E. B., "General Relativity and Radio Interferometry," (*The Physics Teacher*, vol. 14, no. 6, 1976).

Fomalont, E. B., and Sramek, R. A., "A Confirmation of Einstein's General Theory of Relativity By Measuring the Bending of Microwave Radiation in the Gravitational Field of the Sun," (*The Astrophysical Journal*, vol. 199, 1975).

Gold, T., "Rotating Neutron Stars as the Origin of the Pulsating Radio Sources," (*Nature*, vol. 218, 1968).

Goldsmith, Donald, "SETI: The Search Heats Up," (*Sky and Telescope*, vol. 75, no. 2, 1988).

Gordon, Mark A., "New Surface for an Old Telescope," (*Sky and Telescope*, vol. 67, no. 4, 1984).

Gordon, Mark A., "VLBA—A Continent-Size Radio Telescope," (*Sky and Telescope*, vol. 69, no. 6, 1985).

Greenstein, J. L., "Washington Conference on Radio Astronomy—1954, ed. by J. L. Greenstein," (*Journal of Geophysical Research*, vol. 59: 149, 1954).

Gregory, P. C., Kronberg, P. P., Seaquist, E. R., Hughes, V. A.,

Woodsworth, A., Viner, M. R., and Retallack, D., "Detection of a Radio Burst in Cygnus X-3," (*Nature*, vol. 239, 1972).

Gregory, P. C., Kronberg, P. P., Seaquist, E. R., Hughes, V. A., Woodsworth, A., Viner, M. R., Retallack, D., and Balick, M. R., "The Nature of the First Cygnus X-3 Radio Outburst," (*Nature Physical Science*, vol. 239, 1972).

Hagen, John, "Radio Astronomy Conference," (*Science*, vol. 119, April 1954).

Harrison, E. R., "Electrified Black Holes," (*Nature*, vol. 264 no. 5586, 1976).

Harrison, E. R., "Acceleration of Supermassive Compact Objects by Emission of Asymmetric Radiation," (*The Astrophysical Journal*, vol. 213, 1977).

Harten, R. H., "Beam Characteristics of the 300-ft Telescope," (*The Astronomical Journal*, vol. 78, no. 7, 1973).

Heeschen, D. S., "Observations of Radio Sources at Four Frequencies," (*Astrophysical Journal*, vol. 133, January 1961).

Heeschen, D. S., "Radio Galaxies," (*Scientific American*, vol. 206, no. 3, 1962).

Heeschen, D. S., and Dieter, N. H., "Extragalactic 21-cm line Studies," (*Proceedings of the IRE*, vol. 46, no. 1, 1958).

Heeschen, D. S., and Wade, C. M., "A Radio Survey of Galaxies," (*The Astronomical Journal*, vol. 69, no. 4, 1964). Hefland, David J., Taylor, J. H., Backus, P. R., and Cordes, J. M., "Pulsar Timing: Observations From 1970 to 1978," (*The Astrophysical Journal*, vol. 273, April 1, 1980).

Heiles, Carl, and Hoffman, Wilson, "The Beam Shape of the NRAO 300-ft Telescope and Its Influence on 21-cm Line Measurements," (*The Astronomical Journal*, vol. 73, no. 6, 1968).

Hewish, A., Bell, S. J., Pilkington, J. D., Scott, P. F., and Collins, R. A., "Observations of a Rapidly Pulsating Radio Source," (Nature, vol. 217, February 1968).

Hey, J. S., "Reports on the Progress of Astronomy: Radio Astronomy," (*Monthly Notices of the Royal Astronomical Society*, vol. 109, 1949).

Hey, J. S., Phillips, J. W., and Parsons, S. J., "Cosmic Radiation at 5 Meters Wavelength," (*Nature*, vol. 157, 1946). Hoerner, Sebastian von, "Astronomical Aspects of Interstellar Communication," (*Astronautica Acta*, vol. 18, 1973).

Hoerner, Sebastian von, and Wong, Woo-Yin, "Improved Efficiency with a Mechanically Deformable Subreflector," (*IEE Transactions on Antennas and Propagation*, vol. A P-27, no. 5, September 1979).

Hogbom and Shakeshaft, "Secular Variation of the Flux Density of the Radio Source Cassiopeia A," (*Nature*, vol. 190, May 20, 1961).

Hogg, David E., "Radio Observations of the Galaxy and

Extragalactic Sources," (*R. A. S. C. Journal,* vol 58, no. 5, 1962).

Höglund, B., and Mezger, P. G., "Radio Recombination Lines: A New Observational Tool in Astrophysics," (NRAO Reprint Series A, 1960).

Howard, William E., "From the Director's Office," (*The Observer,* vol. 10, no. 2, March 1970).

Hjellming, R. M. "An Astronomical Puzzle Called Cygnus X-3," (*Science,* vol. 182, 1973).

Hjellming, R. M., and Bignell, R. C., "Radio Astronomy with the Very Large Array," (*Science,* vol. 126, no. 4552, 1982).

Hjellming, R. M., and Wade, C. M., "Radio Novae," (*The Astrophysical Journal,* vol. 162, October 1970).

Hjellming, R. M., and Wade, C. M., "Detection of Radio Emission from Antares," (*The Astrophysical Journal,* vol. 163, February 1971).

Hjellming, R. M., and Wade, C. M., "Radio Stars," (*Science,* vol. 173, no. 4002, September 1971).

Jansky, K. G., "Directional Studies of Atmospherics at High Frequencies," (*Proceedings of the IRE,* vol. 20, no. 13, 1932).

Jansky, K. G., "Electrical Disturbances Apparently of Extraterrestrial Origin," (*Proceedings of the IRE,* vol. 21, no. 1, 1932).

Jansky, K. G., "Electrical Phenomena that Apparently Are of Interstellar Origin," (*Popular Astronomy,* vol. 41, 1933).

Jansky, K. G., "Radio Waves from Outside the Solar System," (*Nature,* vol. 132, 1933).

Jansky, K. G., "The Beginnings of Radio Astronomy," (*American Scientist,* vol. 45, 1957).

Kahn, F. D. "Sound Waves Trapped in the Solar Atmosphere II," (*The Astrophysical Journal,* vol. 135, 1962).

Kaifu, N., et al., "Detection of Interstellar Methylamine," (*The Astrophysical Journal,* vol. 191, L135, 1974).

Kellermann, K. I., "Thermal Radio Emission from the Major Planets," (*Radio Science,* vol. 5, no. 2, 1970).

Kellermann, K. I., "Extragalactic Radio Sources," (*Physics Today,* vol. 26, no. 10, 1973).

Kellermann, K. I., Shafer, D. B., Clark, B. G., and Geldzahler, B. J., "The Small Radio Source at the Center of the Galaxy," (*The Astrophysical Journal,* vol. 214, L61, 1977).

Kellermann, K. I., and Thompson, A. R., "The Very Long Baseline Array," (*Science,* vol. 229, 1985).

Kevles. Daniel J., "Book Review: *Jodrell Bank and Cambridge: The Emergence of Radio Astronomy in Britain,* D. Edge and M. J. Mulkay," (*Science,* vol. 196, 1979).

Kiernan, Vincent, "How Far to the Galaxies," (*Astronomy,* vol. 17, no. 6, 1989).

Kourganoff, Vladimir, "Otto Struve: Scientist and Humanist," (*Sky and Telescope,* vol. 75, no. 4, 1988).

Kraus, J. D., "Radio Telescope Antennas of Large Aperture," (*Proceedings of the IRE,* vol. 46, 1958).

Kraus, J. D., "Recent Advances in Radio Astronomy," (*IEEE Spectrum,* vol. 1, 1964).

Kraus, J. D., "The First 50 Years of Radio Astronomy Part 1: Karl Jansky and His Discovery of Radio Waves from Our Galaxy," (*Cosmic Search,* vol. 3, 1981).

Kraus, J. D., "The First 50 Years of Radio Astronomy Part 2: Grote Reber and the First Radio Maps of the Sky," (*Cosmic Search,* vol. 4, no. 14, 1982).

Kraus, J. D., "The First 50 Years of Radio Astronomy Part 3: Post-War Radio Astronomy," (*AstroSearch,* vol. 1, 1983).

Kraus, J. D., and Ko, H. C., "Radio Radiation from the Supergalaxy," (*Nature,* vol. 172, 1953).

Kuiper, T. B. H., Zuckerman, B., Kakar, R. K., and Kuiper, Eva N., "Detection of 2.6 Millimeter Radiation Probably Due to Nitrogen Sulfide," (*The Astrophysical Journal,* vol. 200, L151, 1975).

Lancaster, Jack, "VLA Dedicated- October 10, 1980," (*The Observer (NRAO),* vol. 22, No. 1, 1981).

Little, C. G., and Lovell, A. C. B., "Origin of the Fluctuations in the Intensity of Radio Waves from Galactic Sources—Jodrell Bank Observations," (*Nature,* vol. 165, 1950).

Lovell, A. C. B., "The New Science of Radio Astronomy," (*Nature,* vol. 167, 1951).

Lovell, A. C. B., "The Jodrell Bank Radio Telescope," *Nature,* vol. 180, 1957).

Lovell, A. C. B., "The Emergence of Radio Astronomy After WWII," (*Quarterly Journal of the Royal Astronomical Society,* vol. 28, 1978).

Malphrus, Benjamin, and Bradley, Richard, "The NRAO 40-foot Radio Telescope Operators' Manual," (NRAO Internal Report, edited by the NRAO Electronics Department, July 1987).

Maslowski, J., "A Green Bank Sky Survey in Search of Radio Sources at 1400 MHz," (*Astronomy and Astrophysics,* vol. 16, 1972).

Menon, R. C., and Albaugh, N. P., "Cooled Loads as Calibration Noise Standards for the mm-Wavelength Range," (*Proceedings of the IEEE,* vol. 54, no. 10, 1966).

Mezger, P. G., "Observations of the Hydrogen Recombination Line n_{110}-n_{109} Emitted from Galactic HII Regions," (*The Astronomical Journal,* vol. 71, no. 3, 1966).

Muller, C. A., and Oort, J. H., "The Interstellar Hydrogen Line at 1,420 Mc./sec., and an Estimate of Galactic Rotation," (*Nature,* vol. 168, 1951).

Napier, Peter J., Thompson, Richard, and Ekers, Ronald D.,

"The VLA," (*Proceedings of the IEEE,* vol. 71, no. 11, 1983).

Needell, Allan, "Lloyd Berkner, Merle Tuve, and the Federal Role in Radio Astronomy," (*Osiris,* 2nd series, vol. 3, 1987).

Norberg, Arthur L., "Book Review: *Astronomy Transformed: The Emergence of Radio Astronomy in Great Britain,* D. Edge and M. J. Mulkay," (*ISIS* vol. 70, no. 4, 1979).

Norris, Ray, "Cosmic Masers," (*Sky and Telescope,* vol. 71, no. 3, 1986).

Papagainnis, Michael D., "Bioastronomy: The Search for Extraterrestrial Life," (*Sky and Telescope,* vol. 67, no. 1, 1984).

Pater, Imke De, and Dickel, John R., "VLA Observations of Saturn at 1.3, 2, and 6 cm," (*Icarus,* vol. 50, 1982).

Pauliny-Toth, I. I. K., and Kellermann, K. I., "Measurements of the Flux Density and Spectra of Discrete Radio Sources at Centimeter Wavelengths. II. The Observations at 5 GHz (6 cm)," (*Astronomical Journal,* vol. 73, no. 10, part 1, 1968).

Pawsey, J. L., "Radio Astronomy in Australia," (*Journal of the Royal Astronomical Society of Canada,* vol. 47, 1953).

Penzias, A. A., Wannier, P. G., Wilson, R. W., and Linke, R. A. "Deuterium in the Galaxy," (*The Astrophysical Journal,* vol. 211, 1977).

"Proceedings of Washington Conference on Radio Astronomy, January 4–6, 1954," (*Journal of Geophysical Research,* vol. 59, 1954).

Ralston, John, "The 45-foot Portable Antenna," (*The Observer,* vol. 13, no. 5, October 1972).

Reber, Grote, "Cosmic Static," (*Proceedings of the IRE,* vol. 28, 1940).

Reber, Grote, "Radio Emissions from the Milky Way," (*The Astrophysical Journal,* vol. 100, 1944).

Reber, Grote, "Radio Astronomy," (*Scientific American,* vol. 181, 1949).

Reber, Grote, "Early Radio Astronomy at Wheaton, Illinois," (*Proceedings of the IRE,* vol. 46, 1958).

Reber, Grote, and Greenstein, J. L., "Radio Frequency Investigations of Astronomical Interest," (*Observatory,* vol. 67, 1947).

Reid, Mark J., and Muhleman, Duane, *et al.,* "The Structure of Stellar Hydroxyl Masers," *The Astrophysical Journal,* vol. 214, 1977).

Reifenstein, Edward C. III, Staelin, David H., and Brundage, William D., "Crab Nebula Pulsar NPO527," (*Physical Review Letters,* vol. 22, no. 7, 1969).

Roberts, Morton S., "Recent Discoveries in Radio Astronomy," (*Physics Today,* February 1985).

Robinson, Leif J., ed., "New Class of Radio Sources," (*Sky and Telescope,* vol. 69, no. 6, 1985).

Robinson, Leif J., ed., "Blowing Bubbles in the Center of M51," (*Sky and Telescope,* vol. 70, no. 6, 1985).

Robinson, Leif J., ed., "Of Whirlpools, Warps, Bubbles and Jets," (*Sky and Telescope,* vol. 75, no.1, 1988).

Rothman, Tony, "In Memoriam: The 300-foot Radio Telescope Has Collapsed," (*Scientific American,* February 1989).

Ryle, M., "Evidence for the Stellar Origin of Cosmic Rays," (*Proceedings of the Physical Society,* vol. 62A, 1949).

Ryle, M., "A New Interferometer and Its Application to the Observation of Weak Radio Stars," (*Procedures of the Royal Society,* A211, 1952).

Ryle, M., "The New Cambridge Radio Telescope," (*Nature,* vol. 194, 1962).

Ryle, M., and Elsmore, B., "A Search for Long-period Variations in the Intensity of Radio Stars," (*Nature,* vol. 168, 1951).

Ryle, M., Elsmore, B., and Neville, A. C., "High-Resolution Observations of the Radio Sources in Cygnus and Cassiopeia," *Nature,* vol. 205, 1965).

Ryle, M., and Smith, F. G., "A New Intense Source of Radio Frequency Radiation in the Constellation of Cassiopeia," (*Nature,* vol. 162, 1948).

Saslaw, William C., and Mauri, J. Valtonen, "The Gravitational Slingshot and the Structure of Extragalactic Radio Sources," (*The Astrophysical Journal,* vol. 190, 1974).

Schorn, Ronald A., "The Extragalactic Zoo- I," (*Sky and Telescope,* vol. 75, no. 1, 1988).

Schorn, Ronald A., "The Extragalactic Zoo- II," (*Sky and Telescope,* vol. 75, no. 4, 1988).

Shalloway, A. M., Mauzy, R., Greenlaugh, J., and Weinreb, S., "The NRAO 416-Channel Autocorrelation Spectrometer," (NRAO Electronics Division Internal Report, 124, 1972).

Sinha, R. P., "Survey of Neutral Hydrogen in the Galactic Center Region," (*Astrophysical Journal Supplement,* vol. 37, November 17, 1978).

Small, Maxwell M., "The New 140-foot Radio Telescope," (*Sky and Telescope,* vol. 30 no. 6, 1965).

Smith, David H., "Merlin: A Wizard of a Telescope," (*Sky and Telescope,* vol. 67, no. 2, 1985).

Smith, F. Graham, "Martin Ryle, Pioneer Radio Astronomer," (*Sky and Telescope,* vol. 69, No. 1, 1984).

Snyder, Lewis E., et al., "Radio Detection of Interstellar DCO+," (*The Astrophysical Journal,* vol. 209, L83, 1976).

Snyder, Lewis E., and Buhl, David, "Detection of New Stellar Sources of Vibrationally Excited Silicon Monoxide

Maser Emission at 6.95 Millimeters," (*The Astrophysical Journal,* vol. 197, L31, 1974).

Snyder, Lewis E., and Buhl, David, "Detection of Possible Maser Emission Near 3.48 Millimeters From an Unidentified Molecular Species in Orion," (*The Astrophysical Journal,* vol. 189, 329, 1975).

Snyder, Lewis E., and Buhl, David, et al., "Microwave Detection of Interstellar Formaldehyde," (*Physical Review Letters,* vol. 22, no. 13, 1969).

Snyder, Lewis E., Hollis, J. M., Lovas, F. J., and Ulich, B. L. "Detection, Identification, and Observations of Interstellar $H_{13}CO^+$," (*The Astrophysical Journal,* vol. 209, no. 67, 1976).

Snyder, Lewis E., Hollis, J. M., and Ulich, B. L., "Radio Detection of the Interstellar Formyl Radical," (*The Astrophysical Journal,* vol. 208, L91, 1976).

Snyder, Lewis E., Hollis, J. M., Ulich, B. L., and Buhl, David, "Radio Detection of Interstellar Sulfur Dioxide," (*The Astrophysical Journal,* vol. 198, L81, 1975).

Spradley, Joseph L., "The First True Radio Telescope," (*Sky and Telescope,* July 1988).

Staelin, David H., and Reifenstein, Edward C., III, "Pulsating Radio Sources Near the Crab Nebuls," (*Science,* vol. 162, 1481, 1968).

Struve, Otto, "Progress in Radio Astronomy- I," (*Sky and Telescope,* vol. 9, no. 27, 1949).

Struve, Otto, "Progress in Radio Astronomy- II," (*Sky and Telescope,* vol. 9, no. 55, 1950).

Struve, Otto, Emberson, R. M., and Findlay, J. W., "The 140-foot Radio Telescope of the National Radio Astronomy Observatory," (*Publications of the Astronomical Society of the Pacific,* vol. 72, no. 429, December 1960).

Sullivan, W. T., III, "Radio Astronomy's Golden Anniversary," (*Sky and Telescope,* vol. 64, 1982).

Swarup, G., "Radio Astronomy," Commission 40 Report covering surveys of radio sources, NRAO Reprint Series B, 534, 1981.

Swenson, G. W., and Kellermann, K. I., "An Intercontinental Array—A Next-Generation Radio Telescope," (*Science,* vol. 188, 1263, 1975).

Swenson, G. W., and Mathur, N. C., "The Interferometer in Radio Astronomy," (*Proceedings of the IEEE,* vol. 56, 2114, 1968).

Thompson, A. R., Clark, B. G., Wade, C. M., and Napier, P. J., "The Very Large Array," (*The Astrophysical Journal Supplement Series,* vol. 44, October 1980).

Thomsen, Dietrick, E., "Gravitational Refractions," (*Science News,* vol. 125, 1984).

Townes, C. H., "Microwave Spectra of Astrophysical Interest," (*Journal of Geophysical Research,* vol. 59, 1954).

Tucker, K. D., et al., 'The Ethynyl Radical C_2H- A New Interstellar Molecule," (*The Astrophysical Journal,* vol. 193, L115, 1974).

Tully, J. Brent, and Fisher, J. Richard, "A New Method of Determining Distances to Galaxies," (*Astronomy and Astrophysics,* vol. 54, 1977).

Turner, Barry, "Anomalous OH Emission from New Types of Galactic Objects," (*Astrophysical Letters,* vol. 8, 1971).

Turner, Barry, "Interstellar Molecules," (*Readings from the Scientific American,* Reprints, 1973).

Turner, Barry, "Interstellar Molecules- A Review of Recent Developments," (*Journal of the Royal Astronomical Society of Canada,* vol. 68, no. 2, January 1974).

Turner, Barry, "Detection of OH at 18-Centimeter Wavelength in Comet Kohoutek (1973)," (*The Astrophysical Journal,* vol. 189, L137, 1974).

Turner, Barry, "A New Interstellar Line With Quadrupole Hyperfine Splitting," (*The Astrophysical Journal,* vol. 193: L73, 1974).

Turner, Barry, "Microwave Detection of Interstellar Ketene," (*The Astrophysical Journal,* vol. 213, L75, 1977).

Turner, Barry, "On the Identification of U-Lines in the 3 mm Region of the Interstellar Spectrum," (*Astrophysical Letters,* vol. 23, 1983).

Turner, Barry, et al., "Observations of Interstellar Water Vapor," (*Astronomy and Astrophysics,* vol. 4, 1970).

Turner, Barry, Kislyakov, A. G., Liszt, H. S., and Kaifu, N., "Microwave Detection of Interstellar Cyanamide," (*The Astrophysical Journal,* vol. 201, L149, 1975).

Turner, Barry, and Zuckerman, B., "Microwave Detection of Interstellar CH," (*The Astrophysical Journal,* vol. 187, L59, 1974).

Turner, Barry, Zuckerman, B., Fourikis, N., Morris, M., and Palmer, Patrick, "Microwave Detection of Interstellar HDO," (*The Astrophysical Journal,* vol. 198, L125, 1975).

Turner, Barry, Zuckerman, B., Morris, M., Gilmore, W., and Palmer, Patrick, "Detection of Interstellar SiS and a Study of the IRC +10216 Molecular Envelope," (*The Astrophysical Journal,* vol. 199, L47, 1975).

Turner, Barry, Zuckerman, B., Morris, M., and Palmer, Patrick, "Cyanoacetylene in Dense Interstellar Clouds," (*The Astrophysical Journal,* vol. 205, 1976).

Turner, Barry, Zuckerman, B., Morris, M., Palmer, Patrick, and Rickard, L. J., "Observations of Extragalactic Molecules HCN and CS," (*The Astrophysical Journal,* vol. 214, 1977).

Tuve, M. A., "Symposium on Radio Astronomy, delivered before the autumn meeting of the National Academy of Sciences, Indiana University, Bloomington, November 17, 1959, Chairman, Merle A. Tuve," (*Proceedings of the National Academy of Sciences,* Special Reprint, 1960).

Tyler, W. C., Hogg, D. E., and Wade, C. M., "First Results with the National Radio Astronomy Observatory Interferometer," (*American Institute of Physics for the American Astronomical Society Abstracts*, presented at the 118th meeting of the American Astronomical Society held 14–17 March 1965).

Verschuur, Gerrit L., "Positive Determination of an Interstellar Magnetic Field by Measurement of the Zeeman Splitting of the 21-cm Hydrogen Line," (*Physical Review Letters,* vol. 21, no. 11, 1968).

Verschuur, Gerrit L., "An Intermediate Velocity Cloud Showing a Velocity Bridge to Local Matter," (*Astronomy and Astrophysics*, vol. 3, 1969).

Verschuur, Gerrit L., "Further Measurements of Magnetic Fields in Interstellar Clouds of Neutral Hydrogen," (*Nature*, vol. 223 no. 5202, 1969).

Verschuur, Gerrit L., "High Resolution Observations of a High Latitude Neutral Hydrogen Concentration," (*Astronomy and Astrophysics*, vol 1, no. 4, 1969).

Verschuur, Gerrit L., "Some Very Cold HI Clouds Found in Emission," (*Astrophysical Letters*, vol. 4, 1969).

Verschuur, Gerrit L., "Interesting Neutral Hydrogen Feature in the Direction of the North Polar Spur," (*Astrophysical Letters*, vol. 6, 1970).

Verschuur, Gerrit L., "High Velocity Clouds and 'Normal' Galactic Structure," (*Astronomy and Astrophysics*, vol. 22, 1973).

Verschuur, Gerrit L., "Molecules Between the Stars," (*Mercury*, May-June, 66, 1987).

Verschuur, Gerrit L., "A New 'Yardstick' for the Universe," (*Astronomy*, vol. 16, no. 11, 1988).

Wade, C. M., "A Possible New Radio Galaxy in the Virgo Cluster," (*The Observatory*, vol. 80, no. 919, 1959).

Wade, C. M., "Fine Structure of the Radio Source Cygnus A," (*The Physical Review Letters,* vol. 17, no. 20, 1966).

Wade, C. M., and Miley, G. K., "Positions of Unidentified Radio Sources," (*The Astronomical Journal,* vol. 76, no. 2, 1971).

Waldorp, Mitchell M., "Collapse of a Radio Giant," (*Science,* vol. 242, no. 1120, 1988).

Waldorp, Mitchell M., "The Farthest Galaxies: A New Champion," (*Research News*, August 1988).

Weinreb, S., Barrett, A. H., Meeks, M. L., and Henry, J. C., "Radio Observations of OH in the Interstellar Medium," (*Nature,* vol. 200, no. 829, 1963).

Wesseling, Karel H., "A Single-Sideband-Double-Sideband Interferometer Receiver for Radio Astronomy," (*IEEE Transactions on Antennas and Propagation,* vol. AP-15, no. 2, 1967).

Will, Clifford M. "Testing General Relativity: 20 Years of Progress," (*Sky and Telescope*, vol. 66, no.4, 1983).

Williams, D. R. W., and Davies, R. D., "A Method for the Measurement of the Distance to Radio Stars," (*Nature,* vol. 173, no. 1182, 1954).

Wright, M. C. H., Clark, B. G., Moore, C. H., and Coe, J., "Hydrogen-line Aperture Synthesis at the National Radio Astronomy Observatory: Techniques and Data Reduction," (*Radio Science,* vol. 8, no. 8, 1973).

NRAO Internal Documents/Archives

J. W. Findlay, "Large Radio Telescopes—1950 to 1989," NRAO Internal Publication, August 20, 1989.

The Green Bank Telescope: A Radio Telescope for the Twenty-First Century, Final Proposal to the National Science Foundation, June 1989, submitted by Associated Universities, Inc.

"Historical Telescopes at the National Radio Astronomy Observatory in Green Bank West Virginia," Operated by the Associated Universities, Inc. under contract with the National Science Foundation.

Interferometer Memo Book One.

Interferometer Memos Book Two.

Lancaster, John, "Recollections of Brookhaven National Laboratory and the NRAO," 1981.

The Making of the 140-foot. NRAO Video. 1965.

"NRAO and AUI: The First 9,131 Meridian Crossings 25 at the Research Frontier," 1987.

National Radio Astronomy Observatory Brochure, Green Bank, West Virginia, 1984.

NRAO Annual Reports, (Bulletin of the American Astronomical Society, (Reprints), (1969–1986).

NRAO Annual Reports, 1975–1982.

NRAO Computer Division Internal Reports, 1966–1980.

NRAO 85-foot Radio Telescope Log Book, 1958–1980.

NRAO Electronics Division Internal Reports, 1962–1980.

NRAO Forty-foot Radio Telescope Log Book, 1958–1962; 1987–1988.

NRAO Newsletter, 1981–1990.

NRAO Observing Summmary Statistics, 1973–1987.

NRAO 140-foot Radio Telescope Log Book, 1962–1980.

NRAO Quarterly Reports, 1963–1990.

NRAO Reprints, 1957–1990.

NRAO 300-foot Radio Telescope Log Book, 1962–1980.

The Observer, Newsletter, 1961–1981.

140-foot Computer Assisted Observing Manual Editions 1–3.

140-foot Construction Records 1960–1965.

140-foot K Band Performance, Engineering Memo.

140-foot Observer's Log Books, 1965–present.

140-foot Radio Telescope Dedication, October 13, 1965.

Oref, Wally, "Radio Astronomy and the NRAO" 1981. Unpublished NRAO Document.

"Planning Document for the Establishment of a Radio Astronomy Observatory," Prepared for the National Science Foundation by Associated Universities, Inc., New York, 1956.

"Recollections and Humor from the First Twenty-Five Years with the 300-foot Telescope" NRAO Internal Publication, 1987.

Report of the Technical Assessment Panel, "Collapse of the 300-foot Radio Telescope," including "National Radio Astronomy Observatory 300-foot Telescope Finite Element Analysis, Volume I," prepared by Henry Ayvazyan, Joel Stahmer, and Joseph Vellozi of Amman and Whitney, Consulting Engineers, New York, New York, February 1989.

"Science Highlights with the 300-foot Telescope: The First Quarter Century," NRAO Internal publication included in an internal document for the 25 year anniversary ceremony of the 300-foot telescope, 1987.

Seielstad, George A., "A Radio Telescope Larger than Earth" Testimony to House of Representatives Space Science and Applications Subcommittee, February 9, 1989.

"Technological Developments Fostered By Radio Astronomy" National Radio Astronomy Observatory, April 1987.

300-foot Construction Records, 1961–1962.

300-foot Observer's Log Books, 1962–1988.

300-foot Radio Telescope Book, Appendix 1, 2, 2A, 1962.

"300-foot Telescope Technical Evolution, 1962–1987: A Quarter Century of Science on the Meridian," NRAO Electronics Division Internal Report, 1987.

Interviews and Lectures

Crews, J. F., Interview with B. Malphrus, NRAO, Green Bank, West Virginia, March 3, 1989.

Crews, J. F., Lecture "The Collapse of the 300-Foot Telescope," NRAO, Green Bank, West Virginia, June 19, 1989.

Fisher, J. Richard, Interview with B. Malphrus, NRAO, Green Bank, West Virginia, March 3, 1989.

Fleming, Richard, Interview with B. Malphrus, NRAO, Green Bank, West Virginia, March 31, 1989.

Lockman, Jay, Lecture, "The History of the 300-foot Radio Telescope, or 'Astropolitics'," NRAO, Green Bank, West Virginia, July 20, 1988.

Reber, Grote, Lecture, "The Early Years of Radio Astronomy." NRAO, Green Bank, West Virginia, February 25, 1988.

Other Sources

CSIRO and Astronomy. CSIRO information Service. Leaflet #2. January 1978.

CSIRO in Brief. CSIRO Sciences Communication Unit. Canberra, Australia. April 1986.

Keller, K. S., and Coleman, H. P., U.S. Naval Report 4088, 1948.

"The NASA Microwave Observing Project for the Search for Extraterrestrial Intelligence (SETI)," Publication of the SETI Institute, February 16, 1989.

Name Index

Subject Index